HANDBOOK OF ORGANIZATIONAL AND ENTREPRENEURIAL INGENUITY

Handbook of Organizational and Entrepreneurial Ingenuity

Edited by

Benson Honig

McMaster University, Hamilton, Ontario, Canada

Joseph Lampel

Cass Business School, City University London, UK

Israel Drori

College of Management, Academic Studies, Rishon LeZion, Israel

Edward Elgar

Cheltenham, UK • Northampton, MA, USA

Published by
Edward Elgar Publishing Limited
The Lypiatts
15 Lansdown Road
Cheltenham
Glos GL50 2JA
UK

Edward Elgar Publishing, Inc.
William Pratt House
9 Dewey Court
Northampton
Massachusetts 01060
USA

A catalogue record for this book
is available from the British Library

Library of Congress Control Number: 2013949872

This book is available electronically in the ElgarOnline.com
Business Subject Collection, E-ISBN 978 1 78254 904 8

ISBN 978 1 78254 903 1

Typeset by Servis Filmsetting Ltd, Stockport, Cheshire
Printed and bound in Great Britain by T.J. International Ltd, Padstow

Contents

Contributors

Aneesh Banerjee is a doctoral researcher at the Cass Business School, London. His research investigates the triggers of qualitatively superior innovations in organizations.

Stewart Clegg is Research Professor at the University of Technology, Sydney, and Director of the Centre for Management and Organisation Studies Research, and a Visiting Professor at Nova School of Business and Economics. His research is driven by a fascination with power and theorizing.

Larry W. Cox is Associate Professor of Entrepreneurship at Pepperdine University in the United States. He holds a PhD in Strategic Management from the University of Nebraska. His publications appear in the *Journal of Enterprising Culture*, *Journal of International Business and Entrepreneurship*, and *Entrepreneurship, Innovation and Change*.

Miguel Pina e Cunha is Professor of Organization Studies at Nova School of Business and Economics, Portugal. His research interests include: process-based views of organizations; the paradoxes of organizing; virtuous and toxic leadership; and the unfolding of positive and genocidal forms of organization.

Israel Drori (PhD, UCLA) is Professor of Management, School of Business, College of Management, Academic Studies, Israel and Visiting Professor, Recanatti School of Management, Tel Aviv University. His research interests include organizational ethnography, evolution of industries and transnational and high-tech entrepreneurship. He has published seven books and in journals such as: *American Sociological Review*, *Organization Science* and *Organizational Studies*, among others.

Gregory Gorse holds an MBA in Sustainability from Duquesne University, and a BBA in Marketing from the University of North Texas, USA. His team won third place in the 2013 Aspen Institute MBA Case Competition. His publication appears in the *Handbook of Research on Organizational Ingenuity and Creative Institutional Entrepreneurship*.

Peter Groenewegen is Professor of Organization Sciences at the Department of Organization Sciences, Faculty of Social Sciences, VU University Amsterdam. His research concerns the networked character of organizing and entrepreneurship taking place in emergency management organizations, health care and online communities.

Benson Honig (PhD, Stanford University) is the Teresa Cascioli Chair in Entrepreneurial Leadership, DeGroote School of Business, McMaster University, Hamilton, Ontario, Canada. Studying entrepreneurship worldwide (including environments of transition), his research includes organizational theory, ethics in scholarship, social capital, social entrepreneurship, business planning, nascent entrepreneurship and transnational entrepreneurship.

Jochen Koch is a Professor of Management and Organization at the European University Viadrina, Frankfurt/Oder, Germany. He received his doctorate from the Freie Universität Berlin. His current research interests include modern and postmodern organization theory, organizational routines and practices, strategic management and creativity, and the theory of organizational and strategic path dependence.

Janne M. Korhonen (MSc in Engineering from Helsinki University of Technology) is with the Aalto University School of Business, Department of Management and International Business, Helsinki, Finland. He is a PhD student studying the interplay of constraints and technological evolution.

Joseph Lampel is Professor of Strategy and Innovation at Cass Business School, City University London. He is the co-author and co-editor of five books, and more than 50 scholarly and practitioner articles. His main areas of research are strategy in creative industries, innovation processes, and management of project-based organizations.

Sandra R.H. Mariano is Associate Professor at Universidade Federal Fluminense, Brazil. She holds a PhD in Systems and Computer Engineering from Universidade Federal do Rio de Janeiro. She published two books in the area of entrepreneurship and articles in academic journals such as *Espacios (Caracas)* and *RAUSP*.

Francesca Masciarelli is Assistant Professor at the University G. d'Annunzio (Italy). She received her doctorate from the University of Trento. Her research interests include social capital, strategy and management of innovation and international business, with particular emphasis on the implications of social capital on firms' competitiveness.

Judy Matthews is a Senior Lecturer in the QUT Business School, Australia, engaged in researching and teaching innovation, creative problem solving and design thinking to senior managers and executives. Judy investigates innovation management, entrepreneurial strategies and organizational strategic renewal, presenting and publishing widely on design thinking and design led innovation.

David T. Methé is a Full Professor at the Institute of Business and Accounting, Kwansei Gakuin University, Kobe, Japan. His research focuses on innovation in organizations. He has held faculty positions at Sophia University in Tokyo, Japan, Kobe University's Research Institute for Business and Economics (RIEB) and at the University of Michigan Ross Business School.

Joysi Moraes is Adjunct Professor at Universidade Federal Fluminense, Brazil. She holds a PhD in Management from Universidade Federal do Rio Grande do Sul. Her publications appear in *Ephemera: Theory and Politics in Organization, Organizacoes e Sociedade, Cadernos EBAPE (FGV), Revista de Administracao FACES Journal*, and *Espacios.*

Pedro Neves is Assistant Professor at Nova School of Business and Economics, Portugal. His research has been published in journals such as the *Journal of Applied Psychology*, *The Leadership Quarterly*, *Human Performance* and *Group & Organization Management*. His research interests include leadership and interpersonal relationships in the workplace, entrepreneurship, improvisation/risk taking, and change management.

Pedro Oliveira is an Associate Professor at Católica Lisbon School of Business and Economics. His research focuses on innovation and technology management, and particularly on the role of users in developing new services. His research has appeared in *Research Policy*, *Production and Operations Management*, *Journal of Product Innovation Management*, *Organizational Research Methods*, and *International Journal of Operations and Production Management*, among others.

Andrea Prencipe (PhD, SPRU, University of Sussex) is a Professor of Economics and Management Innovation at LUISS Guido Carli University (Italy). His research interests include: division and coordination of knowledge and labour; routines in project-based contexts; and the relationships between social capital and innovation.

Arménio Rego is Associate Professor at the Universidade de Aveiro, Portugal. He has published in journals such as *Journal of Business Ethics*, *Journal of Business Research*, *Journal of Occupational Health Psychology*, and *The Leadership Quarterly*. His research deals mainly with positive organizational scholarship.

Wasko Rothmann is a post-doctoral research assistant at the European University Viadrina, Frankfurt/Oder, Germany. His current research interests include creativity in strategic management, organizational and strategic change, entrepreneurship and innovation processes.

Tamar Sagiv is a Lecturer of Organizational Behavior at the Recanati Graduate School of Business Administration at Tel Aviv University. Her current research projects include studies on knowledge evolution in various industries; creativity and authenticity in economic transactions; innovation and entrepreneurship in creative industries; and the effect of the interplay between individual agents and institutional mechanisms on the formation of new organizational forms.

Ninja Natalie Senf is a doctoral research assistant at the European University Viadrina, Frankfurt/Oder, Germany. Her current research interests include organizational aspiration level, strategic reference points, performance feedback, creativity, as well as feelings and measures of success.

John G. Shearer is Practitioner Faculty of Entrepreneurship at Pepperdine University in the United States. He holds an MBA from the University of Colorado. He is an award-winning inventor and serial entrepreneur. He holds multiple patents and received the 2008 Carnegie Science Award for Start-Up Entrepreneur.

Ana Cristina O. Siqueira is Assistant Professor of Management at Duquesne University, United States. She holds a PhD in Management from the University of Cambridge. Her articles appear in the *International Journal of Innovation Management*, *Academy of Management Best Paper Proceedings*, *Technology Analysis and Strategic Management*, and *IEEE Transactions on Engineering Management*.

A.M.C. Eveline Stam is a PhD candidate at the Department of Organization Sciences, Faculty of Social Sciences and lecturer at the Amsterdam Centre for Entrepreneurship, Faculty of Economics and Business Administration, both at the VU University Amsterdam. Her research focuses on entrepreneurship, network dynamics and new venture legitimation in Dutch health care.

Liisa Välikangas is Professor of Innovation Management at Aalto University and Hanken School of Economics in Helsinki, Finland. Her related book *The Resilient Organization: How Adaptive Organizations Thrive Even When Strategy Fails* was published by McGraw-Hill in 2010.

Ingrid A.M. Wakkee is Associate Professor at the Faculty of Social Sciences, and Business Developer at the Technology Transfer Office both at the VU University, Amsterdam. Her research focuses on the entrepreneurial process, failure and recovery, the role of social networks in entrepreneurship as well as university–industry collaboration.

David B. Zoogah is currently Associate Professor of Management in the Earl Graves School of Business and Management at Morgan State University, Baltimore, Maryland, USA. His research interests centre on micro issues (OB, HRM, Psychology) related to environmental management, Employee Development and Training, Team Effectiveness, Alliance Management.

Acknowledgements

As we hope this book describes, ingenious organizational activity (for which we aspire this book represents) requires the 'buy in' of many organizational actors. We would like to thank the following individuals and institutions for making this book possible: Social Science Humanities Research Council, as well as DeGroote School of Business, for providing the resources for our first conference on organizational ingenuity at McMaster; the team at the Halbert Center at Hebrew University, for providing a fellowship and support allowing for the completion of this book; Alan Sturmer, acquisition editor for Edward Elgar, for believing in this project from the very beginning; Teresa Cascioli, for supporting this initiative with her time and resources; Ruth Sutherland, for her unending and essential editorial and organizational assistance; Yervant Terzian and Lucy Djelaian, for making our first conference a success; Bill Starbuck, for his intellectual support and incisive keynote during the McMaster conference; all the many scholars involved in our project too numerous to name; and of course, in our personal lives, our wives and families who put up with our late nights, weekend vigils, and occasional absences from hearth and home.

Organizational ingenuity: insights and overview

Joseph Lampel, Benson Honig and Israel Drori

The construction of creativity in organizations continues to puzzle scholars. Creativity is often associated with organizational spaces where rules and boundaries are weak to non-existent, and by the same token is considered to decline where they are strong and pervasive. This dichotomy, between organizational freedom and necessity, results in opposing images of creativity and its absence. On the one hand, we have romantic images of entrepreneurs asserting creativity by breaking rules and transcending boundaries and, on the other, institutionalized managers constrained by rules and prevented by boundaries from exercising creativity.

The interplay of freedom and necessity in organizations, however, is more nuanced. Most organizational actors have to operate both within and through existing rules and boundaries to effect creative solutions to existing problems. To meet the challenge of the situation these actors develop a set of skills, social tactics, and mental orientation that express 'organizational ingenuity': *the ability to create innovative solutions within structural constraints using limited resources and imaginative problem solving.*

This book is an exploration of the many facets of organizational ingenuity. When we extended an invitation for contributions to this volume we gave authors the definition of organizational ingenuity provided above, and a general framework that we challenged them to develop. The final selection more than lived up to our expectations: it is rich and varied, and yet relatively coherent. When it came to writing the opening chapter our challenge was to give you, the reader, the perspective needed to appreciate the readings. Reading the contributions it became clear that a single fixed perspective is not sufficient to explore our central construct while at the same time respecting the variety and richness of the chapters in this book. Organizational ingenuity has a number of facets that flow from the original definition, and the contributions in this book focus on one or several of these facets, building and developing them in different directions.

After some discussion we decided to employ a device that is often used in literature, theatre, and cinema: convey key insights about organizational ingenuity by telling a story that is well known, in our case the story of Jeff Bezos and Amazon.com.

Amazon.com today is a US$61 billion colossus that spans the globe. Founded in 1994 by Jeff Bezos, Amazon has become the largest online retailer, which sells everything from books, to jewelry, to toys, apparel and sporting goods. Over the years the rise and evolution of Amazon has been covered in the press and media, and in book length publications. Our intention here is not to recount Amazon.com's story, but to take episodes that to our mind capture key features of organizational ingenuity and, at the same time, provide background for highlighting how these features are discussed in one or more chapters in this book.

RESOURCE CONSTRAINTS

Organizational ingenuity is the creative attempt by individuals and organizations to solve problems in situations where resources are limited. The paradox of organizational ingenuity is that the innovative solutions that emerge would probably not have occurred if the individuals and organizations involved had all the resources they needed at their disposal. Organizational ingenuity happens when actors refuse to abide by constraints, and instead search for a superior solution sometimes in spite, but often also because, of the constraints. An episode from the history of Amazon.com highlights this point.

Amazon.com is the archetypical garage start-up. Working out of a garage, Bezos and his team were able to grow to five employees, where they developed the necessary software routines to develop the new product. Unfortunately, the circuit breaker in the garage they were working on was insufficient. As a result, the early Amazon.com was developed by using extension cords running from other rooms to provide power. Eventually, they were siphoning off the entire house's electric supply, which was still insufficient, and they had to move to a slightly larger location (Spector, 2000).

Rather than hire consultants or a marketing firm to evaluate their design, they asked friends and family to enter 'fake' orders, under strict secrecy. Once the site was de-bugged and ready to go live, they sent out notices to all their beta testers announcing the launch and encouraging them to spread the word (Spector, 2000).

Frugality, and ingenuity, were embedded into the corporate culture at Amazon.com from the very beginning. Bezos built his first desk out of an exterior door that weighed 80 pounds. These 'door desks' became iconic – a measure of how Amazon kept their prices and costs low. They were widely displayed during the first television commercials, and became an integral component of corporate culture. Amazon eventually built several thousand door desks for packing purposes:

> In those formative years, Bezos repeatedly reminded employees that Amazon.com's customers didn't care what the offices or the desks look like; he continually reinforced the virtue of frugality. All the furniture – with the exception of the door desks – was purchased from garage sales or auction, or wherever they could find it. (Spector, 2000, p. 1471)

The same spirit of ingeniously overcoming constraints pervaded Amazon.com's work in almost every domain. One of the major constraints of the period was internet browser loading time and modem speed. Many dot.com and bricks-and-mortar companies that went online accepted the constraints, focusing their efforts on the look of the website. As a result, their sites looked great but took too long to load and use. Amazon.com looked at the problem differently. It recognized that website graphics and looks should take second place to functionality. What was important was to build an efficient website that loaded quickly and was as usable. As one book publisher put it:

> a lot of people were spending a lot of time and energy making sites that were hard to use because they had all kinds of fancy graphics. Amazon.com was so stripped down. They realized that what they were building was not a brochure; they were building an application. (Spector, 2000, p. 1162)

Constraints are not always technical or financial. Sometimes they are created by industry conventions that are entrenched by decades of use. In the early days, Amazon.com, like most start-ups, wanted to conserve cash by only ordering from wholesalers the minimum number of books required. The major book wholesalers insisted on a ten-book minimum order. They would not budge, even when offered incentives. Bezos found an ingenious way around this constraint. All Amazon.com had to do was order the one book it needed, and add nine out-of-stock books which the wholesaler could not supply.

Many of the conceptual contributions to this volume emphasize the issue of constraints. For example, Aneesh Banerjee's chapter, 'The roots of organizational ingenuity: how do qualitatively superior ideas come about?', develops a theory of constraints, providing a framework for examining how constraints impact resource slack, bricolage and organizational slack. Cunha, Rego, Clegg, Neves, and Oliveira's chapter 'Unpacking the concept of organizational ingenuity: learning from scarcity' examines organizational ingenuity when time, resources, or affluent customers are missing. In their framework, improvisation, bricolage, and frugal innovation are organizational responses to this scarcity. Thus, scarcity triggers the need to adopt some ingenious approach to a given problem.

The empirical contributions to this volume also report the influence of resource constraints on ingenuity. For example, Stam, Wakkee, and Groenewegen, in their chapter 'Acting ingeniously: opportunity development through institutional work' examine the institutional environment of home nursing in the Netherlands. They study an entrepreneur who cleverly denounces and restructures institutional constraints. In this study, they utilize institutional work to highlight a case where there is a gap between the current institutional setting and an ingenious entrepreneurial solution. Judy Matthews's chapter 'Stimulating organizational ingenuity with design methods' examines a design method that stimulates organizational ingenuity, where constraints are used to create possibilities for new ways of working in a company that is a market leader. Tamar Sagiv's chapter suggests that constraints are not always external to the actors, a byproduct of the system in which they operate. In her study of the evolution of modern dance in Israel she argues that innovative choreographers and dancers are often constrained by their need to preserve and achieve authenticity in their art. The paradox that these choreographers and dancers confront is that, unlike classical ballet which is highly restrictive, modern dance vocabulary is quite open-ended and eclectic. It would be quite easy for these individuals to generate variety and claim differentiation, but their inner integrity constrains their creation of new styles. Sagiv's chapter points to an elusive factor in creative problem solving: the inner dialogue that often leads innovators to aspire to an ingenious solution even when an adequate solution is readily available.

Ana Cristina Siqueira, Sandra Mariano, Joysi Moraes, and Gregory Gorse explore a very different context than Sagiv, but nevertheless their chapter also looks at the influence of leadership style on ingenuity. Their chapter examines organizational ingenuity by addressing the context of Brazilian community banks providing microfinance in areas with severe economic constraints. The chapter provides a comparative view of two leadership styles that shape the level of impact of these microfinance institutions. They find that leaders' involvement in local communities and perception of social disparities can increase the likelihood of organizational ingenuity, and that leaders' organizational ingenuity can increase the likelihood of creative solutions with an impact at the community and national levels.

Ingenious solutions are often obvious after the fact, and for that reason they seem the least risky of the alternatives. In reality, until they prove their value, ingenious solutions are risky – probably more risky than conventional solutions. The risk of ingenious solutions is amplified if they require additional resources. David Methé's paper is concerned with how Japanese managers view risk in starting up new projects. Risk is perceived as inherent in the management task and process and all the managers believed that they had to deal with risk by employing strategies of action that would minimize it. Methé maintains that minimizing risk requires entrepreneurial ingenuity and takes the form of re-configuring the sources of risk to provide a solution. The ingenious solution is not a stand-alone action, but usually an ongoing configuring of resources used to solve an evolving problem that may pose a risk to the organization.

INDIVIDUAL INGENUITY

Organizational ingenuity is impossible without individual ingenuity. But arguably within organizations individual ingenuity will not be effective without organizational support. Many organizations do not always support innovation, or welcome it, but they nevertheless depend for their success on ingenious individuals that solve problems caused by the organization's own constraints and rigidities. This raises the interesting question of whether such individuals are ingenious to begin with, or become ingenious because they simply refuse to accept the status quo to which the rest of the organization has become accustomed.

Jeff Bezos and the story of Amazon.com suggest that the answer may be both. Growing up, Bezos would spend his summers on his grandfather's farm in Texas. His grandfather, a retired scientist and former regional manager of the Atomic Energy Commission, actively encouraged him to tinker with different projects both around the farm and in the adjoining garage. Bezos busied himself with projects as diverse as windmills, amateur radios, and hovercraft, using whatever resources he could scrounge. His mother referred to his self sufficiency in this way: 'One of the things that [Jeff] learned is that there really aren't any problems without solutions. Obstacles are only obstacles if you think they're obstacles. Otherwise, they're opportunities' (Spector, 2000, p. 5).

Many entrepreneurs avoid established organizations with their routines and suffocating structures. Many who are forced to seek employment in established organizations see a start-up as an escape from unbearable constraints. Jeff Bezos's corporate career demonstrates that there is no contradiction between entrepreneurial ability and corporate success. After graduating from Princeton in 1986 with summa cum laude in electrical engineering and computer science, Bezos took up employment at a start-up that managed international telecommunications for financial transactions. He next moved to Bankers Trust, where he became vice president of global fiduciary services, the youngest VP in the history of the Bank.

Bezos developed considerable experience in moving past conventional barriers when he worked at Bankers Trust. One of his early assignments was to computerize a bank client's investment performance. While up to the minute financial reporting is commonplace today, before the advent of the internet investors relied upon monthly and quarterly reports, carefully managed and produced by investment institutions. Many of the established bankers were uncomfortable with clients receiving up-to-date daily investment statistics. As Harvey Hirsch, his boss claimed:

> Jeff has a way of stripping away the extraneous and focusing on what's really important. He sees different ways of doing things and better ways of doing things. He told the naysayers 'I believe in this new technology and I'm going to show you how it's going to work' – and he did. At the end of the day, he proved them wrong. He has no trouble puncturing someone's balloon if he thinks that they're proposing to do something the wrong way or in an inappropriate way. (Spector, 2000, p. 14)

In their study of the development of flash smelting in the copper industry Janne M. Korhonen and Liisa Välikangas show how individuals can decisively change the innovation dynamics in an organization that is struggling to survive. Devastated by war, Finland had to pay reparations and cede large parts of its hydroelectric capacity to the Soviet Union. In this context, Eero Mäkinen, the CEO of Outokumpu, a small Finnish mining company, was determined to restore the firm's refining operations in spite of a ruinous hike in electricity rates which made such a project uneconomical. He empowered a small team of engineers, headed by Petri Bryk, to search for a smelting method that radically reduced energy demand. The team's ingenious design not only met the requirement laid out by Mäkinen, it also raised the possibility of developing a smelting process that was energy neutral, i.e. sufficient energy being generated during the smelting process to make it practically self-sustaining.

Outokumpu's process eventually became a world standard, and is today used by 60 per cent of all copper smelting. In their chapter, Korhonen and Välikangas weave together the multiple levels of analysis of how this came about. The research and experimentation that led to the development of flash smelting came from multiple sources, some dating as far back as the middle of the 19th century. But Outokumpu's success owed everything to Mäkinen's determination and Bryk's ingenuity. The context was not propitious, but the individuals that drove the innovation forward knew how to work the system to their advantage.

Whereas Korhonen and Välikangas examine the interaction between individuals and their organization, Masciarelli and Prencipe examine the interaction between individuals and their geographic regions. In their chapter titled 'Connecting regional ingenuity to firm innovation: the role of social capital', Masciarelli and Prencipe analyze the influence within a region of creative actors that possess the ability to create innovative solutions within structural constraints, i.e. regional ingenuity. They find that local social capital facilitates the exchange of information among actors and influences the relationship between regional ingenuity and firm innovation. In effect, they show that both individual and organizational attributes may be agglomerated in the case of social capital.

DEVELOPING ORGANIZATIONAL INGENUITY

Individual ingenuity may be the essential building block of organizational ingenuity. Many successful entrepreneurs are ingenious, but this does not translate into organizational ingenuity. In many instances, this can be traced to entrepreneurial temperament and approach to doing business. Many entrepreneurs see their organizations as a tool designed to accomplish their ideas. They are loath to delegate, and in general prefer subordinates who are enthusiastic when it comes to finding ways of implementing plans drafted at the top, provided these subordinates do not put forward creative ideas of their own. The result is ingenuity at the center, usually confined to the entrepreneur founder, and an organization that is conventional when it comes to solving problems.

Jeff Bezos and Amazon.com are an exception to this tendency. It is likely that Bezos's experience as a manager taught him the peril of centralizing creativity at the top. It is also probable he looked ahead and understood that developing creative teams is essential for survival in a dynamic environment. One of Bezos's first tasks when building Amazon.com was to identify good programmers. Rather than choosing individuals who had experience in business or retail software development, he instead chose programmers that were well known for being both creative and interdisciplinary. He thus demonstrated that he was more committed to hiring the best and the brightest, regardless of their previous experience (Spector, 2000).

Amazon employees were carefully selected not only to be highly creative and intelligent, but also to maintain innovation as the highest priority. All employees were interviewed by several Amazon employees, with the goal of raising standards with each entry (*Wall Street Journal*, 4 May 1999). Applicants were given on the spot puzzle type questions, and were carefully examined for a range of metrics that were an attempt to focus on hiring only the most intelligent staff: 'Every hire required a unanimous vote, and fantastically enough, Jeff sometimes allowed himself to be overruled by the majority. Given these stringent procedures, it now strikes me as amazing that we hired anyone at all' (Marcus, 2004, p. 42).

Organizational institutional characteristics, norms and structures are likely to be an important component leading to ingenuity. For example, Cox, Siqueira, and Shearer in their chapter 'Organizational ingenuity in the commercialization of early-stage technological innovations' provide a very interesting case of two institutions – one idea 'rich', and one idea 'poor', collaborating in order to advance opportunities for both. They report on a pilot test of an 'Ideation Team Model for IP commercialization', suggesting that specific institutional structures and procedures may be modified to increase ingenuity.

When innovation and novelty are important, ingenuity is a function of the field in which actors solve problems. In their chapter, Ninja Natalie Senf, Jochen Koch and Wasko Rothmann explore the institutional context of the field of haute cuisine in Germany. The authors claim that chefs are taking an active role in shaping the context in which they work. This spills over to how they develop creative ideas. By resorting to ingenuity, chefs are able to produce an innovative and creative kitchen through working on an authentic and flexible profile that expresses the mannerisms and characteristics of haute cuisine.

Companies such as Amazon.com are not democracies: ultimately Jeff Bezos and outside investors have financial control. From the outset, however, Bezos understood that allowing financial control to become managerial control can easily undermine organizational ingenuity. An early employee of Amazon.com put it this way: 'In most companies, having the boss join your group would have prompted an orgy of shameless sycophancy. Here he was just one of us.' (Marcus, 2004, p. 52).

Bezos may be willing to be overruled at times, and he works hard to bridge the gap between boss and employees, but this does not mean he eschews decisive leadership when this is required. In fact, one could argue that staying close to his employees and gaining his trust gives him exceptional insight into their mood and how they see the obstacles they confront. When the young Amazon.com was confronting a move by Barnes and Noble into online book merchandizing, Bezos called together his 125 employees to a meeting to calm their fears. Yes, he told his employees, Amazon.com may be small next to Barnes and Noble with its 30 000 employees and $3 billion in sales, but your thoughts should not be on how Barnes and Noble or other competitors can hurt us, but how our customers see us:

> We can't be thinking about how Barnes and Noble has so much more in the way of resources than we do. I told everyone, Yes, you should wake up every morning terrified with your sheets drenched in sweat, but not because you're

> afraid of our competitors. Be afraid of our customers, because those are the folks who have the money. Our competitors are never going to send us money. (Kirby and Stewart, 2007, p. 80)

Ingenious solutions often come from seeing problems differently than others who are tackling the same problem. For Bezos the problem is not keeping Barnes and Noble at bay, but getting customers to fork over their hard won cash. In his chapter titled 'Ingenuity spirals and corporate environmental sustainability', David Zoogah argues that ingenious solutions to environmental problems usually begin with seeing the problem clearly in contexts where perception is often stymied by complexity. But seeing the problem is only the first step. For Zoogah ingenuity is a process that is centered on the problem but expands to mobilize resources during analysis and solution. To develop corporate sustainability it is important for the organization to engage in ingenious analysis, to turn 'the table upside down, looking in crevices that would typically not be looked into'. In effect, for Zoogah (and arguably for Bezos) organizations should not be satisfied with ingenious solutions, but should aspire to ingenious problem solving processes that defeat the routines and biases that normally constrain individuals and organizations from creatively solving problems.

CONCLUSION

As our technological, political, and economic systems become increasingly more complex and interconnected, the need for institutional ingenuity increases (Homer-Dixon, 2002). However, ingenuity is not always a generic solution to market or human problems. As the various contributions to this book show, there is no general formula when it comes to ingenuity. The way that organizational members deal with resource constraints, and whether this leads to ingenuity, is ultimately context sensitive. Accordingly, we propose that ingenuity is an intentional event or action through which an organization's members are able to enact effective problem solving techniques that are viewed as positive and important by others in the organization, and often beyond, because they are surprising and non-conventional. Ingenuity is thus an interplay between those who come up with creative solutions and the organization in which he or she works.

As we have shown in our Amazon case, an important contextual factor that influences organizational ingenuity is what some have described as the organization's 'DNA' characteristics, including its values, practices, routines, knowledge, and their managerial blueprint or human relations practices (Burton et al., 2002). In this vein, the organization's history is also an important factor in creating a culture of ingenuity. Stinchcombe's (1965) imprinting hypothesis (see also Marquis and Tilscik, 2013) claims that the environment and its characteristics, such as social relations, institutions or legal systems, 'are historically contingent and imprint an organization with the characteristics of the era when it was founded' (Stinchcombe, 1965, p. 142). In other words, the initial conditions at the time of founding, which include the technological, social, political, economic and cultural characteristics, exert enduring effects on future organizational development. Specifically, the imprinting hypothesis encompasses two major aspects that may help us to understand the prevalence of organizational ingenuity. First, elements which form the initial conditions of founding are resilient and remain in the organization during its life cycle. Numerous studies demonstrate the resilience of initial conditions. Burton and Beckman (2007) and Baron and Hannan (2005), using data from the Stanford Project on Emerging Companies (SPEC), examining firm history, demonstrate how initial conditions constrain subsequent organizational outcomes, such as turnover rate of successors who later occupy certain positions, or the composition and the preferred background of the top management team. This implies that values and practices, including those associated with identifying and implementing ingenious actions, are embedded in the organization.

Second, organizational elements and characteristics may be selectively reproduced across time within the organization (Johnson, 2007). Thus, the elements of the initial conditions of founding members may also shape organizational characteristics that could enhance a culture of ingenuity, serving as 'blueprints' of human resources practices (Baron et al., 1999) or cultural policies (Johnson, 2007). Consequently, we can explain activation and prospects of organizational ingenuity by the nature and formation of the organization's core values and practices. Thus, ingenuity is not always an ad-hoc opportunity of exploitation providing surprise and effective solution against obvious odds, but may also reflect deeply ingrained norms in the organizational history. At the same time, the initial founding conditions constrain subsequent capabilities or outcomes, including the propensity for organizational ingenuity.

Although the legacy of organizations exerts lasting effect, the changing context and environment may constrain or encourage organizational ingenuity. Furthermore, when practices and values are perceived as worthy, their preservation serves as 'templates' for the new members (DiMaggio and Powell, 1983). In this way, ingenious members of organizations may engage in practices that encourage creative solutions under resource constraints (Lampel et al., 2011). Note that ingenious solutions could take different forms and actions, including creating, maintaining or disrupting an existing order (Lawrence et al., 2009). Thus, ingenuity relates to the way the organization is reconstructing experience through designing and implementing non-conventional solutions.

REFERENCES

Baron, J.N., M.D. Burton, and M.T. Hannan (1999), 'Building the iron cage: determinants of managerial intensity in the early years of organizations', *American Sociological Review*, **64**(4), 527–47.

Baron, J.N. and M.T. Hannan (2005), 'The economic sociology of organizational entrepreneurship: lessons from the Stanford project on emerging companies', in V. Nee and R. Swedberg (eds), *The Economic Sociology of Capitalism*, Princeton University Press, pp. 168–203.

Burton, M.D. and C.M. Beckman (2007), 'Leaving a legacy: position imprints and successor turnover in young firms', *American Sociological Review*, **72**, 239–67.

Burton, M.D., J.B. Sorensen and C.M. Beckman (2002), 'Coming from good stock: career histories and new venture formation', in Michael Lounsbury and Marc J. Ventresca (eds), *Research in the Sociology of Organizations*, Oxford: Elsevier, pp. 229–62.

DiMaggio, P.J. and W.W. Powell (1983), 'The iron cage revisited: institutional isomorphism and the collective rationality in organizational fields', *American Sociological Review*, **48**(2), 147–60.

Homer-Dixon, T. (2002), *The Ingenuity Gap: Facing the Economic, Environmental, and Other Challenges of an Increasingly Complex and Unpredictable Future*. New York: Vintage.

Johnson, V. (2007), 'What is organizational imprinting? Cultural entrepreneurship in the founding of Paris opera', *American Journal of Sociology*, **113**(1), 97–127.

Kirby, J. and T. Stewart (2007), 'The HBR interview; Jeff Bezos: the institutional yes', *Harvard Business Review*, **85**(10), 74–82.

Lampel, J., B. Honig and I. Drori (2011), 'Discovering creativity in necessity: organizational ingenuity under institutional constraints', *Organization Studies*, **32**, 584.

Lawrence, T., R. Suddaby and B. Leca (eds) (2009), *Institutional Work*, Cambridge: Cambridge University Press.

Marcus, J. (2004), *Amazonia: Five Years at the Epicenter of the dot.com Juggernaut*. New York: The New Press.

Marquis, C. and A. Tilcsik (2013), 'Imprinting: toward a multilevel theory', *The Academy of Management Annals*, **7**(1), 193–243.

Spector, R. (2000), *Amazon.com: Get Big Fast*. New York: HarperInformation.

Stinchcombe, A.L. (1965), 'Social structure and organization', in J.G. March (ed.), *Handbook of Organizations*, Chicago, IL: Rand McNally, pp. 142–93.

Wall Street Journal (2000), 'Boss talk: taming the out-of-control in-box', February 4, p. B1.

PART I

UNDERSTANDING INGENUITY

1 The roots of organizational ingenuity: how do qualitatively superior ideas come about?

Aneesh Banerjee

INTRODUCTION

The ability to continuously innovate has long been cited as the cornerstone of growth and regeneration (Burgelman, 1983). It is also clear that some firms are able to innovate better than others, for example larger firms (Damanpour, 1992), new entrants (Foster, 1986), or firms with access to greater resources and capabilities (Methe et al., 1997). Furthermore there have been studies to understand the impact of internal organizational processes on innovation, for example researchers have studied the impact of culture (Tellis et al., 2009), leadership (Elenkov et al., 2005), learning (Ahuja and Lampert, 2001) and knowledge (Zhang and Li, 2010) on innovation.

In contrast to the rich research on these perspectives, little research exists that examines the relationship between constraints and innovation, especially the qualitative nature of the outcome. Why are some organizations able to create innovative solutions within structural constraints using limited resources and imaginative problem solving (Lampel et al., 2011), while others succumb to sub-optimal performance? It would seem that the organizations that make a break from the sub-optimal are often characterized by a 'can-do' ambition in the face of these constraints coupled with 'out-of-box' thinking. It is this rare combination that delivers transformational leaps rather than incremental progress. At the heart of this combination are ingenious ideas that represent a qualitative jump in the nature of the solution and have often changed the course of industries not just the organizations that show ingenuity. These ideas result in a quantum leap in the design of products, services or business models and become the new standard. Consider the history of the evolution of jet engines, which reveals many episodes where ingenious ideas had a profound impact on shaping the industry.

Engine Overheating

> Towards the end of the Second World War the Germans and the Americans were competing to build more powerful jet engines for their aircrafts. The problem facing both parties was material fatigue due to overheating. The Americans, under relatively fewer resource constraints, pursed a significantly costly approach to develop a more durable alloy for the engine. This was targeted at solving the problem of material fatigue. While the Germans, under significant financial constraints and restrictions relating to trade of metals, pursued a cost effective approach to solve the overheating problem by designing a more efficient cooling system. The war ended before the Germans could implement their design, but the 'cooling' solution was so elegant that it is still used in jet engines. (Adapted from Gibbert and Scranton, 2009)

For us, ingenuity is reactive creativity in problem solving. It is reactive to the constraints that do not allow the pursuit of linear problem solving methods. Ingenious solutions are also specific to a problem and to that extent their impact cannot be compared across cases. For example, even a small problem can be solved ingeniously.

Ice on the Nose of the Engine

> As jet engines became more powerful and could now take aircrafts to higher altitudes, it was found that the nose of a jet engine often froze, which led to the accumulation of ice on its tip. Sometimes chunks of ice would break free from the tip and crash into the blades of the engine. Even though these hits were not critical, they still had a long-term effect on the maintenance and longevity of the blades.
>
> When this problem was put to the engineers at a reputed aeronautical firm, the solution was obvious. The jet engine produces tremendous heat and so they needed to find a way to divert some of this heat back to the nose, i.e. remove the ice by heating or melting it. The 'heating' solution required some design modifications and would have been expensive to implement but nevertheless met the required objective. However, working under cost constraints, an alternate solution was proposed. It was to use a rubber tip on the nose which would not allow ice to form on it. The 'rubber nose' solution was significantly cheaper and also qualitatively superior to the other option. All jet engines now have a rubber tip on the nose to prevent the formation of ice on it. (Adapted from Dewulf and Baillie, 1999)

These examples show that some organizations are able to create innovative solutions within structural constraints using limited resources and imaginative problem solving (Lampel et al., 2011). The examples also show that solutions that were triggered by the constraints were in fact qualitatively better than the ones designed without constraints. Constraints play a central role in triggering the generation of these ingenious solutions and we define them as any inhibiting condition that is beyond the control of the organization (Campbell and Pritchard, 1976, p. 65). It is always easy to notice an ingenious solution with the benefit of hindsight, but can we find some common characteristics of all ingenious solutions?

CHARACTERISTICS OF INGENIOUS SOLUTIONS

Before we turn our attention to reviewing the theoretical frames, it is worth defining what makes a solution ingenious. Based on examples of ingenuity, we identify the conditions that a solution must satisfy to be ingenious.

Let us assume a case where a problem P has n known solutions, which belong to the stock of solutions $S_{known} = \{S_1, S_2, S_3,..,S_n\}$ and that the solutions can be ordered by the amount of resources needed to apply them in order to solve P. However none of solutions from S_{known} can be applied due to structural constraints or non-availability of resources, then a new solution S_i would be an *ingenious solution* if it satisfies the following conditions:

a. Novelty condition: $S_i \notin S_{known}$
b. Efficiency condition: S_i consumes significantly less resources compared to the known solutions S_{known}
c. Specific solution condition: S_i is generated to solve a particular predefined problem P
d. Usefulness condition: S_i can be applied to solve P within the structural constraints and available resources
e. Dominant solution condition: Over time S_i becomes the dominantly accepted solution to the problem P

When we apply the above conditions on the two examples of ingenuity, we find that both the cooling solution to solve the engine overheating problem and the rubber nose solution to prevent the formation of ice on the nose of the engine meet the criteria of being ingenious solutions.

Cases of ingenuity are clearly a subset of the literature on innovation. The Schumpeterian view of innovation (i.e. a new good or a new quality of a good; a new method of production; a new market; a new source of supply or a new organizational structure (Scherer, 1986)) is largely accepted as the foundation of innovation studies. From this definition it is evident that all ingenious solutions are also innovations, but not all innovations are ingenious. From a process perspective, what sets apart an ingenious solution from any novel and useful innovation is that structural constraints exist and in fact are at the heart of triggering the generation of the solutions. From the perspective of the outcome, we find that ingenious solutions adapt to structural constraints and are qualitatively superior to other solutions, thereby often setting a new industry standard. Therefore, within the innovation literature, it is useful to look for relevant theoretical frameworks that may explain the emergence of ingenious solutions.

In our search for answers, we review three frameworks that address innovation under constraints. We start with the theory of constraints, which proposes ways to optimize a solution under constraints. We then review the literature on organizational slack, which studies the relationship between the availability of resources and the generation of new ideas or performance, and conclude with a review of bricolage, which studies value creation under resource constraints. We argue that none of these frameworks adequately explain ingenuity and propose that researchers need to pay closer attention to the aspirations behind the pursuit of ingenious solutions. In particular we focus on the conceptualization of aspirations in the behavioural theory of the firm and propose an interaction effect with constraints. Our contribution in this chapter to the emerging field of ingenuity is to propose the interaction of high aspirations and constraints. To further that agenda we conclude by proposing future directions of research for scholars studying ingenuity.

THREE FRAMEWORKS DEALING WITH INNOVATION UNDER CONSTRAINTS

Theory of Constraints

Pioneered by E.M. Goldratt (Goldratt, 1987) in the 1980s, the theory of constraints looks at how constraints limit the ability of achieving higher levels of performance relative to the goal (Aryanezhad et al., 2010). Rooted in the operations management literature, it builds on the principles of continuous improvement, but its point of departure from such theories is that it takes a systems perspective (Dettmer, 1997). For instance, a standard continuous improvement methodology would prescribe that all components of a process must be optimized to their full potential to achieve the best performance, whereas the theory of constraints would highlight the interdependence of the processes and their links with constraints to prescribe ways to exploit constraints, i.e. get the most out of the system as a whole under constraints. Goldratt's central premise is that organizations exist as systems of interacting and not independent processes.

The theory of constraints is not so much a management theory devised to explain creative problem solving under constraints as a theory aimed at the optimization of a solution in an iterative process. As Dettmer (1997) notes, it is a collection of 'system principles and tools, or methods for solving the problem of improving overall system performance' (p. xxi). Since its introduction, the theory has been steadily enriched by a wide range of tools and techniques applicable in diverse settings, from accounting to operations research. The theory is fairly broad in its consideration of constraints like equipment, people and policy.

Its main limitation with regards to studying the creation of ingenious solutions is that it does not provide a theoretical basis to understand how the solutions arise in the first place. In our definition of ingenious solutions, the theory of constraints does not explain the first condition i.e. the novelty condition. For example in the case of the 'cooling' solution a theory of constraints approach can help optimize the solution but does not comment on the triggers to creative solution ideation.

Moving away from an 'optimization' perspective to understand if the lack of resources can 'trigger' the creative problem solving process, we turn our attention to two other concepts, namely organizational slack and bricolage. In both these concepts the generation of novel solutions is central to the argument and therefore meets our first criterion, the novelty condition.

Organizational Slack

Researchers have often used organizational slack to understand the effect of availability of resources on innovation. Nohria and Gulati (1996) define slack as resources that are in excess of the necessary minimum amount required to run the operations of a firm. While it is recognized that by nature slack resources can be diverted or redeployed for the achievement of organizational goals (George, 2005), scholars have also noted that some types of slack resources are more easily redeployed than others. Therefore, the slack construct is often studied as a contrast between slack that is easy to recover (i.e. high-discretion or unabsorbed slack) and slack that is not easy to recover (i.e. low-discretion or absorbed slack) (see Nohria and Gulati, 1996; Sharfman et al., 1988; Singh, 1986 for further details on the type of slack). In our discussion, we are interested to understand what may be the effect of availability of slack (i.e. no resource constraints) and non-availability of slack (i.e. resource constraints) on innovation outcomes.

It has been theorized that the presence of recoverable slack in an organization acts as a buffer. Scholars have argued that the presence of such a resource buffer can have a positive as well as a negative effect on performance outcomes. For instance, organizational theorists who draw parallels between the firm and an organism view the ultimate goal of organizations as survival and growth (Cyert and March, 1963; Salancik and Pfeffer, 1978). In that context, while the organization theorists recognize the cost of slack to the firm in the short term, they propose that it is necessary for the survival of the firm in the long term. They argue that the presence of slack resources buffers the core of the firm in times of distress (Cyert and March, 1963; Levinthal and March, 1981) and from environmental shocks (Meyer, 1982) thereby impacting the long-term performance. This 'buffer' also enables the firm to take risks and experiment with innovations. With more experimentation, it is argued, the firm is more likely to produce more innovation (Nohria and Gulati, 1996) and better performance (Bromiley, 1991; Singh, 1986). Consistent with organizational theory, the presence of slack facilitates multiple other functions in the organization (Tan and Peng, 2003) and therefore the definition of slack from an organization theory perspective as proposed by Bourgeois (1981, p. 30) is:

> A cushion of actual or potential resources which allow an organization to adapt successfully to internal pressures for adjustment or to external pressures for change in policy, as well as to initiate changes in strategy with respect to the external environment.

Therefore, according to organization theorists, slack translates into innovation outcomes because organizational actors inherently seek innovations. In contrast to this view, agency theorists consider the firm as a nexus of contracts between principals and agents (Fama, 1980). Therefore, agency theory explicitly rejects the notion of the firm as an organization and in the words of Davis and Stout (1992) turns the organization theory perspective 'upside down'. The agency argument is that managers acting as agents inherently have a set of goals that are not always aligned with the principal (for example, pursuit of power, prestige, money, and job security). Managers may use slack to engage in excessive diversification, empire-building, and on-the-job shirking (Tan and Peng, 2003). These agency theorists go on to claim that slack is in fact the source of the agent problem, i.e. firms are inefficient in allocating resources termed as 'X-inefficiency' (Leibenstein, 1980). From this perspective, the presence of excess slack resources has also been found to diminish competitiveness in organizations (Davis and Stout, 1992).

Building on this tension between organizational theory and agency theory, researchers have therefore proposed a curvilinear, i.e. an inverted U, relationship between organizational slack and innovation outcome (Nohria and Gulati, 1996).

To summarize the implications of slack, we find that while resources are necessary for innovation, too little or too many resources are not conducive to produce new solutions. Therefore, we can infer that depending on the level of slack in the organization, an increase in constraints may indeed improve performance. As a next step we should therefore have a closer look at how the variables slack and performance are operationalized in research. Table 1.1 summarizes the highly cited literature in this space and the corresponding variables used.

As we see from the table, the measurement of organization slack is often a combination of various financial metrics capturing different types of slack. For example, Bromiley's (1991) highly cited study takes three different combinations of financial metrics to define available, recoverable and potential slack as predictors of the firm's risk taking behaviour and performance measured in terms of financial performance as return on total assets (ROA), return on equity (ROE) and return on sales (ROS). For the purpose of our discussion it is important to note that the empirical evidence on the theory of slack considers the outcome variables as the financial return or the incidence of innovation over a period of time as opposed to the quality of each innovation or solution.

Going back to our conditions of ingenious solution, while slack can potentially explain the novelty condition, it does not shed light on the other conditions, most notably the efficiency condition. For example, in the case of jet engine development, we find that the theory of slack can explain the incidence of innovations, i.e. why innovation would drop with too few or too many resources. In 1971, while developing the next generation jet engines for Lockheed, Rolls-Royce went into bankruptcy due to the costs of the project. RB 221, the first three spool turbofan engine could only be completed after the British government nationalized the company and restructured it (divesting the automotive unit). The new nationalized company kept funding the RB 221 project, which on completion catapulted Rolls-Royce to the big league in the airline industry. But slack does not shed light on how some firms are able to make a qualitative leap in their solutions – only that it provides enough resources to ensure the completion of a project.

Bricolage

Another body of literature cited in this context of innovation under resource constraints is that of bricolage. Originally proposed by Claude Levi-Strauss in his seminal book in 1962, *La pensée sauvage* (English version published in 1966 as *The Savage Mind*) in the field of anthropology, it later gained popularity in management literature in various contexts like innovation research and organization theory (Duymedjian and Rüling, 2010). This emergence in management theory is fairly recent; as Boxenbaum and Rouleau (2011) note, of all the papers published in

Table 1.1 Theories and corresponding variable to understand the effect of slack

Primary theoretical lens	Study	Measure of slack	Measure of outcome variable	Outcome
Organization theory	Singh, 1986	Absorbed slack (selling, general, and administrative expenses and working capital) and unabsorbed slack (cash and securities)	Performance measured as a composite measure of financial performance and top executive subjective response to questionnaire	A high level of absorbed and unabsorbed slack is related to good performance
Organization theory	Hambrick and D'Aveni, 1988	Unabsorbed slack (equity-to-debt ratio and working capital as a percentage of sales)	Performance measured as financial bankruptcy	Bankrupt companies have substantially less slack than surviving companies
Organization theory	Bromiley, 1991	Available slack (current ratio), recoverable slack (selling, general, and administrative expenses divided by sales), and potential slack (debt-to-equity ratio)	Performance measures as return on total assets (ROA), return on equity (ROE) and return on sales (ROS)	Slack, particularly available and potential slack, increases performance
Organization theory	Miller and Leiblein, 1996	Recoverable slack (accounts receivable/sales, inventory/sales, and selling, general, administrative expenses/sales)	Performance as measured by ROA	Firm performance is strengthened by the presence of slack
Organization theory	Reuer and Leiblein, 2000	Recoverable slack (accounts receivable/sales, inventory/sales, and selling, general, administrative expenses/sales)	Downside risk is a probability weighted function of below-target performance outcomes. Performance measured by ROA and ROE	Slack is negatively related to firms' downside risk

Agency theory	Davis and Stout, 1992	Cash flow	Performance measured by ROE. To the extent that takeovers are meant to discipline underperforming firms, those that are earning higher returns should be subject to less risk of takeover	Greater cash flow increases the risk of being taken over
Inverted U relationship	Nohria and Gulati, 1996	A single composite measure of slack based on two questionnaire items	Performance measured by subjective responses from top executives.	There is an inverse U-shaped relationship between slack and innovation: both too little and too much slack may be detrimental to innovation
Prior performance	Greenley and Oktemgil, 1998	Generated slack (6 measures) Cash flow/investment, debt to equity, earnings before interest and taxes (EBIT)/interest cover, market to book value, current assets/current liabilities, sales per employee Invested slack (4 measures) Administration costs/sales, dividend pay-out, sales/total assets, working capital/sales	Performance (5 measures) Sales revenue, return on investment (ROI), return on net assets (RONA), ROS, ROE	A positive relationship between slack and performance exists only for high-performance firms; it does not exist for low-performance ones

the database ABI/INFORM between 1992 and 2009 with the keyword bricolage, 87 per cent were published after 2000.

Our interest is to understand how the concept of bricolage is used in the context of innovation. We find that there are two ways to look at this literature: from the perspective of the central actor called the *bricoleur* (for example, the entrepreneur or the artist), or from the process perspective (for example, resource mobilization). These two ways do intersect and it is at those points of intersection that we find examples of bricolage but our primary interest is more on the process side of the literature.

In Levi-Strauss's original conceptualization, the artisan or *bricoleur* plays a central role in bringing together seemingly redundant artefacts in order to compose something meaningful. Therefore, it is not surprising that many scholars, such as Miner et al. (2001) and Garud and Karnøe (2003), have highlighted the characteristics of the involved actors, most notably their resourcefulness and ability to improvise. Here the ability to improvise is used as a personal quality of the bricoleur. Improvisation as a process is a construct that is closely related to bricolage – but it does not necessarily imply bricolage. The distinction between improvisation and bricolage is an important one and will be made clearer from a process perspective.

Bricolage is understood as a process of resource mobilization when the usual resources to meet an objective are not available (Desa, 2012). Such resource mobilization can lead to novel solutions and entrepreneurial ventures as noted in Baker and Nelson's (2005) definition, 'making do by applying combinations of resources already at hand to new problems and opportunities' (p. 333). This view is close to 'improvisation' and therefore the two constructs have often been studied in close association (Moorman and Miner, 1998; Weick, 1998). However, as Baker notes 'while bricolage may imply improvisation, bricolage also occurs in the absence of improvisation, and that it is therefore important to recognize that they are separate constructs' (Baker, 2007, p. 698). We therefore briefly pursue the literature on improvisation to understand how it is different from bricolage.

In the study of improvisation, scholars like Weick (1993) and Baker et al. (2003) take an 'amalgamated' view in which there is significant overlap and spontaneity in the design and execution of novel activities. Furthermore, in the improvisation construct there is no separation between the creator and the interpreter, and between the design and production of the result (Weick, 1993). Therefore, in improvisation thinking and doing unfold simultaneously and is marked by retrospective sense making (Weick, 1996), implying that a way to reconcile these two constructs could be to think that 'bricolage may be a cause of improvisation' (Baker, 2007, p. 698). While improvisation consists of assembling elements based on simple rules in order to yield an original composition (Duymedjian and Rüling, 2010), it does not necessarily depend on resource constraints. For example, improvisation has been consistently observed in many creative disciplines where constraints are not observed, such as musical improvisation (Chase and Portney-Chase, 1988; Kernfeld, 1997), theatre (Knapp, 1985; Spolin, 1999), and sports (Bjurwill, 1993). In summary, we find that the central difference between bricolage and improvisation is that improvisation can happen without constraints whereas the central theme of bricolage is 'scavenging' for resources by the bricoleur.

Within the innovation literature, bricolage has been used in two contexts which are known to have resource constraints. First in the context of entrepreneurial ventures often termed as entrepreneurial bricolage (Baker and Nelson, 2005) and second to understand social ventures, termed as social bricolage (Di Domenico et al., 2010). Table 1.2 lists the central theme of some of the cited studies on bricolage.

So far we find that, while bricolage is a powerful theory, it however looks at value creation in general, i.e. creating something from nothing (Baker and Nelson, 2005), as opposed to a specific case of problem solving or comparison between solutions to solve the same problem. Therefore, while the theory of bricolage at best provides a basis to understand value creation in resource constrained environments (often termed as 'frugal innovation' or *Jugaad* – Radjou et al., 2012) it does not address goal oriented problem solving and the qualitative nature of one innovation outcome over other outcomes. Going back to the criteria of ingenious solutions that we defined earlier, bricolage is limited in explaining the dominant solution and specific solution conditions.

When we look at the examples of the cases presented at the beginning of the chapter through the bricolage lens, we find that while the element of creativity under resource constraints is common, ingenious innovations diverge from the example of bricolage in their qualitative superiority over existing solutions. For example the 'cooling' solution or the 'rubber nose' solution are not make-do solutions but clearly a quantum jump in the design: triggered by constraints, such that they have set the new standard in solving that particular problem.

From the brief review of the three frameworks that are commonly cited in the context of innovation under constraints (i.e. theory of constraints, organizational slack and bricolage) we found only a partial explanation for the emergence of qualitatively superior solutions. We found that the theory of constraints, rooted in production management, takes a systems perspective to optimize solutions in an iterative process. Its main

Table 1.2 Summary of cited studies on bricolage

Study	Focus	Sample	Measure of bricolage	Key finding
Garud and Karnøe, 2003	Bricolage as a process	Case study on the emergence of wind turbines in Denmark and the United States	As a process that could harness the inputs of distributed actors who are embedded in accumulating artefacts, tools, practices, rules and knowledge.	A process of bricolage has been more successful than a process aiming to generate 'breakthrough' innovation
Baker and Nelson, 2005	Bricolage as a process Entrepreneurial bricolage	29 resource-constrained firms	Refusal to enact the limitations imposed by dominant definitions of resource environments	Demonstrates the socially constructed nature of resource environments and the role of bricolage in this construction
Di Domenico et al., 2010	Bricolage as a process Social bricolage	8 social enterprises	Make do with resources at hand	Key element of successful social bricolage are make do, refusal to be constrained by limitations and improvisation
Desa, 2012	Bricolage as a process Social bricolage	202 technology social ventures from 45 countries	Reconfiguring existing resources to meet institutional demands	Social entrepreneurs who adopt a process of bricolage are better at succeeding in the face of institutional constraints
Boxenbaum and Rouleau, 2011	Bricolage as a process Theory building	Theoretical frame for epistemic scripts of knowledge production	Assembly of different knowledge elements that are readily available to the researcher to create new knowledge	
Banerjee and Campbell, 2009	Bricoleur characteristics	197 firms in Life Science Diagnostics	Inventor bricolage measured as the construction of technological capabilities through recombining the knowledge of inventors on hand to address opportunities.	Inventors with less assimilative capacity and more creative capacity in teams where there is relevant experience will promote inventor bricolage

limitation is that it focuses on optimization and does not comment on how the solutions come up in the first place. A review of the literature on organizational slack revealed that the focus has been on the quantitative aspects of the generation of solutions but it does not shed light on the qualitative comparison between solutions i.e. if one solution is better than another. Bricolage explains the process of scavenging for resources to create value from seemingly nothing, but once again does not address qualitatively superior problem solving and also how a solution becomes the industry standard to solve the problem.

As discussed in the introduction to this chapter, one of the common elements of the examples of ingenuity, apart from the existence of structural constraints, is the high level of aspiration of the teams working on the innovation project. These aspirations can be set by the strategic intent of the organization (Hamel and Prahalad, 1989), past performance or the performance of peers. While an assessment of the aspirations of the actors may have been out of the scope of these perspectives, it plays a central role in triggering ingenious solutions. We therefore propose to take a closer look at a possible interaction effect between high aspirations and constraints on the innovation outcome.

DO ASPIRATIONS CONDITION THE EFFECT OF CONSTRAINTS?

Since Cyert and March's publication of *A Behavioural Theory of the Firm* in 1963, it has had a profound impact on both organization theory and strategic management. A full-scale review of the behavioural theory of the firm would be beyond the scope of this chapter but for a detailed review of research on the behavioural theory of the firm see Argote and Greve (2007) and Gavetti et al. (2012). Our interest is in using the foundations of the behavioural theory of the firm to understand if there would be an interaction between high aspirations and constraints on the qualitative nature of the innovation outcome. More precisely, is there an effect on the search behaviour of such an organization?

We know from the central postulates that make up the cognitive foundations that individuals do not maximize, they *satisfice*. This implies that as the knowledge of all possible and optimum solutions is limited; decision makers (within the limited knowledge) are likely to seek a solution that meets their *aspirations* – and not necessarily the optimum solution. These aspirations are set as a result of past (historical) and social (peer) performances. Aspirations can also be set by the strategic intent (Hamel and Prahalad, 1989) of the organization.

Simultaneously, processes in an organization are governed by *rules or standard operating procedures* which are built on past experiences. However, if performance falls below aspirations, it triggers changes in these rules such as the organization's search behaviour. This leads to *problemistic search* – implying that the search for new solutions is motivated by the objective to achieve goals when performance falls below aspirations. As a result of the cycle of search and change, the *organization learns* by adapting the rules and procedures.

Our interest is focused on understanding if the behavioural theory can shed some light on how ingenuity is triggered. The central concepts that we would like to focus on are aspirations, defined as 'the smallest outcome that would be deemed satisfactory by the decision maker' (Schneider, 1992, p. 1053), and problemistic search – triggered when performance falls below aspirations. There is strong empirical evidence to show that firms indeed trigger problemistic search when performance falls below aspirations; for example, we know that firms tend to invest more in R&D (Antonelli, 1989; Hundley et al., 1996), as well as seek new R&D processes (Bolton, 1993), when their performance is below aspiration levels. Therefore, when performance falls below aspirations an organization triggers problemistic search processes.

Note that a behavioural theory based explanation of quality of innovation from problemistic search does not preclude the role of slack based search or from acquiring solutions available in the environment. Therefore, a firm may build a solution stock using slack resources or by acquiring knowledge from the environment (Greve, 2003).

In such a context we propose that the introduction of constraints would have a moderating effect on the applicability of the existing stock of solutions in meeting the performance targets. If the performance falls below the aspirations we expect one or a combination of the following:

a. Lowering of aspiration levels of the decision maker. Given the limitations in resources, the decision maker may satisfice by accepting that the current performance is the best possible under the circumstances and therefore adjusts the aspiration level. In the case of extreme resource constraints, the decision maker may simply 'give up'.
b. Generation of a novel solution that meets the performance expectations within the resource constraints. This approach builds on an interactive process (Slappendel, 1996), wherein innovations are created by the interaction of individual traits and structural influences. In this possibility the firm may be able to produce a solution that meets the conditions of being an ingenious solution.

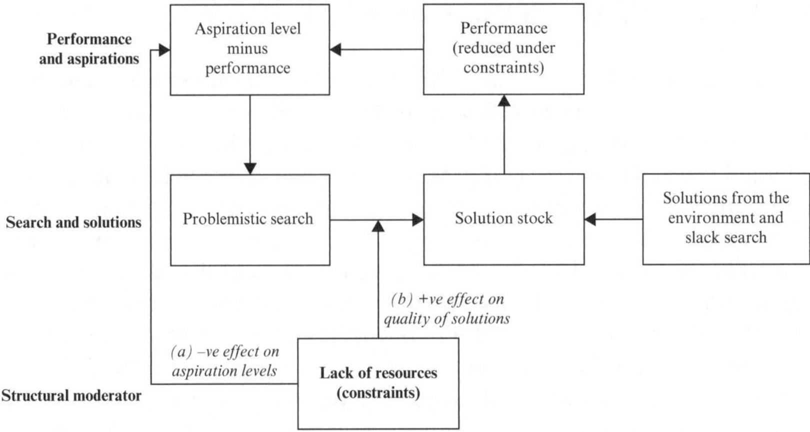

Source: Inspired by Greve (2003).

Figure 1.1 Theoretical model: proposed moderating effect of the lack of resources on the generation of the solution stock

This leads us to propose that:

When faced with resource constrains a firm either reduces aspiration levels and meets the lower performance level or a firm maintains high aspiration levels and pursues ingenious solutions that have high performance even under constraints.

Figure 1.1 shows the theoretical model for these effects.

Considering the conditions and examples of ingenuity that we had introduced earlier, we find that an explanation based on the cognitive foundations of the behavioural theory of the firm may provide a more comprehensive explanation of the qualitative nature of the outcome. For example, in the 'engine overheating' case, when the Germans and Americans were competing to build a stable jet engine, the presence of resource constraints on the German side forced them to search for solutions that would still work within their resource constraints, ultimately leading to a better quality solution. Whereas the Americans pursued what would appear to be a slack solution to solve the same problem. Similarly, in the 'ice on the nose' case, the solution only emerged as the engineers were forced to search for cost effective solutions. In either case, the teams could have settled to lower their aspiration levels and chosen a solution from the existing stock of solutions – instead they pursued the path of developing ingenious solutions.

FUTURE DIRECTIONS FOR RESEARCH

In our pursuit to find the roots of organizational ingenuity we found only partial explanation for the effect of structural constraints on the quality of innovation outcomes from the theory of constraints, bricolage and organizational slack. However, going back a step and introducing the moderating effect of constraints on an explanation based on the behavioural theory of the firm provides exciting opportunities for further investigation. This proposal clearly does not preclude the individualistic perspective of innovation which focuses on individual level constructs (Amabile et al., 1996; Scott and Bruce, 1994) but instead calls for multi-level research to understand the phenomenon.

Questions such as what influences the organization's ability to create qualitatively superior solutions within structural constraints using limited resources and imaginative problem solving can only be answered through creative and robust research across different dimensions of innovation (e.g. product/process, technological/administrative). Furthermore, research is needed to understand why only a few firms produce hugely influential products or services regularly (e.g. Apple, Dyson) or innovations in environments that by nature have structural constraints. It is important to highlight that structural constraints need not be limited to resources but can arise from other sources such as conditions set by a management team or culture.

While measuring innovation outcome, researchers have also shown a bias towards measuring productivity in terms of the commercial success of products or incidence of innovation. For example, number of patents filed has been the preferred measure of innovation output for many decades. However, this clearly does not capture the quality of the innovation. New variables that aim to capture the quality of innovation should be encouraged in mainstream innovation research. For example, recent studies by Joshi and Nerkar (2011) and Lahiri (2010) used a patent citation index to measure the quality of the innovation. While these variables are not error free they still enable us to derive a more balanced view of innovation quality.

And finally this enquiry into ingenious problem solving would not be complete without extending it to other units of analysis like teams and individuals that work under constraints. We need to develop a better understanding of how these units perceive and react to constraints. After all, 'creative' is an anagram of 'reactive', reactive to constraints.

ACKNOWLEDGEMENT

The author would like to thank Joseph Lampel and Ajay Bhalla for their inputs in developing the concepts discussed in the chapter.

REFERENCES

Ahuja, G. and Lampert, M.C. 2001. Entrepreneurship in the large corporation: a longitudinal study of how established firms create breakthrough inventions. *Strategic Management Journal*, **22**: 521–543.

Amabile, T.M., Conti, R., Coon, H., Lazenby, J. and Herron, M. 1996. Assessing the work environment for creativity. *Academy of Management Journal*, **39**(5): 1154–1184.

Antonelli, C. 1989. A failure-inducement model of research and development expenditure: Italian evidence from the early 1980s. *Journal of Economic Behavior & Organization*, **12**(2): 159–180.

Argote, L. and Greve, H.R. 2007. A behavioral theory of the firm – 40 years and counting: introduction and impact. *Organization Science*, **18**(3): 337–349.

Aryanezhad, M.B., Badri, S.A. and Komijan, A.R. 2010. Threshold-based method for elevating the system's constraint under theory of constraints. *International Journal of Production Research*, **48**(17): 5075–5087.

Baker, T. 2007. Resources in play: bricolage in the toy store(y). *Journal of Business Venturing*, **22**(5): 694–711.

Baker, T., Miner, A.S. and Eesley, D.T. 2003. Improvising firms: bricolage, account giving and improvisational competencies in the founding process. *Research Policy*, **32**(2): 255.

Baker, T. and Nelson, R.E. 2005. Creating something from nothing: resource construction through entrepreneurial bricolage. *Administrative Science Quarterly*, **50**(3): 329–366.

Banerjee, P.M. and Campbell, B.A. 2009. Inventor bricolage and firm technology research and development. *R&D Management*, **39**: 473–487.

Bjurwill, C. 1993. Read and react: the football formula. *Perceptual and Motor Skills*, **76**(3c): 1383–1386.

Bolton, M.K. 1993. Organizational innovation and substandard performance: when is necessity the mother of innovation? *Organization Science*, **4**(1): 57–75.

Bourgeois, L.J. 1981. On the measurement of organizational slack. *Academy of Management Review*, **6**(1): 29–39.

Boxenbaum, E. and Rouleau, L. 2011. New knowledge products as bricolage: metaphors and scripts in organizational theory. *Academy of Management Review*, **36**(2): 272–296.

Bromiley, P. 1991. Testing a causal model of corporate risk taking and performance. *Academy of Management Journal*, **34**(1): 37–59.

Burgelman, R.A. 1983. A process model of internal corporate venturing in the diversified major firm. *Administrative Science Quarterly*, **28**(2): 223–244.

Campbell, J.P. and Pritchard, R.D. 1976. Motivation theory in industrial and organizational psychology. In M. Dunnette (ed.), *Handbook of Industrial and Organizational Psychology*, Chicago: Rand McNally.

Chase, M.P. and Portney-Chase, M. 1988. *Improvisation: Music from the Inside Out*. Creative Arts Book Company.

Cyert, R.M. and March, J.G. 1963. *A Behavioral Theory of the Firm*. Englewood Cliffs, NJ: Prentice-Hall.

Damanpour, F. 1992. Organizational size and innovation. *Organization Studies*, **13**(3): 375–402.

Davis, G.F. and Stout, S.K. 1992. Organization theory and the market for corporate control: a dynamic analysis of the characteristics of large takeover targets, 1980–1990, *Administrative Science Quarterly*, **37**: 605–633.

Desa, G. 2012. Resource mobilization in international social entrepreneurship: bricolage as a mechanism of institutional transformation. *Entrepreneurship: Theory and Practice*, **36**(4): 727–751.

Dettmer, H.W. 1997. *Goldratt's Theory of Constraints: A Systems Approach to Continuous Improvement*. Milwaukee, WI: ASQ Quality Press.

Dewulf, S. and Baillie, C. 1999. *CASE: Creativity in Art, Science and Engineering: How to Foster Creativity*. London: Department for Education and Employment.

Di Domenico, M., Haugh, H., and Tracey, P. 2010. Social bricolage: theorizing social value creation in social enterprises. *Entrepreneurship: Theory and Practice*, **34**(4): 681–703.

Duymedjian, R. and Rüling, C-C. 2010. Towards a foundation of bricolage in organization and management theory. *Organization Studies*, **31**(2): 133–151.

Elenkov, D.S., Judge, W. and Wright, P. 2005. Strategic leadership and executive innovation influence: an international multi-cluster comparative study. *Strategic Management Journal*, **26**(7): 665–682.

Fama, E.F. 1980. Agency problems and the theory of the firm. *Journal of Political Economy*, **88**(2): 288–307.

Foster, R.N. 1986. *Innovation: The Attacker's Advantage*. New York: Summit Books.

Garud, R. and Karnøe, P. 2003. Bricolage versus breakthrough: distributed and embedded agency in technology entrepreneurship. *Research Policy*, **32**(2): 277–300.

Gavetti, G., Greve, H.R., Levinthal, D.A. and Ocasio, W. 2012. The behavioral theory of the firm: assessment and prospects. *Academy of Management Annals*, **6**(1): 1–40.

George, G. 2005. Slack resources and the performance of privately held firms. *Academy of Management Journal*, **48**(4): 661–676.

Gibbert, M. and Scranton, P. 2009. Constraints as sources of radical innovation? Insights from jet propulsion development. *Management & Organizational History*, **4**(4): 385–399.

Goldratt, E.M. 1987. *Essays on the Theory of Constraints*. Great Barrington, MA: North River Press.

Greenley, G.E. and Oktemgil, M. 1998. A comparison of slack resources in high and low performing British companies. *Journal of Management Studies*, **35**: 473–487.

Greve, H.R. 2003. A behavioral theory of R&D expenditures and innovations: evidence from shipbuilding. *Academy of Management Journal*, **46**(6): 685–702.

Hambrick, D.C. and D'Aveni, R.A. 1988. Large corporate failures as downward spirals. *Administrative Science Quarterly*, **33**(1): 1–23.

Hamel, G. and Prahalad, C.K. 1989. Strategic intent. *Harvard Business Review*, **67**(3): 63–78.

Hundley, G., Jacobson, C.K. and Park, S.H. 1996. Effects of profitability and liquidity on R&D intensity: Japanese and US companies compared. *Academy of Management Journal*, **39**(6): 1659–1674.

Joshi, A.M. and Nerkar, A. 2011. When do strategic alliances inhibit innovation by firms? Evidence from patent pools in the global optical disc industry. *Strategic Management Journal*, **32**(11): 1139–1160.

Kernfeld, B. 1997. *What to Listen for in Jazz*. New Haven, CT: Yale University Press.

Knapp, B.L. 1985. *Machine, Metaphor and the Writer*. Philadelphia: Penn State University Press.

Lahiri, N. 2010. Geographic distribution of R&D activity: how does it affect innovation quality? *Academy of Management Journal*, **53**(5): 1194–1209.

Lampel, J., Honig, B. and Drori, I. 2011. Discovering creativity in necessity: organizational ingenuity under institutional constraints. *Organization Studies*, **32**(4): 584–586.

Leibenstein, H. 1980. *Inflation, Income Distribution, and X-efficiency Theory: A Study Prepared for the International Labour Office Within the Framework of the World Employment Programme*. London: Croom Helm.

Levinthal, D. and March, J.G. 1981. A model of adaptive organizational search. *Journal of Economic Behavior & Organization*, **2**(4): 307–333.

Methe, D., Swaminathan, A., Mitchell, W. and Toyama, R. 1997. The underemphasized role of diversifying entrants and industry incumbents as the sources of major innovations. *Strategic Management Society: 15th International Conference*, pp. 99–116.

Meyer, A.D. 1982. Adapting to environmental jolts. *Administrative Science Quarterly*, **27**(4): 515–537.

Miller, K.D. and Leiblein, M.J. 1996. Corporate risk-return relations: returns variability versus downside risk. *Academy of Management Journal*, **39**(1): 91–122.

Miner, A.S., Bassoff, P. and Moorman, C. 2001. Organizational improvisation and learning: a field study. *Administrative Science Quarterly*, **46**(2): 304–337.

Moorman, C. and Miner, A.S. 1998. Organizational improvisation and organizational memory. *Academy of Management Review*, **23**(4): 698–723.

Nohria, N. and Gulati, R. 1996. Is slack good or bad for innovation? *Academy of Management Journal*, **39**(5): 1245–1264.

Radjou, N., Prabhu, J. and Ahuja, S. 2012. *Jugaad Innovation: Think Frugal, Be Flexible, Generate Breakthrough Growth*. San Francisco, CA: Jossey-Bass.

Reuer, J.J. and Leiblein, M.J. 2000. Downside risk implications of multinationality and international joint ventures. *Academy of Management Journal*, **43**(2): 203–214.

Salancik, G.R. and Pfeffer, J. 1978. A social information processing approach to job attitudes and task design. *Administrative Science Quarterly*, **23**: 224–253.

Scherer, F.M. 1986. The world productivity growth slump. In R. Wolff (ed.), *Organizing Industrial Development*, New York: Walter de Druyter, pp. 15–27.

Schneider, S.L. 1992. Framing and conflict: aspiration level contingency, the status quo, and current theories of risky choice. *Journal of Experimental Psychology: Learning, Memory, and Cognition*, **18**(5): 1040–1057.

Scott, S.G. and Bruce, R.A. 1994. Determinants of innovative behavior: a path model of individual innovation in the workplace. *Academy of Management Journal*, **37**(3): 580–607.

Sharfman, M.P., Wolf, G., Chase, R.B. and Tansik, D.A. 1988. Antecedents of organizational slack. *Academy of Management Review*, **13**(4): 601–614.

Singh, J.V. 1986. Performance, slack, and risk taking in organizational decision making. *Academy of Management Journal*, **29**: 562–585.

Slappendel, C. 1996. Perspectives on innovation in organization. *Organization Studies*, **17**(1): 107.

Spolin, V. 1999. *Improvisation for the Theater 3E: A Handbook of Teaching and Directing Techniques*. Evanston, IL: Northwestern University Press.

Tan, J. and Peng, M.W. 2003. Organizational slack and firm performance during economic transitions: two studies from an emerging economy. *Strategic Management Journal*, **24**: 1249–1263.

Tellis, G.J., Prabhu, J.C. and Chandy, R.K. 2009. Radical innovation across nations: the preeminence of corporate culture. *Journal of Marketing*, **73**(1): 3–23.

Weick, K.E. 1993. Organizational redesign as improvisation. In G.P. Huber and W.H. Glick (eds), *Organizational Change and Redesign: Ideas and Insights for Improving Performance*, New York: Oxford University Press, pp. 346–379.

Weick, K.E. 1996. Drop your tools: an allegory for organizational studies. *Administrative Science Quarterly*, **41**(2): 301–313.

Weick, K.E. 1998. Improvisation as a mindset for organizational analysis. *Organization Science*, **9**(5): 543–555.

Zhang, Y. and Li, H. 2010. Innovation search of new ventures in a technology cluster: the role of ties with service intermediaries. *Strategic Management Journal*, **31**(1): 88–109.

2 Unpacking the concept of organizational ingenuity: learning from scarcity

Miguel Pina e Cunha, Arménio Rego, Stewart Clegg, Pedro Neves and Pedro Oliveira

INTRODUCTION

Organizational theorists have recently become particularly interested in the processes and the management of organizational ingenuity (e.g., Harhoff and Hoisl, 2007). Ingenuity in organizations refers to the ability to create innovative solutions within the context of structural constraints, making use of those resources that are at hand, via imaginative problem solving (Lampel et al., 2011; Nalebuff and Ayres, 2003). In a world of scarcity, ingenuity may be a crucial organizational skill.

We present ingenuity, the capacity of showing skill or inventiveness, as an umbrella concept (Hirsch and Levin, 1999) that articulates several dimensions and processes that have previously been explored independently. We suggest that these distinct processes may be articulated because they share some features but require different managerial responses. We consider three potential constraints that stimulate ingenious organizational approaches (lack of three factors: time, resources, and affluent customers) as well as three ingenious responses to those forms of scarcity (improvisation, bricolage, and frugality). We extend previous work on the effects of scarcity as constraints for organizational action (Cunha et al., forthcoming) and unpack the concept of ingenuity by reference to constraints resulting from some form of scarcity that triggers the need to adopt some ingenious approach to a given problem.

In the face of scarcity, ingenuity is essential. The resource-based view, in both classical and modern versions (Penrose, 1959; Wernerfelt, 1984), argues that companies obtain competitive advantage only by cultivating and using unique resources. Resources are thus, to some extent, 'plastic' and malleable, in the sense that the same resources can render different service, depending on how ingeniously they are used. Different organizations will make different use of resources in conditions of scarcity according to this view, with some even able to use them to build dynamic capabilities when addressing weaknesses in the institutional and market environments in which they operate (Tashman and Marano, 2010).

Cultivated ingenuity can turn into a source of competitive advantage when it allows organizations to be speedier, more innovative, or affords greater sensitivity to changing environments and institutional fields.

With the above in mind, we ask three questions in this chapter. Our first question asks *what* is ingenuity in a world of scarcity? In this section, we contrast different assumptions about ingenuity. The second question asks *when* does ingenuity occur and *how* does it manifest itself? We contrast three forms or expressions of ingenuity that have been previously identified but not articulated: bricolage, improvisation, and the base of the pyramid (BOP). Finally we ask *why* does ingenuity matter for organizations? These questions are relevant for exploring forms of ingenuity and how they interact with their contexts and among themselves. These are relevant questions because ingenuity has mostly been approached obliquely via its manifestations. Less obliquely, ingenuity may be an umbrella concept, articulating processes that have been insufficiently theorized, awaiting rendering as something more explicit and open to investigation.

WHAT INGENUITY MEANS IN PRACTICE IN A WORLD OF SCARCITY

Scarcity is too complex a concept to be captured by any uni-dimensional approach. In this section we explore complexity by contrasting two perspectives: (1) scarcity as stubborn facticity versus scarcity as enactment; and (2) scarcity as threat versus scarcity as opportunity (see Figure 2.1). These perspectives, in turn, help to illuminate the part that scarcity plays in organizational ingenuity, which will be the theme of the following section. Explicitly questioning one's assumptions about the nature of resources is necessary to understand ingenuity because the way resources are enacted influences the way they are used: re-enacting resources frees them for ingeniously rendering new services.

The relation between scarcity, need and ingenuity should be problematized. Jared Diamond (1998) offers a relevant cue in a chapter of *Guns, Germs and Steel*, 'Necessity's mother'. Consider the following passage from that chapter:

> many or most inventions were developed by people driven by curiosity or by a love of tinkering, in the absence of any initial demand for the product they had in mind. Once a device had been invented, the inventor then had to find an application for it. (p. 242)

In other words, sometimes invention is the mother of necessity. As argued by Steve Jobs, 'customers don't know what they want until we've

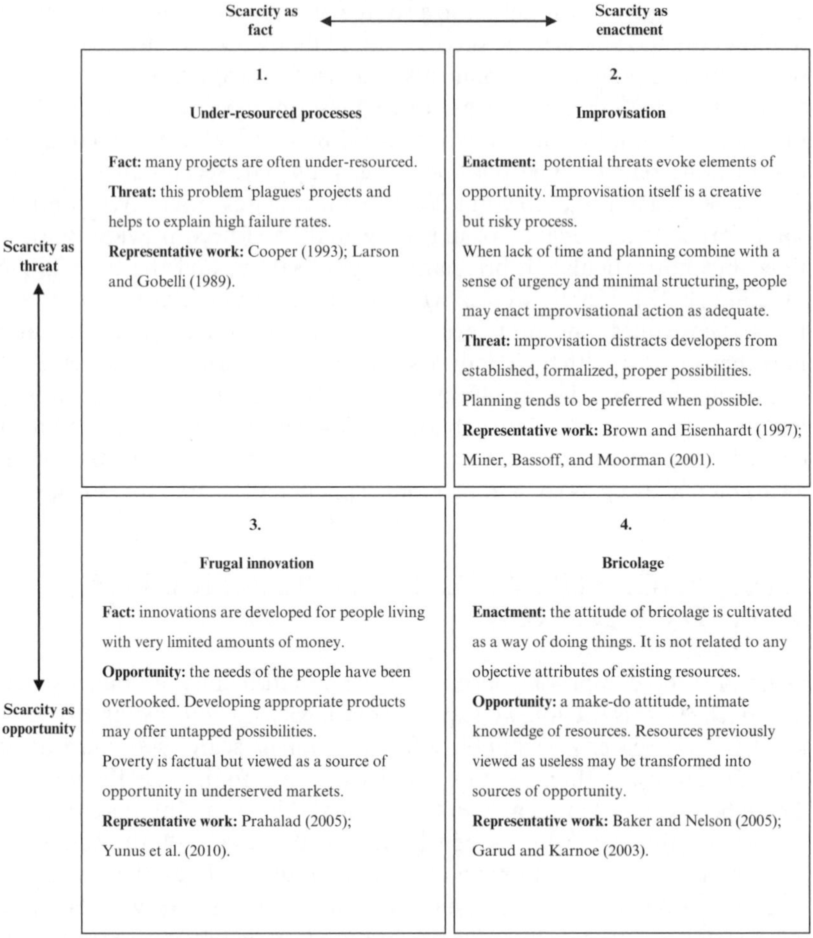

Scarcity as fact ⟷ **Scarcity as enactment**

Scarcity as threat ↕ **Scarcity as opportunity**

	Scarcity as fact	Scarcity as enactment
Scarcity as threat	**1.** **Under-resourced processes** **Fact:** many projects are often under-resourced. **Threat:** this problem 'plagues' projects and helps to explain high failure rates. **Representative work:** Cooper (1993); Larson and Gobelli (1989).	**2.** **Improvisation** **Enactment:** potential threats evoke elements of opportunity. Improvisation itself is a creative but risky process. When lack of time and planning combine with a sense of urgency and minimal structuring, people may enact improvisational action as adequate. **Threat:** improvisation distracts developers from established, formalized, proper possibilities. Planning tends to be preferred when possible. **Representative work:** Brown and Eisenhardt (1997); Miner, Bassoff, and Moorman (2001).
Scarcity as opportunity	**3.** **Frugal innovation** **Fact:** innovations are developed for people living with very limited amounts of money. **Opportunity:** the needs of the people have been overlooked. Developing appropriate products may offer untapped possibilities. Poverty is factual but viewed as a source of opportunity in underserved markets. **Representative work:** Prahalad (2005); Yunus et al. (2010).	**4.** **Bricolage** **Enactment:** the attitude of bricolage is cultivated as a way of doing things. It is not related to any objective attributes of existing resources. **Opportunity:** a make-do attitude, intimate knowledge of resources. Resources previously viewed as useless may be transformed into sources of opportunity. **Representative work:** Baker and Nelson (2005); Garud and Karnoe (2003).

Figure 2.1 Four ingenuity modes: perspectives on scarcity

shown them' (in Isaacson, 2012, p. 97). As Henry Ford once remarked, instead of automobiles initial market conditions were such that customers wanted better horses. The 'necessity leads to invention' perspective may thus be complemented with an 'invention leads to necessity perspective'. The role of ingenuity and resources in the two processes is fundamentally different.

In the 'invention leads to necessity perspective', lack of the 'adequate' resources may not constitute a constraint because resources may reveal unexpected valences when there is deep knowledge about them and an attitude of ingenuity. As pointed out by Ciborra (1996, p. 116), stable pools of 'junk', badly organized resources, may actually provide platforms for responding innovatively to surprising competitive moves. Tinkering with resources may lead to inventions that then have to find a need to fulfill. Research in management has systematically downplayed the role of tinkering in the development of new products and processes. New products and processes, however, can simply emerge out of experiments with resources. There are marginal references to the process in the innovation literature (e.g., Garud and Karnoe, 2003) but the bulk of research accepts the superiority of a mechanistic, engineering mode, where need is fulfilled by invention rather than the other way around. In a traditional marketing perspective, innovation is viewed as a process that is managed-cum-engineered to satisfy some identified need, whose development requires appropriate resources be properly executed.

A story of Australian innovation demonstrates this approach. Each year, after Year 12 exams are completed, hundreds of thousands of High School students leave for the coast to engage in the kind of pursuits that 17 and 18 year old teenagers, let off the lease, enjoy: sun, sand, surf, etc. 'Schoolies' generally occurs between November 24 and December 17. In the past, the major destinations were places such as Queensland's Gold Coast. A travel entrepreneur, founder of a company called Unleashed Travel, realized that:

> many domestic 'schoolies' offerings weren't offering commissions to travel agents. So he decided to research the possibility of taking teenagers overseas for schoolies week, selling the product through travel agents. He ensured that the agents could earn a commission, thus incentivising them to promote the product. (http://www.smh.com.au/small-business/managing/blogs/enterprise/its-party-time-for-schoolies-entrepreneur-20121025-286sn.html#ixzz2AMKAQ5lC)

The business model is simple: a resort is booked out for the period; trained staff are hired with a firm policy on acceptable behavior policed by application of rules and the schoolies have a ball, entirely with people like themselves, without the hassles that interaction with the broader public and authorities might occasion. It is now an AUD$6 million business after just four years. Ingenuity consisted of seeing 'schoolies' as a market and creating a resource-based experiment, which was taking the initial risk of booking out a resort in Fiji to see if the idea would work.

Perspective 1: Stubborn Facticity vs. Enactment

Some environments are objectively richer than others (Diamond, 1998). In this sense, scarcity is a fact. Such a matter of factual perspective is adopted, for example, by such influential theories as resource dependence (Pfeffer and Salancik, 1978) and population ecology (Hannan and Freeman, 1989). As argued by Baker and Nelson (2005), the population ecology perspective tends to assume that the nature of resources is relatively unproblematic or taken for granted. Consider the following passage of Hannan and Freeman (1989, p. 94):

> The larger environment fluctuates between two conditions. When resources are abundant, inconsistencies among the various demands, placed on these organizations are 'solved' by creating special pools of resources for each problem and encouraging the target organizations to create specialized programs for each demand. When resources are scarce, public organizations are required to justify programs and document the links between claims and practices. The strategy of institutional isomorphism may be a good one when resources are rising; but it may not be so good when resources decline.

In this view, resources are either scarce or abundant and organizational strategies should be adopted depending on the levels of resource munificence. The same view of resources as objective is present in Pfeffer and Salancik's (1978) resource dependence theory. For these authors 'the key to organizational survival is the ability to acquire and maintain resources' (p. 2) but resources are taken to consist of materials endowed with known characteristics and functionalities. The process of acquisition and maintenance is at the core of resource dependence but the nature of resources in this macro view is not viewed as problematic: resources are taken as objective facts of the organization's ecology. Baker and Nelson (2005, p. 331) describe this perspective as imagining markets to be marked by a 'stubborn facticity': a resource is a tangible something or other. However, interpretation of what are or are not resources has first to be collectively built, shared and then taken for granted.

An alternative interpretation suggests that there is a crucial element of enactment in the process of resource utilization, such that resources are better seen from a process perspective as a matter of resourcing. Travel Unleashed did not use a tangible resource – it enacted the use of tangible resources by networking disparate actors and actants: Schoolies as an institution; Year 12 graduates and their families who pay for the vacation; travel agents who were not gaining commissions from flats booked online; Fijian resorts whose facilities were under-utilized in a period outside of high season. Creating the actor network created the resource. Such organizations interpret their resources in ways that differ from generalized interpretations, triggering ingenious experimentations. In this case, organizations may discover possibilities that were not previously activated because of the way resources were interpreted. Managers may assume that resource pools can be enhanced (Heynoski and Quinn, 2012). Organizations such as Travel Unleashed deal with scarcity by changing 'poison', in the form of the often adverse press coverage of out of control teenagers on the Gold Coast into 'medicine', the total institutional and supervised resort experience, transforming a crisis experience 'into the development of new, innovative practices, products or services' (Clair and Dufresne, 2007, p. 68).

In a similar way, some entrepreneurs extract valuable services 'out of nothing', i.e., from dormant resources, materials that were viewed as useless by the majority (Baker and Nelson, 2005). In this enactment perspective, supported by a 'why not?' disposition (Nalebuff and Ayres, 2003), organizations may create interpretations about their resources that allow them to extract new services with ingenuity (e.g., Haier modified existing washing machines to allow poor, rural Chinese to use them for washing clothes *and* vegetables; Anderson and Markides, 2007). The range of potential services rendered by a resource should then be viewed as characterized by plastic possibility rather than by stubborn facticity. As Penrose (1959, p. 86) put it, 'no firm ever perceives the complete range of services available from any resource'. As a remark on the untapped potential of 'any' resource this suggests that the facticity perspective provides an important but incomplete understanding of the relationship between organizations and the resources they control and offers an opportunity to reframe the nature of organizational ingenuity.

Perspective 2: Threat vs. Opportunity

The representation of scarcity as a problem is present in many discussions of management and organizations. Inadequate resourcing will lead, for example, to inferior new product development (NPD) processes and therefore increases the probability of organizational failure:

> the degree to which sufficient resources have been made available to complete the project is likely to affect success. Projects with inadequate resources are likely to be doomed to begin with regardless of the project structure used. (Larson and Gobeli, 1989, p. 120)

Lack of resources comes in several forms. Sometimes they are induced through managerial practices: 'managers constantly indicated that they had done a rushed or sloppy job on many critical activities and that there was a need for more time, care, and effort' (Cooper, 1993, p. 37). Lack of resources 'plagues' (Cooper, 1993, p. 44) too many projects, leading to poor execution and provoking serious consequences for organizational processes. In another example, the scarcity of tantalum during the 1990s slowed the growth of information and communication technologies (Hilty, 2005).

Resource scarcity can be a problem but research suggests that abundance is not, in itself, a source of competitive advantage. On the contrary, it can lead to inertia and to a sense of invulnerability that may actually damage the organization's competitiveness, stimulating inefficiency and a 'fat and happy' attitude (Miller, 1990; Boulding and Staelin, 1993). As pointed out by Debruyne et al. (2010, p. 175), 'resources may sometimes be a double-edged sword, both stimulating and inhibiting competitive reaction'. In this sense, resource abundance or scarcity are not in and of themselves sources of competitive advantage or disadvantage. It is the way they are deployed that makes the difference, which means that companies should use the 'weapons' they have (Debruyne et al., 2010). Resource abundance may breed complacency as much as opportunity and resource scarcity can be debilitating or a source of profound lateral innovation that seeks to substitute for the scarcity of resources. Resource munificence may be a source of advantage by giving companies more muscle to invest in innovative activities (Daniel et al., 2004), while scarcity may be stimulating if it gives managers and employees the opportunity to re-examine frames of thinking, product portfolios and processes and routines. Scarcities can create preparedness and ingenuity for action, even in the face of resource constraints.

Japan introduced several miniaturization innovations (notably in the electronic and architecture/building industries) due to space scarcity (Karan, 2005). Japan has few natural energy resources, leading McMillan and Wesson (1998, p. 73) to state 'Japan's advantages (. . .) include a century's experience of managing resource scarcity'. The scarcity of land for waste disposal helps to explain why the Japanese cement industry now consumes approximately 6 per cent of the country's waste and why the nation reports the smallest energy consumption per ton of cement produced among developed countries (Hirose, 2004; Morikawa, 2000). Scarcity also explains why Japan responded more efficiently than the US and other countries to the oil shocks of the 1970s (through innovations, for example, in industrial energy efficiency and automobile fuel consumption; Ikenberry, 1986). Gladwell (2008) also argued that the lack of scale of Chinese rice cultivation centered on peasant smallholdings and the correspondent resource scarcity explains the attitudes developed during centuries by Chinese people, attitudes nourishing a culture of persistency that has positive consequences, even for present generations.

The case of India is particularly relevant in regard to the reconsideration of scarcity as either a threat or an opportunity. India is an obvious example of a penurious environment, where millions of people live at the base of the economic pyramid (Prahalad, 2005). Given an absence of fuel for cooking and toilet facilities in rural communities, dried human and animal excrement is used as a bio-fuel. The National Innovation Foundation of India recognizes the important role that people at the grassroots can play in developing new products and services and have created numerous activities aimed at unleashing the creative potential of innovators at grassroots level, as well as a database of about 10 000 innovations under various categories, including utilities and general machinery, farm implements, energy devices, agricultural and traditional knowledge practices, livestock management tools, herbal remedies, and biodiversity (Yadav, 2010). Companies from different fields have developed initiatives aimed at capturing the ingenuity expressed in such environments and eventually transferring it to developed countries.

In addition to rural practices this country is also the source of a number of challenges to Western management thinking. The management innovations produced in the subcontinent are so notoriously idiosyncratic that they have been presented under the notion of an 'India way' (Cappelli et al., 2010). One of the major lessons of the Indian way is that innovation pursued under conditions of scarcity requires new ways of thinking about innovation itself. It is not only a matter of improving the efficiency of existing models but rather the reinvention of innovation models for penurious environments. In other words, mindsets and models that were developed for contexts of relative resource abundance are not necessarily adequate for contexts of scarcity.

An example of this comes from the work of an unusual anthropologist, Jan Chipchase, who worked for seven years for Nokia as a 'human-behavior researcher' and 'user anthropologist' (Corbett, 2008), travelling the globe to explore how people used mobile devices. An issue that intrigued many producers of telecom equipment was the widespread use of mobile or cell phones in areas where there was no electricity to recharge batteries. In-depth inquiry revealed that users developed ingenious ways of charging their cell phones. Using a car battery they were able to build a cell phone charger that served large communities of cell phone users in areas that were not served by electrical power. Moreover, it is commonplace in African markets to see female entrepreneurs equipped only with a cell phone wired to a car battery that they rent to users to make calls more cheaply than with landlines, one of many examples of telecom products being ingeniously adapted in contexts of scarcity. William Kamkwamba, 'The [Malawian] boy who harnessed the wind' (Kamkwamba and Mealer, 2009), built an electricity-producing windmill from spare parts and scrap, in another emblematic demonstration of how scarcity is the mother of peculiar innovations – likely unthinkable within affluent contexts. According to Al Gore, former USA vice president and Nobel laureate, 'William Kamkwamba's achievements with wind energy should serve as a model of what one person, with an inspiring idea, can do to tackle the crisis we face' (Kamkwamba and Mealer, 2009, back cover).

Summary

There is more in resource-poor environments than problems and threats. In the next section, we explore three ingenious responses to scarcity constraints. Before that, we briefly summarize the main aspects of the previous discussion that favor ingenious organizational responses. First, some environments are objectively more penurious than others but different organizations may react distinctly to penury by enacting the value of existing resources in different ways. For some, penury may be a justification to inaction. For others, it is a context with specific and idiosyncratic needs that require ingenious organizational responses. Second, for the former, scarcity is more likely to be represented as a threat, whereas for the latter it is another difficulty that needs to be faced. Finally, organization theories tend to see needs as the origins of new products or services: products are created to respond to needs. But in some cases, however, the product results from familiarity with resources, including non-valuable ones. Familiarity with resources combined with cultivated ingenuity may lead to organizational responses that will potentially satisfy some need.

WHEN AND HOW IS INGENUITY PRACTISED?

Lack of Resources, Lack of Time, and Lack of Affluent Customers as Sources of Ingenuity

In this section we ask: when does ingenuity occur? How does it occur? To answer this question, we distinguish three different factors that lead to ingenuity: lack of resources, lack of time, and lack of affluent customers. These are relevant because they play important roles in organizational contexts. Time has been presented as a critical vector of competitive advantage (Stalk and Hout, 1990), and insufficient access to customers and the resources they provide are potentially lethal for an organization's survival (Hannan and Freeman, 1989; Pfeffer and Salancik, 1978). Organizations, however, may counter these obstacles via bricolage, improvisation and frugal innovation (for a brief consideration of the three streams, see Table 2.1).

Table 2.1 Comparing three literature streams

Concepts	Improvisation	Bricolage	Frugal innovation
Key idea	Success via rapid response to unexpected events	Success via familiarity with resources	Success via ultra-low cost innovation
Definition	Convergence of planning with execution drawing on the available resources t = time	Creative bundling of scarce resources r = resources	Application of a clean sheet approach to organizational problems q = quanity
Critical scarce resource	Time	Material stuff	Affluent customers
Theory	Metaphor-based (jazz, performing arts)	Anthropology-based (Levi Strauss, in particular)	Mostly a-theoretical
Research methods	Inductive	Inductive	Anecdotal evidence, case studies
Ingenuity skills	Capacity to work without plans around a minimal structure	Imagining new uses for known resources	Stripping products of features formerly viewed as critical
Market	Fast, unpredictable	'Red oceans' with strong competition for resources	Un-served
Performance (dependent variable)	Speed of reaction	'Resource-less' innovations	Efficiency
Managerial implications	Cultivating minimal structures, a culture of proactivity, an action orientation and an aesthetics of imperfection	Hands-on approach, familiarity with resources, resourcing	An appreciation for frugality, a 'less is more' attitude, a sustainability mindset
Limitations	Improvisation may produce drift and lack of response consistency	Imperfect resources may produce imperfect solutions	Low cost may be perceived as low quality
Representative authors	Brown and Eisenhardt (1997); Moorman and Miner (1998)	Baker and Nelson (2005); Ciborra (1996)	Hart and Christensen (2002); Prahalad (2005)

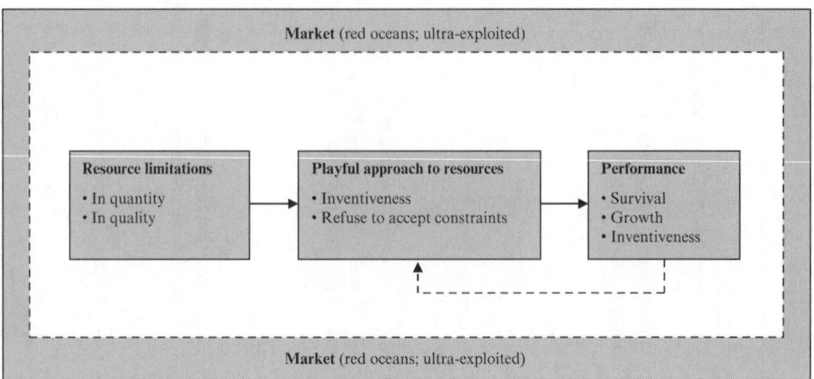

Figure 2.2 The process of bricolage

Bricolage: When Resources are Imperfect

Bricolage refers to the 'creative bundling of scarce resources' (Halme et al., 2012, p. 1), the ingenious application of existing resources to new problems or opportunities (Baker, 2007). Bricoleurs 'play with possibilities' and use available resources in search of solutions (Nohria and Berkley, 1994, p. 133). Bricolage refers to doing what has to be done with the resources available at a given moment. These are unlikely to be 'optimal'. If resource allocation involves the creation and distribution of adequate resources, bricolage is about making the best out of the available, often simple and general, resources (see Duymedjian and Ruling, 2010, for a discussion of bricolage in organization theory, and Figure 2.2 for an overview of the process).

The activities of bricolage occur when people ingeniously explore or exploit existing resources in novel ways, not necessarily under time pressure, as in the case of improvisation. As Baker and Nelson (2005) pointed out, a bricolage approach involves a bias against received limitations. In bricolage it is the accumulation of familiarity with resources that makes possible the revelation of potential services waiting to be extracted. Ingenious acts of bricolage can have evolutionary potential, developing end products that can be effective and diverse (Duboule and Wilkins, 1998). Companies with limited resources in red ocean hypercompetitive markets need to extract new rents from the resources at hand (Kim and Mauborgne, 2005). When inventiveness proceeding from a playful approach to resources is combined with a refusal to accept constraints as an inevitable condition, the result is the attitude of the bricoleur.

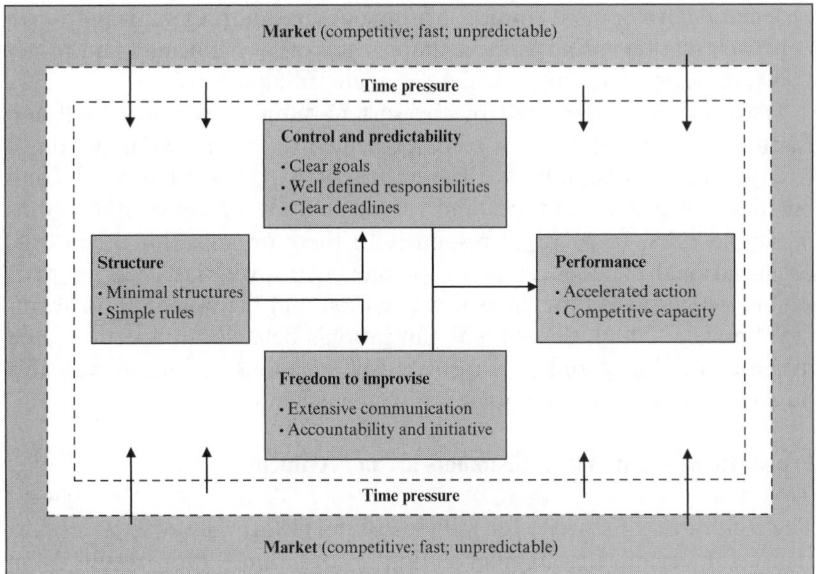

Figure 2.3 The process of improvisation

Improvisation: When Time is Scarce

Improvisation has been defined as the convergence of planning and execution, while action unfolds, drawing on available resources (Cunha et al., 1999). Improvisational ingenuity occurs when there is time pressure. In improvisation, planning and execution cannot be separated – they take place in tandem, often combining formal and informal conducts (Cunha et al., 2012). Such ingenuity consists of responding to opportunities, some of them serendipitous (Cunha et al., 2010), in the face of scarce time and possibly with less than adequate resources (Miner et al., 2001). Figure 2.3 offers a general overview of the process.

Sometimes innovations result from the combination of a pressing urgency with a lack of a systematic plan. When such conditions occur, innovators have to improvise. Improvisation may thus be viewed as a form of short-term, real-time learning, where learning must happen before the outcomes of these improvisations are known (Miner et al., 2001).

When facing unexpected and relevant issues organizational members may thus need to improvise solutions, which means that they have to act deliberately and unexpectedly (Moorman and Miner, 1998) and resort to whatever materials they have, precisely because there is no time to search for more adequate resources, be they market studies, time to follow a systematic development routine, or some other factor. Organizations can approach markets using a planned approach or, as Moorman and Miner (1998) indicated, they may plan and execute simultaneously.

The process is supported by the sort of simple structures (Sull and Eisenhardt, 2012) that favor complex behaviors via the combination of freedom and accountability. These structures, designed around combinations of goals, deadlines and responsibilities, or some other forms of simple rules, favor rapid response. In these organizational contexts, organizational members respond to challenges ingeniously via improvisation, which saves time, increases attention and the potential for learning (Vendelo, 2009), flexes the organization's behavioral repertoire, and increases flexibility and a propensity for ingenious action, all valuable outcomes in fast-changing organizational environments.

Frugal Innovation: When Customers are not Affluent

The conditions of scarcity for millions of people are viewed as an opportunity for organizational intervention. Frugal innovation results from an overarching philosophy that enables a radical clean sheet approach to organizational processes. In this philosophy, cost discipline is a core element of the process but organizational frugality is more than cost cutting: it aims to avoid unnecessary costs (Sehgal et al., 2010).

Frugal innovation aims to respond with extreme efficiency to some essential need of BOP consumers. BOP is 'the largest remaining global market frontier for businesses' (Nakata and Weidner, 2012, p. 21). It represents 4 billion people that spend more than \$2 trillion each year, being the target of initiatives of companies such as Hewlett-Packard, Unilever, Ericsson, Cemex, Motorola, DuPont, Citigroup, Monsanto, Johnson and Johnson, Novartis, Philips, and Danone (Elaydi and Harrison, 2010; Gabel, 2004; Hart and Christensen, 2002; Tashman and Marano, 2010; Yunus et al., 2010).

Rather than departing from existing products to adapt them as low cost versions, frugal innovation aims to provide the essential functions people try to satisfy with a given product. In this sense, because scarcity is a fact on the demand side, some companies have decided to approach their markets with a clean sheet approach, closely observing the behavior of their clients, identifying their needs, and responding with new, ingenious solutions. Successful innovation requires taking into account four product or service factors (Anderson and Markides, 2007): affordability, acceptability, availability, and awareness.

The process of frugal innovation departs from four conditions, as represented in Figure 2.4: (1) appreciating the high growth potential of the

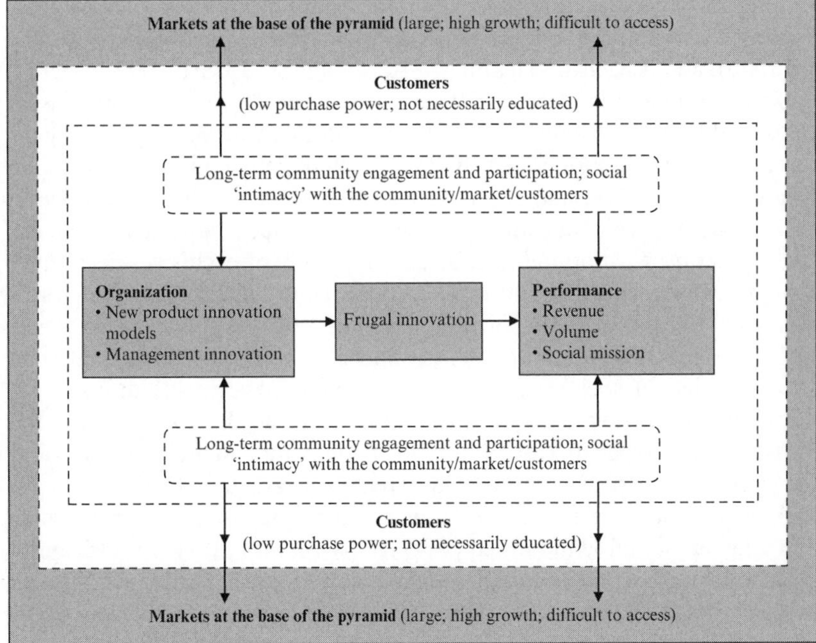

Figure 2.4 The process of frugal innovation

difficulty to access billions of people at the base of the economic pyramid; (2) organizing consideration of the needs of these poor and not necessarily well-educated consumers; (3) developing a willingness to develop new organizational ways of working; and (4) nurturing 'social embeddedness' practices that establish 'social intimacy' with these communities, in order to understand community idiosyncrasies and foster loyalty among them to the company's social mission (Elaydi and Harrison, 2010; London and Hart, 2004; Simanis and Hart, 2009).

Social intimacy is crucial to success in subsistence markets because in them people 'survive on a relational, reciprocation based system' (Elaydi and Harrison, 2010, p. 654) and tend to distrust outsiders whose behaviors may seem exploitive and uncaring (Viswanathan et al., 2009). As Simanis and Hart (2009) argued, Grameen's innovative banking did not start because Muhammad Yunus conducted grassroots market research or consulted a global positioning system for the best target markets. Instead, Yunus' vision came into focus as he spent time with villagers as a community, working together, learning from one another (Yunus himself had already lent money to poor people; Yunus et al., 2010). Other examples of this kind of embedded innovation, which contrasts with the traditional 'structural innovation', can be found in BOP protocol projects such as S.C. Johnson (Kenya), DuPont/Solae (India), The Water Initiative (Mexico), and Ascension Health (USA; Simanis and Hart, 2009).

Social intimacy is also crucial because, contrary to what happens in those industrialized countries in which values of autonomy and egalitarianism prevail, individuals in many low-income populations derive meaning largely from participating in a shared way of life and pursuing collective goals (Nakata and Weidner, 2012). Such collective emphasis enables people to adopt new products not mainly or exclusively as a function of personal needs but rather as a consequence of the welfare and preferences of the collective to which they belong. Without social intimacy and embeddedness it is difficult to identify and satisfy local needs innovatively. It was a kind of social intimacy and embeddedness (including informal talks with rural consumers to learn and understand the aspirations of the masses) that allowed TATA Motors to create Nano, an ultra-low-cost and sub-compact car combining two apparent contradictory features: (1) the ability to manoeuvre in crowded towns and cities, through narrow roads and alleyways, and with limited parking spaces; and (2) being sufficiently large to fit typical Indian families and accommodate adequate seating and luggage space (Ray and Pradeep, 2011). The market sought was occupied not by existing automobiles but scooters and light motorbikes that would often be loaded with more people and goods than one would think possible.

When the above-mentioned four conditions are articulated, organizations may engage in frugal innovation. The expected results of this approach are significant revenues, an increase in the firm's volume, and an accomplished social mission of responding to the needs of people that were circumvented by traditional business models (Anderson and Markides, 2007; Elaydi and Harrison, 2010; Hart and Christensen, 2002; London and Hart, 2004; Tashman and Marano, 2010; Viswanathan et al., 2009; Yunus, 1999).

While in developing countries, the BOP represents a huge amount of relatively demographically young impoverished people, in developed and demographically old countries elderly people have progressively widened the top of the age pyramid, turning pyramids into mushrooms. The purchasing power of aged people reliant on pensions tends to decrease after retirement, a problem exacerbated by higher health-care needs. Moreover, state budget deficits and structural problems with financing social security systems (the ratio of active/inactive has been increasing) render states less able to deal effectively with the elderly challenge, a combination that requires ingenious approaches to emerging societal problems. Surprisingly, this penurious environment has been neglected by companies, keeping millions of people away from needed benefits for their health and wellbeing.

INGENUITY PROCESSES: LINES OF CONTINUITY

We previously discussed how scarcity in different forms requires ingenious responses from organizations. They can be discussed separately because they respond to different issues, on the basis of different materials, with different actions, envisioning different outcomes. The literature, however, revealed interesting overlaps and continuities between these approaches, as depicted in Figure 2.5. These areas of overlapping and continuity are still to explore because their mention was not followed by rich theoretical elaboration. We contribute to this exploration via the integration of the three streams discussed above. This conceptual analysis will subsequently require empirical testing.

Before elaborating the point, we recapitulate previous notes. Bricolage has been studied as a relevant process *per se* in several literatures, including the organizational. The concept has been used to show how familiarity with a given material resource (e.g., a technology) may reveal opportunities for organizational action using inferior material capacity. Another body of research discussed how improvisation may be necessary under

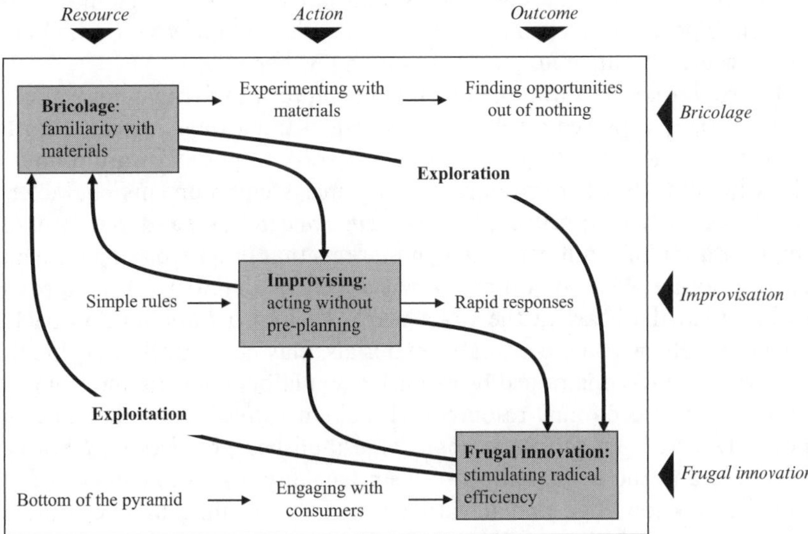

Figure 2.5 Articulating forms of ingenuity

time pressure to respond to pending problems. When time is scarce and issues are relevant, organizations may have no opportunity to plan their responses. One practical implication of improvisation, therefore, is that improvisers improvise with the existing resources. They need to develop bricolaging skills, hence the reference to bricolage in some definitions of improvisation (e.g., Cunha et al., 1999).

Interestingly, improvisational skills and bricolaging have been associated with the discipline of frugal innovation, namely in India. Research suggests that the three processes of bricolage, improvisation and frugal innovation may be embedded in one another, i.e., one process sometimes activating another. The embeddedness approach is supported by preliminary anecdotal evidence that we now develop.

A common form of scarcity resides in the absence of adequate materials for organizational processes (time, money, people, and technologies). Some organizations, however, develop a competence for bricolage. They stimulate an intimate relationship with materials, in such a way that people can ingeniously extract uses from those resources that are not visible for developers with a less ingrained understanding of resource plasticity. This familiarity, in turn, can be a trigger for improvisation: improvisers make do with the available materials, hence the association between improvisation and bricolage (Cunha et al., 1999). In fact, because improvisers are pressed to act by the situation, they need to use whatever resources they have at hand. In this sense, they have to express the ingenious attitude of the bricoleur. The tinkering process associated with bricolage may thus stimulate people to feel comfortable in the role of improvisers, who plan and execute simultaneously (Moorman and Miner, 1998).

Improvisation, in turn, may be involved in forms of frugal innovation. For example, it played a role in the so-called India way. *Jugaad*, a form of improvisation, has been said to be necessary to push forward innovations in contexts where resource penury meets environments characterized by red tape (Cappelli et al., 2010). *Jugaad* can be viewed as a form of improvisation of significant value in markets that differ from the affluent societies of the West. As observed by Cappelli et al. (2010, p. 4), improvisation is 'at the heart of the India way'. An improvisational approach, including reliance on the available materials, may be particularly valuable in environments constrained by extensive regulatory controls and stymied by the lack of economic resources. (Brazilian *jeitinho* is another case in point: Duarte 2006.) In these contexts, established practices may not be adequate and the combination of deep frugality, the willingness to test established knowledge, and an extreme capacity to adjust, may be sources of competitive advantage while developing new products.

The articulation of bricolage, improvisation, and frugal innovation captures a process of product ingenuity that is, to a great extent, exploratory and emergent, pervaded by serendipity, reaction, and ingenuity. However, after witnessing the potential benefits of innovation in this emergent mode in resource-poor environments, some organizations seek to reverse the cycle by exploiting and institutionalizing ingenuity. They stabilize and structure improvisational practices and cultivate the practice of ingenuity. This will feed the cycle back. For example, after investing in the exploration of the BOP and taking advantage of the acquired knowledge, organizations may start introducing some formalization and institutionalization in the process, in order to make it more systematic rather than emergent. The development of Tata's Nano (Ray and Pradeep, 2011), for example, incorporated a strong element of goal setting in the process, forcing developers not only to explore but also to exploit existing knowledge, including Indian suppliers with a track record of manufacturing to global standards as well as from foreign partners, such as Bosch Automotive. The challenge of producing a one-*lakh* (the clearly defined price tag) automobile focused attention and introduced a significant element of exploitation and efficiency orientation. Finally, the acquired improvisational competences, once framed and institutionalized, may lead to a cultivation of competences of both improvisation and bricolage within established limits and carefully institutionalized processes.

In fact, improvisational and bricolage practices can be trained, stimulated, choreographed, and managed (Brown and Eisenhardt, 1997; Bechky and Okhuysen, 2011; Vera and Crossan, 2004). And the attitude of bricolage, the leveraging of innovation from intimacy with materials, can be cultivated in highly institutionalized organizational systems, such as Toyota. As noted by Takeuchi et al. (2008), employees in this automaker are stimulated continuously to grapple with challenges and problems and are expected to advance ideas to tackle them. The system, in other words, is paradoxical: Taylorism for knowledge workers, or a highly institutionalized system that stimulates local experimentation, a massive company with a frugality culture (see p. 99 in Takeuchi et al., 2008). When people are empowered for innovation, they will potentially use whatever resources they have to approach problems creatively. The more one is empowered within the clear limits provided by minimal structures and organizational heuristics (Bingham and Eisenhardt, 2011), the more one may feel confident to test materials and to solve problems in a creative way. Institutionalization and exploitation provide the structure that guides action and that supports further explorations. Exploration and exploitation can reinforce one another (Farjoun, 2010), and routinization can facilitate ingenuity (Ohly et al., 2006).

Different companies may favor different approaches, but it seems possible to predict that the different approaches may be part of a holistic process of organizational ingenuity. The competences acquired over time, from experience with a process, may be used to gain competences in other types of processes that partly overlap. Knowledge can be transferred between processes, a possibility that explains why some companies are able to show consistent and disciplined ingenuity with a 'less is more' type of attitude, blending improvisational, bricolage, and frugal capabilities. Such companies may have discovered how to develop comfort and familiarity with paradoxical requirements involving structure and freedom, exploration and exploitation, lack and slack, need and scarcity. They will challenge the idea that ingenuity requires slack and that scarcity is a hostile environment for an organization's existence by cultivating ingenuity.

A COUNTER POINT AND RESEARCH DIRECTION: ON INGENUITY AND SLACK

We approached the potentiality of ingenuity under conditions of scarcity. As we discussed, sometimes organizations activate ingenious responses because of scarcity but slack can also stimulate and support ingenuity. As our editors pointed out, many firms in Silicon Valley started in a garage – utilizing 'extra' space attached to a house with no set boundaries or expected uses (beyond storage) that can easily be adapted to other uses. Perhaps cultures without this surplus of space (organizational slack) are at a relative disadvantage? Thus, the question becomes, while scarcity is clearly one avenue, to what extent is surplus another? What are the trade-offs between them?

Bricoleurs can better improvise innovations in 'tolerant organizations' (Halme et al., 2012) with surplus spaces and resources – slack – that allow for experimentation. Of course, they will then need whatever is imagined and created to fulfill users and markets. Improvisers may improvise because they have the free time to engage in creative attempts in cultures favorable to slack and playfulness (Mainemelis, 2010). Entrepreneurs may find that slack is critical for launching new ventures (Vanacker and Collewaert, forthcoming). Slack and abundance, in other words, can be highly favorable environments for the expression of ingenuity. Not all slack equals 'fat' and slack can actually facilitate creative behavior (Bourgeois, 1981). The scarcity/slack relationship with innovation should thus be further explored in future research, namely by analyzing the conditions under which slack and scarcity can be creative or destructive (see, for example, Nohria and Gulati, 1997).

CONCLUSION

We presented organizational ingenuity as an umbrella concept that articulates a number of organizational responses to constraints that have not been previously articulated. Organizations need to be ingenious when time, resources or affluent customers are missing. In response to constraints, organizations can respond by resorting to processes such as improvisation, bricolage or frugal innovation. These processes, in turn, may relate and interact. We offered an integrated view of organizational ingenuity as a rich and multiform process that aims to develop the organization's capacity to respond to problems or opportunities rather than to represent constraints as impediments to organizational action. We contributed to the literature by opening the black box of organizational ingenuity, distinguishing ingenious possibilities and articulating different streams of ingenuity that so far remain unrelated. This way, we answer to Katila and Shane's (2005) 'intriguing' finding 'that resource constraints can be enabling' and that 'the enabling features of scarcity remain mostly unexplored' (p. 825). An approach based on synergistic relationships between improvisation, bricolage and frugal innovation makes 'resource constraints as enabling' more explicable.

ACKNOWLEDGEMENTS

This paper is part of the project CMU-PT/OUT/0014/2009. We are grateful to the editors for their feedback and suggestions.

REFERENCES

Anderson, J. and Markides, C. (2007). Strategic innovation at the base of the pyramid. *MIT Sloan Management Review*, **49**(1), 83–8.

Baker, T. (2007). Resources in play: bricolage in the toy store(y). *Journal of Business Venturing*, **22**, 694–711.

Baker, T. and Nelson, R. (2005). Creating something out of nothing: resource construction through entrepreneurial bricolage. *Administrative Science Quarterly*, **50**, 329–66.

Bechky, B. and Okhuysen, G. (2011). Expecting the unexpected? How SWAT officers and film crews handle surprises. *Academy of Management Journal*, **54**, 239–61.

Bingham, C. and Eisenhardt, K.M. (2011). Rational heuristics: the 'simple rules' that strategists learn from process experience. *Strategic Management Journal*, **32**, 1437–64.

Boulding, W. and Staelin, R. (1993). A look on the cost side: market share and the competitive environment. *Marketing Science*, **12**, 144–66.

Bourgeois, L.J. (1981). On the measurement of organizational slack. *Academy of Management Review*, **6**(1), 29–39.

Brown, S.L. and Eisenhardt, K.M. (1997). The art of continuous change: linking complexity theory and time-paced evolution in relentlessly shifting organizations. *Administrative Science Quarterly*, **42**, 1–34.

Cappelli, P., Singh, H., Singh, J. and Useem, M. (2010). The India way: lessons for the US. *Academy of Management Perspectives*, **24**(2), 6–24.

Ciborra, C. (1996). The platform organization: recombining strategies, structures, and surprises. *Organization Science*, **7**, 103–18.

Clair, J.A. and Dufresne, R.L. (2007). Changing poison into medicine: how companies can experience positive transformation from a crisis. *Organizational Dynamics*, **36**(1), 63–77.

Cooper, R.G. (1993). *Winning at New Products*. Reading, MA: Addison-Wesley.

Corbett, S. (2008). Can the cellphone help end global poverty? *The New York Times*, April 13. Available at: http://www.nytimes.com/2008/04/13/magazine/13anthropology-t.html?pagewanted=all&_r=1& (last accessed 20 September 2013).

Cunha, M.P., Clegg, S.R. and Mendonça, S. (2010). On serendipity and organizing. *European Management Journal*, **28**, 319–30.

Cunha, M.P., Cunha, J.V. and Kamoche, K. (1999). Organizational improvisation: what, when, how and why. *International Journal of Management Reviews*, **1**(3), 299–341.

Cunha, M.P., Neves, P., Clegg, S. and Rego, A. (2012). Improvisation sequences. Unpublished manuscript.

Cunha, M.P., Rego, A., Oliveira, P., Rosado, P. and Habib, N. (forthcoming). Product innovation in resource poor environments: three research streams. *Journal of Product Innovation Management*.

Daniel, F., Lohrke, F.T., Fornaciari, C.J. and Turner, R.A. (2004). Slack resources and firm performance: a meta-analysis. *Journal of Business Research*, **57**, 565–74.

Debruyne, M., Frambach, R. and Monaert, R. (2010). Using the weapons you have: the role of resources and competitor orientation as enablers and inhibitors of competitive reaction to new products. *Journal of Product Innovation Management*, **27**, 161–78.

Diamond, J. (1998). *Guns, Germs and Steel: A Short History of Everybody for the Last 13,000 Years*. London: Vintage.

Duarte, F. (2006). Exploring the interpersonal transaction of the Brazilian jeitinho in bureaucratic contexts. *Organization*, **13**(4): 509–27.

Duboule, D. and Wilkins, A.S. (1998). The evolution of 'bricolage'. *Trends in Genetics*, **14**(2), 54–9.

Duymedjian, R. and Ruling, C.C. (2010). Towards a foundation of bricolage in organization and management theory. *Organization Studies*, **31**, 133–51.

Elaydi, R. and Harrison, C. (2010). Strategic motivations and choice in subsistence markets. *Journal of Business Research*, **63**, 651–5.

Farjoun, M. (2010). Beyond dualism: stability and change as duality. *Academy of Management Review*, **35**, 202–25.

Gabel, M. (2004). Where to find 4 billion new customers: expanding the world's marketplace. *The Futurist*, **38**(4), 28–31.

Garud, R. and Karnoe, P. (2003). Bricolage versus breakthrough: distributed and embedded agency in technology entrepreneurship. *Research Policy*, **32**, 277–300.

Gladwell, M. (2008). *Outliers: The Story of Success*. London: Penguin.

Halme, M., Lindeman, S. and Linna, P. (2012). Innovation for inclusive business: intrapreneurial bricolage in multinational corporations. *Journal of Management Studies*, **49**(4), 743–84.

Hannan, M.T. and Freeman, J. (1989). *Organizational Ecology*. Boston, MA: Harvard University Press.

Harhoff, D. and Hoisl, K. (2007). Institutionalized incentives for ingenuity – Patent value and the German Employee Inventions Act. *Research Policy*, **36**, 1143–62.

Hart, S.L. and Christensen, C.M. (2002). The great leap: driving innovation from the base of the pyramid. *MIT Sloan Management Review*, **44**(1), 51–6.

Heynoski, K. and Quinn, E.R. (2012). Seeing and realizing organizational potential: activating conversations that challenge assumptions. *Organizational Dynamics*, **41**(2), 118–25.

Hilty, L.M. (2005). Editorial: electronic waste – an emerging risk? *Environmental Impact Assessment Review*, **25**, 431–5.

Hirose, S. (2004). Waste management technologies in Japanese cement industry – from manufacturing to ecofactuaring™. *Journal of Water and Environment Technology*, **2**(1), 31–6.

Hirsch, P.M. and Levin, D.Z. (1999). Umbrella advocates versus validity police: a life-cycle model. *Organization Science*, **10**, 199–212.

Ikenberry, G.J. (1986). The irony of state strength: comparative responses to the oil shocks in the 1970s. *International Organization*, **40**(1), 105–37.

Isaacson, W. (2012). The real leadership lessons of Steve Jobs. *Harvard Business Review*, **90**(4), 92–100, 102.

Kamkwamba, W. and Mealer, B. (2009). *The Boy who Harnessed the Wind: Creating Currents of Electricity and Hope*. New York: Harper Collins.

Karan, P.K. (2005). *Japan in the 21st Century: Environment, Economy and Society*. Kentucky: University Press of Kentucky.

Katila, R. and Shane, S. (2005). When does lack of resources make new firms innovative? *Academy of Management Journal*, **48**(5), 814–29.

Kim, W.C. and Mauborgne, R. (2005). *Blue Ocean Strategy*. Boston, MA: Harvard Business School Press.

Lampel, J., Honig, B. and Drori, I. (2011). Call for papers: discovering creativity in necessity: organizational ingenuity under structural constraints. *Organization Studies*, **32**(5), 715–17.

Larson, E.W. and Gobeli, D.H. (1989). Significance of project management structure on development success. *IEEE Transactions on Engineering Management*, **36**(2), 119–25.

London, T. and Hart, S.L. (2004). Reinventing strategies for emerging markets: beyond the transnational model. *Journal of International Business Studies*, **35**, 350–70.

Mainemelis, C. (2010). Stealing fire: creative deviance in the evolution of new ideas. *Academy of Management Review*, **35**, 558–78.

McMillan, C. and Wesson, T. (1998). The production revolution in manufacturing. In P. Heenan (ed.), *The Japan Handbook* (pp. 72–86). London: Fitzroy Dearborn.

Miller, D. (1990). *The Icarus Paradox: How Exceptional Companies Bring About their Own Fall*. New York: HarperCollins.

Miner, A.S., Bassoff, P. and Moorman, C. (2001). Organizational improvisation and learning: a field study. *Administrative Science Quarterly*, **46**, 304–37.

Moorman, C. and Miner, A.S. (1998). The convergence between planning and execution: improvisation in new product development. *Journal of Marketing*, **62**, 1–20.

Morikawa, M. (2000). *Eco-industrial Developments in Japan*. Indigo Development Working Paper No. 11. RPP International, Indigo Development Center, Emeryville, CA.

Nakata, C. and Weidner, K. (2012). Enhancing new product adoption at the base of the pyramid: a contextualized model. *The Journal of Product Innovation Management*, **29**(1), 21–32.

Nalebuff, B. and Ayres, I. (2003). *Why Not?: How to Use Everyday Ingenuity to Solve Problems Big and Small*. Boston, MA: Harvard Business School Press.

Nohria, N. and Berkley, J.D. (1994). Whatever happened to the take-charge manager? *Harvard Business Review*, January–February, 128–37.

Nohria, N. and Gulati, R. (1997). What is the optimum amount of organizational slack? A study of the relationship between slack and innovation in multinational firms. *European Management Journal*, **15**(6), 603–11.

Ohly, S., Sonnentag, S. and Pluntke, F. (2006). Routinization, work characteristics and their relationships with creative and proactive behaviors. *Journal of Organizational Behavior*, **27**(3), 257–79.

Penrose, E. (1959). *The Theory of the Growth of the Firm*. New York: Wiley.

Pfeffer, J. and Salancik, G.R. (1978). *The External Control of Organizations: A Resource Dependence Perspective*. New York: Harper & Row.

Prahalad, C.K. (2005). *The Fortune at the Base of the Pyramid. Eradicating Poverty through Profits*. Philadelphia, PA: Wharton School Publishing.

Ray, S. and Pradeep, K.R. (2011). Product innovation for the people's car in an emerging economy. *Technovation*, **31**(5), 216–27.

Sehgal, V., Dehoff, K. and Panneer, G. (2010). The importance of frugal engineering. *Strategy + Business*, Summer, 20–5.

Simanis, E. and Hart, S. (2009). Innovation from the inside out. *MIT Sloan Management Review*, **50**(4), 77–86.

Stalk, G. and Hout, T.M. (1990). *Competing Against Time: How Time Based Competition is Reshaping Global Markets*. New York: Free Press.

Sull, D. and Eisenhardt, K.M. (2012). Simple rules for a complex world. *Harvard Business Review*, **90**(9), 68–74.

Takeuchi, H., Osono, E. and Shimizu, N. (2008). The contradictions that drive Toyota's success. *Harvard Business Review*, June, 96–104.

Tashman, P. and Marano, V. (2010). Dynamic capabilities and base of the pyramid business strategies. *Journal of Business Ethics*, **89**, 495–514.

Vanacker, T. and Collewaert, V. (forthcoming). The relationship between slack resources and the performance of entrepreneurial firms: the role of venture capital and angel investors. *Journal of Management Studies*.

Vendelo, M.T. (2009). Improvisation and learning in organizations – An opportunity for future empirical research. *Management Learning*, **40**, 449–56.

Vera, D. and Crossan, M.M. (2004). Theatrical improvisation: lessons for organizations. *Organization Studies*, **25**, 727–49.

Viswanathan, M., Seth, A., Gau, R. and Chaturvedi, A. (2009). Ingraining product relevant social good into business processes in subsistence marketplaces: the sustainable market orientation. *Journal of Macromarketing*, **29**(4), 406–25.

Wernerfelt, B. (1984). A resource-based view of the firm. *Strategic Management Journal*, **5**, 171–80.

Yadav, V. (2010). User innovation at grassroots. Paper presented at the 2010 International Innovation Workshop, MIT Sloan, Boston, USA, 2–4 August 2010.

Yunus, M. (1999). *Banker to the Poor, Micro-Lending and the Battle Against World Poverty*. New York: Public Affairs.

Yunus, M., Moingeon, B. and Lehmann-Ortega, L. (2010). Building social business models: lessons from the Grameen experience. *Long Range Planning*, **43**, 308–25.

3 Ingenuity spirals and corporate environmental sustainability

David B. Zoogah

The rapid pace of environmental change is propelling corporations, governments, and academics to identify systems that arrest the increasing degradation of the environment and facilitate enhanced sustainability (see Christmann and Taylor, 2002; Shrivastava and Berger, 2010; United Nations World Commission on Environment and Development, 1987). These views emerge from the realization that the world is trapped in environmental degradation, and recent sustainability attempts still yield gaps (Gardner and Stern, 2002; Hoffman and Haigh, 2012; Metzner, 1999). Individuals, groups, and nations have also been mobilized through such groups as the Aspen Institute, The Brundtland Commission, and the United Nations to identify preventive solutions and promotive mechanisms that facilitate environmental sustainability (Jackson et al., 2011). Corporations in particular are becoming proactive in establishing sustainability systems (González-Benito and González-Benito, 2006). They are striving for ingenuity in their attempts to reduce their carbon footprint or environmental impact.

Academics are also endeavoring to understand and help resolve environmental problems such as deforestation, pollution, erosions, catastrophes, toxicity, desertification, aridity, smog, etc. Studies suggest that environmental sustainability can be promoted through positive deviant behaviors (Hoffman and Haigh, 2012) and preliminary evidence seems to provide some support for ecological transcendence (Zoogah, 2013b). One critical factor that seems to be missing, though, is ecological ingenuity. Briefly, ecological ingenuity refers to environment-centered ingenuity (i.e., ingenious ideas that restore health to the natural environment). It differs from, but is interlinked with, other forms of ingenuity – economic, social, political, technological (Homer-Dixon, 2000) – because environmental problems have economic, social, political, and technological sources (Gardner and Stern, 2002; Homer-Dixon, 2000; Holling et al., 2002a; see also Gunderson and Holling, 2002).

My purpose in this chapter therefore is to discuss ecological ingenuity with an emphasis on ingenuity spirals. I elaborate on the attributes, process, antecedents and outcomes of ecological ingenuity. I also discuss how ingenuity spirals may contribute to the resolution of environmental problems and advancement of sustainability. Ecological ingenuity is conceptualized at the organizational level as the aggregate of individual employees' environment-centered ingenuity. Organizations rely on the ingenuity of employees and groups to effectively conduct environmental management practices (Gardner and Stern, 2002).

As Metzner (1999) observed, there has to be a new way of thinking about the relationship between humans and nature. That way of thinking or paradigm includes mechanisms that, although small initially, can magnify in breadth and depth. Ingenuity spirals within organizations can therefore assist in improving environmental sustainability since they affect multiple and diverse agents, contexts, and systems. Their functions and structures are similar to panarchy (Gunderson and Holling, 2002), albeit on a smaller scale. Ingenuity spirals therefore have potential to help resolve environmental sustainability problems – degradation traps and sustainability gaps. Next I briefly discuss corporate environmental sustainability before focusing on ecological ingenuity and spirals.

ENVIRONMENTAL SUSTAINABILITY

There is no consistent definition of environmental sustainability in the literature (Haden et al., 2009) primarily because of the complexity of ecological phenomena, diversity and global dispersion of environmental problems, conflicting views of practitioners and academics, multiplexity of disciplines vested in environmental issues, and scientific lenses used to examine environmental problems. Nevertheless, attempted definitions range from 'environmentally conscious practices of "green" organizations' (Haden et al., 2009, p. 1046) and 'adopting environmentally friendly practices' (Siegel, 2009, p. 5) to use of policies, philosophies, and practices to promote sustainable use of resources and prevent harm arising from environmental concerns within business organizations (Zoogah, 2011). In this chapter, environmental sustainability is defined as the process by which organizations use systems and practices to facilitate, maintain, and improve the quality of the natural environment in the long term. This definition is distinctive in scope, complexity, and urgency. It is part of the broader concept of sustainability[1] and encompasses air, land, water, etc., and other natural environmental elements.

Sustainability phenomena are complex, multilayered, and often characterized by uncertain interdependencies and nonlinearities (see Gunderson and Holling, 2002). One implication is the existence of alternative frames for the problem, including the rational (economic considerations such as productivity and profitability), natural (environmental sustainability, including preservation of natural resources and mitigation of climate change), and humanist (personal satisfaction and social needs such as fair trade practices and human rights) approaches (Elkington, 1994). The complexity of environmental sustainability suggests a need for ingenious solutions that help resolve not only the degradation problems but also gaps in sustainability (Homer-Dixon, 2006). In order to manage that complexity, organizations develop environmental management (EM) strategies.

Strategies

Generally, a strategy is a fundamental course of action undertaken by an organization to achieve a specific goal. Applied to environmental sustainability, it refers to course of actions taken to achieve environmental sustainability goals. Environmental strategies have been examined extensively resulting in several typologies (Clarkson, 1995; Hart, 1995). The typologies show that corporate EM strategies range from proactive, holistic, strategic, to integrated (Margerum, 1999; Steger, 2000). Other strategies proposed by Hart (1995) are based on the natural environment as a resource for organizations. However, organizations can develop behavioral strategies, two of which are prevention and promotion strategies. Promotion strategies seek to advance environmental sustainability while prevention strategies seek to prevent environmental degradation. Both promotion and prevention strategies are fulfilled through environmental management systems.

Systems

Organizations develop environmental management systems based on the International Standards Organization (ISO) standards for business, government and society (http://www.ISO.org). The latest ISO standard, ISO 14001, which is process-oriented and consistent with systems thinking, focuses explicitly on environmental management. It suggests that organizations establish environmental policies; facilitate environmental planning, controlling and monitoring of the policy; determine environmental impact, significance, priorities and objectives consistent with legislative requirements; establish implementation programs; develop accountability and commitment systems for management and employees; provide resources, including training, to achieve targeted performance levels on an ongoing basis; establish audit and management review processes; and establish communications with relevant internal and external parties. ISO 14001 does not mandate a particular organization's optimum environmental performance level but describes a system to help an organization achieve its own environmental objectives.

Thus, organizations have to be ingenious in their adoption of sustainability systems to improve overall environmental performance through prevention of pollution; saving money on energy, materials, and so on; enhancement of existing compliance efforts related to environmental aspects; reduction or mitigation of risks and liabilities; exhibition of environmental due diligence; increase in efficiency; achievement of environmental awareness within organizations; promotion of corporate images; and qualification for incentive programs. The systems facilitate effective execution of environmental sustainability practices.

Practices

Environmental management practices focus on EM activities of organizations. Ziegler and Nogareda (2009) define EM practice as the establishment of internal standards, goals and policies for environmental performance improvement – without formal certification. Others have organized EM activities into typologies (see Azzone and Bertele, 1994; Hunt and Auster, 1990). Zoogah (2013a) uses the productive (source of resources for production processes and disposal of waste from those processes) and value (value creation and value capture) concerns of organizations to identify four types of practices: consumption, sustainability, destruction and management.

Resource consumption practices focus on how organizations consume resources in a way that creates value. Overconsumption of resources seems detrimental to organizations' production concern and their value concern in the long run. Under-consumption does not also optimize their value creation. Optimal consumption, which lies between overconsumption and under-consumption, is impacted by diffusion practices and systems. For example, the desire of customers for more goods may drive an organization (e.g., Walmart) to overproduce even if it initially decided against going over the optimal level as a result of its concern for the natural environment. Organizations can temper this external pressure by adopting consumption behaviors that do not deplete environmental resources.

Resource preservation practices involve preserving resources for future production to enable organizations appropriate future value. Value appropriation occurs through societal surplus or ecological capital (Gardner and Stern, 2002). Value capture has source and temporal dimensions even though the literature has often emphasized the source dimension (see Lepak et al., 2007). Considering that time is an economic (DeSerpa, 1971) and ecological resource (Kronfeld-Schor and Dayan, 2003), organizations can capture value from it by conserving their resources for future activities. Suspending production programs until sometime in the future enhances preservation of resources.

Waste destruction practices focus on effective disposal of waste in a way that creates value. From a Schumpeterian perspective, it refers to creative destruction or innovative ways of disposing of waste. Creative destruction calls for production processes that minimize waste creation (Holling et al., 2002b). Creative waste disposal can save on costs associated with haphazard disposals such as dumping. It can also minimize current and future liabilities.

Waste management practices center on management activities that enable organizations to appropriate value from their waste. They are future oriented. Organizational waste is inevitable; it is part of the production process. If value cannot be created from it now, at least it can be captured in the future. The future capture of value from organizational waste requires management. That means the establishment of standards on types of, frequency of, resources for and storage mechanisms of the waste.

In addition to the above CES strategies, systems, and practices, management scholars have increased efforts to understand, explain, and predict environmental sustainability on organizational effectiveness. As a result, research has identified determinants, processes, and consequences of environmental degradation and sustainability (Gardner and Stern, 2002; Stein, 2000). One major determinant, which unfortunately has not been examined in the literature, is ecological ingenuity. Ecological ingenuity has been proposed in the economics, sociology, and ecological systems literatures as significant (Holling et al., 2002a; Homer-Dixon, 2000).

ECOLOGICAL INGENUITY

Ingenuity, defined as ideas applied to solve problems or specifically sets of instructions on how to arrange the constituent parts of the physical and social worlds in order to achieve certain goals (Homer-Dixon, 2000), is not only conceptualized at the individual level but also has meso (organizational) and macro (societal) level applications. It is viewed as vital to solve the major and complex economic, social, and environmental problems of the world (Homer-Dixon, 2000; Gunderson and Holling, 2002). Homer-Dixon (2000) distinguishes social ingenuity from technical ingenuity. Social ingenuity relates to the arrangement of human relations and the creation of institutions while technical ingenuity encompasses technological and scientific innovations. In this chapter I focus on ecological ingenuity, a concept that Homer-Dixon (2000) hinted at but did not elaborate.

Ecological ingenuity refers to the systematic application of *ideas* to solve *environmental sustainability* problems. That application includes the use of specific sets of instructions on how to arrange the constituent parts of the natural environment to achieve *ecological goals*. Within organizations, it refers to the collective and systematic process of applying ideas, procedures, and systems to achieve ecological goals.[2] Three characteristics of this definition seem significant: ideas which include knowledge, experience, traits, and skills; specificity which limits the context of ingenuity to sustainability; and goals which include minimization of degradation or optimization of sustainability. As Holling et al. (2002a) observed, natural systems are so complex that they require complex goals. As a result there are subordinate and superordinate ecological goals that may have to be achieved. A subordinate goal for example is reduction in waste destruction while an example of a superordinate goal is replacement of all raw materials with biodegradable substitutes.

Ecological ingenuity has five attributes. First, it is unconventional. It departs from traditional or normative approaches, at least initially. That means an ingenious idea has to be different from what is already known in at least a minimal way. However, that is insufficient. A second attribute is that an ingenious idea has to be novel. Novelty shows the contributory value of the idea. Third, it has to be what I term 'simcoplex'. Simcoplexity refers to the extent to which an idea is simple in its complexity.[3] The notion of simcoplexity is that an ingenious application is complex enough to challenge the cognitive capacity or capabilities of others but simple enough to be implemented or adopted. Simcoplexity serves both as a buffer and an amplifier in that it buffers against destruction but enables others to enhance its potential. The fourth attribute of ecological ingenuity is practicality. The extent to which the idea can be transformed into action suggests that it can be utilized to solve an ecological problem. Ingenious ideas that lack practicality are bereft of solutions because they have no guides to application (Homer-Dixon, 2006). The fifth attribute of ecological ingenuity is shock: it shocks or stirs the view of an observer to consider it as interesting. In other words, it creates an 'aha' moment for the actor and observer. Shock creates an appeal on the part of the observer which enhances its value. Origami designs seem to capture these attributes.

Process

A critical reading of Homer-Dixon (2000) and the sparse literature on ingenuity, as well as the dynamics of environmental sustainability, suggests that ecological ingenuity involves a series of steps that cumulatively constitutes the throughput or mechanism that transforms organizational inputs to specific ecological outcomes. Besides, the definition suggests

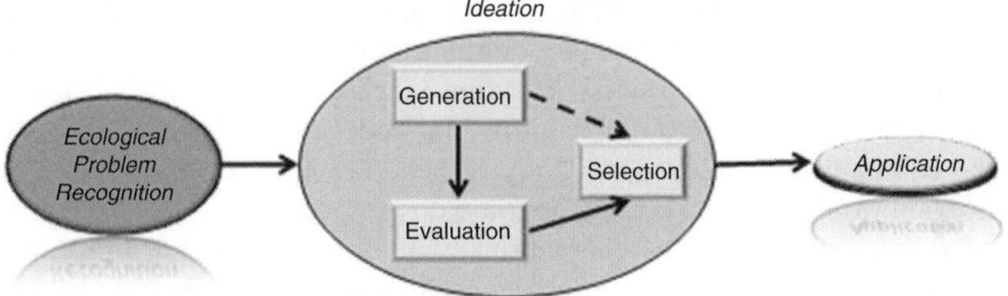

Figure 3.1 Ecological ingenuity process

that organizational ingenuity is a systematic process. Because ingenuity is problem- and goal-oriented, it begins with problem recognition and ends with application. Between these steps is the application of sets of instructions. Figure 3.1 summarizes the process.

Generally, ingenuity addresses a problem, deficit or deprivation. The first stage of the process therefore is recognition of the problem. Recognition is a form of awareness which has been recognized in the environmental management literature as the foundation to sustainability since the 1970s (see Stern, 2000; Goodland, 1995). In environmental sustainability, ecological problems are diverse. As shown in Appendix Table 3A.1, global environmental problems span land, air, and sea, and include pollution, biodiversity reduction, toxicity, and climate changes. Appendix Table 3A.2 also shows the role of corporations in environmental management problems. It lists types of environment issues, problems associated with those issues, and the need for ingenious solutions. The issues center on land, air, and water degradation or pollution. Land problems such as deforestation and desertification require ingenious responses that differ from air problems such as carbon dioxide emissions (i.e., CO_2 pollution) and water problems related to floods and chemical waste of industrial activities (Anderson and Bateman, 2000; Gardner and Stern, 2002). At the organizational level, ecological problems may arise from the consumption of raw materials which invariably derive from the natural environment. They could also center on waste destruction or waste management.

Once the ecological problem is identified, ideas on how to solve it are needed (Homer-Dixon, 2000). Ideation involves generating specific notions or ideas, evaluating them, and selecting one or more ideas that are likely to solve the problem. Idea generation is a deliberative activity, one that involves time, resources, and energies to identify specific *concrete* and *novel* ways of solving the ecological problem (Homer-Dixon, 2000). Sometimes it requires specific task forces. For example, Merck, one of the most successful pharmaceutical companies in the world, attempted a pilot program in Mexico on positive deviance, a transformative technique that uses simple but unconventional approaches. As Pascale, Sternin, and Sternin (2010) observed, 'Merck was engaged in a process that tapped the ingenuity and practical field knowledge of its people (i.e., the individual nodes of intelligence)' (p. 129). The company used groups not only to generate ideas on how the pilot program would succeed, but also to evaluate those ideas.

Evaluation follows idea generation. Evaluation refers to appraisal or assessment of the feasibility, utility, consequence, and viability of an idea. As noted by Pascale et al. (2010) in the case of Merck, 'the group looks in the mirror, decides what makes sense, and determines what to do about it' (p. 129). Sense-making theory (Weick, 1995) suggests that sense-making, sometimes considered an approach to thinking about and implementing communication research and practice, is grounded in identity construction, retrospective, enactive social, ongoing, extracted cues, and driven by plausibility. This form of evaluation is idea-centric: it assesses the potential of ideas to effectively solve an ecological problem in a novel way. It differs from problem evaluation, which appraises an ecological problem. From the perspective of panarchy theory (Holling et al., 2002a), it is embedded in exploitation, conservation, release and reorganization facets because it is an essential requirement to recognize that conditions are needed that occasionally foster novelty and experiment. Evaluation is ingenious to the extent that it focuses on novel combinations and conditions that facilitate creative experimentation. It involves not just the quantity but also the quality of ideas. This is because degradation, for example, is complex and therefore has multiple aspects, all of which require ingenuity. Ingenious evaluation is comprehensive; adopts diverse techniques (quantitative and qualitative) as well as short- and long-term orientations and integrates all, if not, most stakeholders (Homer-Dixon, 2000; Gunderson and Holling, 2002). Such evaluation generates multiple ideas or what Homer-Dixon (2000) calls supply of ingenuity. Similar to awareness, evaluation has been recognized in environmental management by scholars (Stern, 2000) and institutions. The ISO 14001 standard, for example, centralizes evaluation (www.ISO.org/14001) as the novel appraisal of the idea that is critical.

The third step in ideation is selection, the choice of feasible, practical, and efficient ideas. It derives from the meticulous assessment of the ideas. Valuable ideas are selected to be applied to the ecological problem. Selection may include piloting or testing of the idea. Similar to the other steps, ingenious selection follows unconventional steps. Thus, instead of looking for 'excellent' or, in the words of Simon (1982), optimal ideas, one may look for 'satisficing' ideas – those that are good enough in their ingenious potential. As shown in Figure 3.1, the main path goes from generation through evaluation to selection. However, a suboptimal path that goes from idea generation to selection is also possible. This path might lead to process loss unlike the optimal path that leads to process gain.

In sum, ideation involves amassing ideas because the kernel of ingenuity is ideas (Homer-Dixon, 2000). Ideation focuses on practical, feasible, and impactful ideas. Environmental sustainability is invariably intimately linked to, and nested within, interdependent webs of other natural systems (Holling, 2001). Ideation therefore facilitates overcoming the intricate snags or the dynamism and complexity of environmental sustainability. Degradation traps, for example, result from social, cultural, economic, and ecological factors (Holling, 2001; Homer-Dixon, 2000; Gardner and Stern, 2002). Consequently, ingenious ideas have to emerge from those areas. One social factor contributing to degradation in Africa, for example, is social conflicts. Social conflicts lead to destruction of forests and land, natural resources from which organizations derive raw materials. When there is no social conflict, deforestation is unlikely to occur. An organization's worry over deforestation may therefore be reduced. Further, its plans for afforestation may not be exigent (Nyang'oro, 2007). Ideas on cultural transformation that can enhance sustainability are therefore likely to facilitate resolution of degradation traps.

As suggested in the definition of ingenuity (Homer-Dixon, 2000), ingenious ideas have to be applied to ecological problems. That application is part of the ingenuity process. It involves implementing the chosen idea(s). It is how the ideas are applied that makes them ingenious. One attribute is an unconventional approach. However, that is not enough. In addition, the application has to be novel. As discussed above, ingenious application has to be complex enough to challenge the cognitive capacity or capabilities of others but simple enough to be implemented or adopted. Simcoplex ideas can be applied through combinative processes such as the use of both sophisticated and basic technology.

Ingenuity Spirals

Extant conceptualization of ingenuity adopts a static view. Homer-Dixon's (2000) work on the ingenuity gap is an example. Even though he advocates for ingenuity as a mechanism to solve the world's and humanity's challenges, he presents ingenuity as discrete: it is either there or not there. However, like all positive or generative phenomena, ingenuity has the tendency to magnify in scope by increasingly expanding its boundary. Ingenuity appears to broaden an organization's ingenious responses and builds enduring resources in a positively revolving trajectory. In other words, ingenuity spirals are dynamic.

I draw from two theoretical perspectives: broaden and build theory of positive emotions (BBTPE – Fredrickson, 2003) and panarchy (Holling et al., 2002b) to develop ingenuity spirals. Briefly, BBTPE 'describes positive emotions in terms of broadened through-actions repertoires and describes their function in terms of building enduring personal resources' (Fredrickson, 2003, p. 167). It provides perspective on how emotions of individuals evolve meaningfully in an adaptive manner. The boundary of positive emotions expands as if to envelop additional resources. The building occurs when the resources are consolidated or reinforced.

Panarchy theory 'captures the adaptive and evolutionary nature of adaptive cycles that are nested within the other across space and time scales' (Holling et al., 2002b, p. 74). It describes the evolving hierarchical ecosystems which have multiple interrelated elements, and offers a new framework for understanding and resolving the dilemma between change and persistence, between the predictable and unpredictable. Gunderson and Holling (2002) describe panarchy as the structure in which systems, including those of nature (e.g., forests) and of humans (e.g., capitalism), as well as combined human-natural systems (e.g., institutions) that govern natural resource use, are interlinked in continual adaptive cycles of growth, accumulation, restructuring, and renewal. In their model, Gunderson and Holling (2002) describe panarchy as a function of three dimensions: potential (inherent value of accumulated resources of ecosystems), connectedness (degree of connection among controlling variables), and resilience (ability to adapt or rebound from destruction). The model begins with exploitation (r) and moves through conservation (k) circles back to release (Ω) and ends with reorganization (α). Exploitation refers to rapid adoption while conservation refers to slow accumulation and storage of energy or material. Release refers to creative destruction and organization centers on ecological (e.g., soil) processes so as to be available for the next phase of exploitation.

Both theories are integrated to develop ingenuity spirals. The initiation of ingenuity spirals is explained by BBTPE and the cyclical or revolving dynamics is explained by release and reorganization of panarchy. Ingenuity, like creativity, can broaden or expand its boundary in a positive trajectory. Ingenuity spiral therefore refers to the *continuous upward or downward movement in activities or programs caused by interaction with agency, institutions, and resources*. In other words, it is the helical process by which sustainability continuously centers around ingenuity while constantly receding from stasis or approaching resolution. Ingenuity spirals are conceptually positive or upward-bound. In that case they can be applied to promote, advance, or increase positive environmental outcomes. They can also be applied to prevent, arrest, or decrease negative environmental outcomes. Figure 3.2a shows an upward ingenuity spiral.

Homer-Dixon (2000) asked if we need more ingenuity to solve the

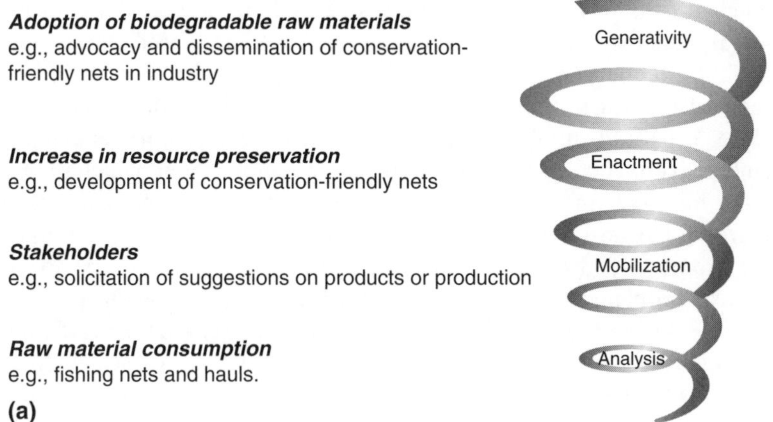

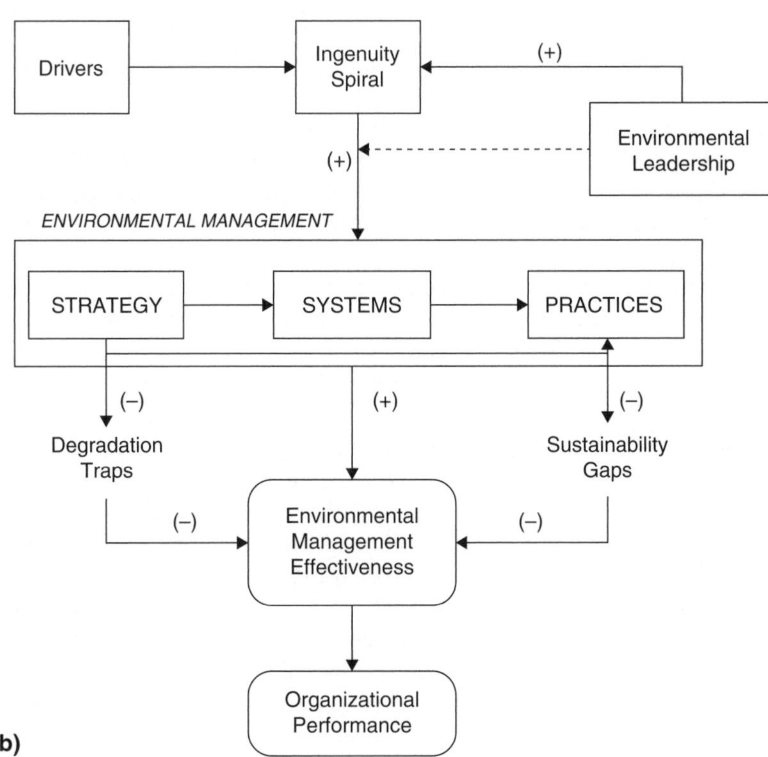

Figure 3.2 Upward ingenuity spiral: (a) process and (b) effects

problems of the future. This question is necessitated by the complexities of social, economic, and environmental phenomena, as well as the vicissitudes of chaotic and turbulent climate changes. Given that 'we live in a world of unknown unknowns' and the imminent potential for a global collapse, it is only supply of ingenuity that can restore hope or balance (Homer-Dixon, 2000, p. 172). Ingenuity is the center around which knowledge, technology, and flourishing of the natural environment can evolve and grow. Upward ingenuity spirals build on this idea. As discussed in the process (see Figure 3.1) upward spirals begin with analysis of an ecological problem necessitating solution. Ingenuity pertains to problem resolution. Using the fishing industry as an example, where overfishing is considered an environmental problem (see Appendix Table 3A.2), an ingenious analysis focuses on the fishing nets and hauls to identify ideas on minimizing the overfishing problem. A fishing organization analyzes the fishing nets to see how they exacerbate the ecological problem before devising ideas to solve that problem. The organization could focus on unconventional appraisal of the ecological problem. For example, instead of following top-down or hierarchical approaches to the problem where executives come up with ideas, a fishing organization may adopt a bottom-up approach by asking local non-commercial fishermen how the problem could be resolved. An ingenious analysis is one that turns the table upside down, looking into crevices that would typically not be looked into. The ingenuity is inherent in the unconventionality, simplicity, and efficacy of the appraisal. Ingenious analysis therefore involves a comprehensive review of how, when, where, why, and who questions so as to determine not only the source and effects, but also potential solutions to the ecological problem. It is ingenious in its cumulative approach: like a snow ball that expands as it rolls, analysis increasingly expands on the understanding of the problem. In the positive deviance literature, problems are analyzed cumulatively beginning with individuals and moving through teams, communities, and villages (see Pascale et al., 2010). They also involve bottom-up rather than top-down approaches.

Following analysis is mobilization of resources. Resources for environmental sustainability are often defined broadly to include psychological, social, cultural, political, and technological resources (Gardner and Stern, 2002; Hart, 1995; Holling et al., 2002b). Mobilization of resources is ingenious to the extent that novel and unconventional approaches are used to garner them. Sometimes it includes identifying agencies that would be dismissed as insignificant (Pascale et al., 2010). In Merck, for example, two sales representatives, dismissed by top executives, turned out to be the linchpins that transformed the performance problem of the company. They exhibited 'extraordinary ingenuity' and achieved 'extraordinary results' (Pascale et al., 2010, p. 131) even though top managers did not think including such employees in the team would provide the answers they were looking for. Psychological, social, political, and cultural resources are forms of capital that, when accumulated, remove potential bottlenecks, or become cornerstones during future implementation or enactment. In the case of fishing, the organization could solicit suggestions from stakeholders such as community fishermen, customers and suppliers on their products and production processes. A crowdsourcing technique could even be adopted to encompass all suppliers not necessarily those in the fishing industry.

Once resources are mobilized, organizations implement the idea(s). The last step in the ideation process is application of the selected ideas (see Figure 3.2a). Action therefore follows ideation. Enactment, ingenious implementation of ideas to resolve degradation traps or sustainability gaps, transforms ideation to desired outcomes. Ideation not only comes up with quality ideas that resolve environmental problems but ideas that mobilize multiple and diverse stakeholders, groups, and institutions. In one sense enactment is the harnessing of human capital resources. To the extent that it is ingenious, enactment minimizes negative externalities. By negative externalities, I mean costs not included in implementation because they are absorbed by systems, human or otherwise, not directly involved in degradation traps or sustainability gaps. To the extent that enactment has positive externalities, it exacerbates degradation traps and therefore does not solve the problem ingeniously. Ingenious enactment leads to resolution of ecological problems. In our overfishing problem, a fishing organization could develop nets that are conservation-friendly so as to minimize the haulage. Ingenious enactment also serves as a basis for generativity.

The term 'generativity' was coined by the psychoanalyst Erik Erikson in 1950 to denote 'a concern for establishing and guiding the next generation'. Generativity is expressed in diverse ways, from stopping a tradition of abuse to starting a new organization. In fact, it includes trying to 'make a difference' in one's community or to the planet. Others define it as 'a system's capacity to produce unanticipated change through unfiltered contributions from broad and varied audiences' (Zittrain, 2006, p. 70). It is viewed as creativity between generations. In the context of ingenuity spirals, generativity describes a self-contained system wherein an organization develops an independent ability to analyze, initiate, generate, or enact novel and unique ways of autonomously solving ecological problems. In that regard, generativity arises out of a sense of optimism about an organization's human potential and capabilities. Generativity thus focuses on the rejuvenation of the natural environment or ingenious nurturance of environmental sustainability. It involves enacting values and beliefs that enable sustainability to flourish.

Further, it involves increased continuous desire and concern for the sustenance of the natural environment. Emerging from ecological actions or behaviors, generativity emphasizes enduring commitment to advance sustainability. Generativity not only resolves degradation traps but establishes mechanisms that impede recidivism or repletion of degradation in future. It embeds resilient mechanisms in environmental systems. In our example, generativity manifests when a fishing organization advocates for and disseminates conservation-friendly nets in the industry. To the extent that it becomes a best practice, the degree of competition is reduced.

To summarize, ingenuity spirals have upward helixes that begin with analysis at the initial stage and revolve upward, mobilizing ingenious factors that lead to actions which not only resolve environmental problems but perpetuate or transmit those solutions to future generations.

INGENUITY SPIRALS AND ENVIRONMENTAL SUSTAINABILITY

The environmental sustainability literature shows two major problems, one of which is degradation traps. Environmental degradation traps refer to the protracted state of degradation that characterizes most countries and regions of the world. In Africa, for example, deforestation has occurred at a very fast rate. 'In less than a generation farmers have seen their wooded environment literally disappear' (Pradervand, 1989, pp. 37–8). This is because of increased exploitation of the forest for fuel (e.g., charcoal), timber (export), real estate, and conflict. Economic deprivation in most African countries, for example, has led to the exploitation of natural resources (land, air, and sea) for income. However, the lack of sophisticated technological capabilities does not allow them to take advantage of their resources and to achieve higher returns. As a consequence, they are driven to greater exploitation, which exacerbates the entrapment. The role of organizations in this entrapment is undeniable (Nyang'oro, 2007). In addition to ecological transcendence, which has been suggested as one means by which societies and organizations can prevent degradation traps (Hoffman and Haigh, 2012; Zoogah, 2013a), I propose ecological ingenuity as another mechanism that can facilitate resolution of degradation traps.

Specifically, ingenuity spirals can help reduce environmental problems. First, reduction occurs when ingenuity spirals are imposed on degradation traps. Ingenious responses to inertia awaken individuals, organizations, and societies to respond positively to deficits in environmental sustainability. Scholars, and indeed the popular press, have recognized inertia as a major impediment to effective environmental sustainability. In other words, inertia contributes to attenuation of sustainability gaps. To the extent that ingenious responses to inertia occur, they help minimize if not eliminate failure in sustainability. Environmental sustainability failure occurs when organizations are unable to achieve desired sustainability goals (Gardner and Stern, 2002; Goodland and Daly, 1996). Through ingenuity organizations are able to minimize or eliminate failure. Besides inability to achieve goals, failure occurs when there is misalignment with economic and social development (Goodland, 1995; Stiglitz et al., 2010). It leads to organizational rigidity in sustainability (i.e., ecological rigidity). Ecological rigidity refers to the tendency of organizations to be rigid in responding to environmental problems. Given that sustainability is now at the center of most organizational missions, very weak sustainability mechanisms represent rigidity. Ecological rigidity tends to be high in developing countries where environmental management is viewed as a constraint imposed by industrial or developed nations. As Collier (2010) observes, developing countries view sustainability as 'hurling up of the development ladder'. Getting out of this mindset requires ingenious responses (Collier, 2010). Ingenious ideas enable organizations to unwind systems that create or exacerbate rigidity which in turn reduces the potential for ecological crisis.

Ecological crisis occurs when the natural environment changes in a way that destabilizes continued survival of the community in which the organization is located. It manifests in the forms of chemical spills, waste deluges, air pollution, and water contamination. The proportionality of an ecological crisis can threaten the survival of an organization because of the potential legal, economic, stakeholder, psychological, and sociocultural consequences. The British Petroleum oil spill off the Gulf Coast continues to affect the company, five years later (Hoffmann and Jennings, 2011; Levy and Gopalakrishnan, 2010). The company is facing legal actions from states, individuals, and other businesses to the point that its stock value has reduced drastically and its profitability is now very low (Smith et al., 2011). Ecological ingenuity is likely to have offered preventive and curative ideas about the crisis and to minimize ecological distress. Ecological distress is defined as suffering that arises from inability to respond ingeniously to environmental problems or sustainability gaps. Ingenuity is likely to help prevent ecological distress (Homer-Dixon, 2000) because ingenuity spirals seem to be able to proffer enduring solutions to environmental problems or permanent removal of sustainability gaps and by extension relatively protracted resolution of distress.

Second, ingenuity spirals can buffer sustainability gaps. Unlike degradation traps that focus on negative consequences of environmental exploitation, sustainability gaps focus on insufficient or inadequate responses to environmental sustainability. This situation seems particularly severe in developing countries. Zoogah (2013a) defines sustainability gaps as the difference between the level of the current environmental state and the desired or optimal preservation state (i.e., the level that maximizes integrated human functioning within the society). Sustainability gaps are wider when degradation is very high (e.g., 95 per cent) and sustainability low (e.g., 5 per cent). In Africa for example:

> afforestation is low relative to deforestation; desertification is high relative to restoration; soil erosion is greater than soil preservation; toxic waste receipt is increasing relative to rejection; environmental institutions are weakened rather than strengthened; wildlife conservation is reducing rather than increasing; carbon capture and storage are lacking rather than prevalent; renewable energy and biomass are low rather than high; and resources for environmental management are minimal rather than profuse. (Zoogah, 2013a: p. 56)

Organizations in Africa can therefore have a sustainability problem. They can solve this problem through ingenuity. Ingenuity spirals can assist them to effectively solve ecological problems related to environmental degradation traps and environmental sustainability gaps. In general, organizations have to cultivate ingenuity spirals in the identification, diagnosis, resolution, and evaluation of those environmental problems. What, then, are the drivers of ingenuity spirals? I turn to that question next.

Drivers

I suggest five drivers of ingenuity spirals. The first driver is intensity of the ecological problem. Ingenuity is problem-oriented. As a result, ecological ingenuity does not manifest when there is no ecological problem to solve. For example, if an organization is judicious in its use of environmental resources and waste generation, ingenious responses to those problems are unlikely to be essential. There are numerous ecological problems with varying degrees of severity and ingenuity-demands (see Appendix Table 3A.1). For example, resource consumption is not so problematic because the amount to consume per production cycle and time (present) is within the control of the organization. In contrast, waste management, which is future oriented (Zoogah, 2013a), is not so much within the control of the organization. The future is unknown and unforeseen conditions can exacerbate or diminish the effects of waste management problems. Ingenuity spirals may therefore be greater in waste management problems than resource consumption problems.

The second driver is organizational (ecological) strategies. As discussed above, strategies are critical to effective environmental sustainability. The sustainability literature has identified diverse strategies. Some of those strategies drive ingenious responses of organizations. Integrated strategies, for example, require ingenuity with regard to analysis, mobilization, and enactment of not only individual employees and managers but also teams, departments and divisions.

A third factor driving ingenuity spirals is the mission of the organization. Some organizations have sustainability-centered missions (e.g., Green Mountain Coffee; Interface, Inc.). As a result, they are driven to adopt ingenious approaches to resource consumption and waste destruction. Not only do they encourage individual ingenuity, they also strive to disseminate ingenuity across teams and departments. They therefore enable ingenuity spirals.

The fourth factor is organizational culture. The organizational culture literature suggests that some organizations have creative cultures (Stevens and Swogger, 2009). The values, beliefs, and norms of organizations can therefore propel individuals to demonstrate ingenuity in environmental sustainability processes and systems. Organizations that value ingenuity are likely to drive employees to be ingenious in their thinking and actions. In fact, they may even establish ingenuity training systems to institutionalize that capability. Such action propels ingenuity spirals. Related to culture is organizational leadership. The fifth driver, leadership, is based on strategic leadership theory (Finkelstein and Hambrick, 1996; Stumpf and Mullen, 1991). It suggests that top management or executives influence organizational processes, structures, and performance. Executives can also influence ingenuity of employees, teams, and departments by supporting and facilitating systems that enable ingenuity to thrive and spread within the organization. Ingenuity is a competence. Strategic leadership theory suggests that executives have responsibility to develop competence of employees, groups, and departments within the organization (Ireland and Hitt, 1999). Organizational leadership can therefore facilitate ingenuity spirals within and across the organization.

Organizational leadership in the context of environmental sustainability is often termed environmental leadership (EL). EL has diverse definitions due to historical and disciplinary foci. In this chapter it refers to the degree to which an organization's purpose, mission, strategy, processes, and systems are oriented toward fulfillment of society's environmental needs while fulfilling shareholder value. This definition extends Kashmania,

Keenan, and Wells' (2010) definition. As shown in Figure 3.2b, EL has two major effects. First, EL directly influences ingenuity spirals. It affects the decisions of managers and behaviors of employees toward the natural environment as well as corporate practices. Consequently, high leadership is likely to lead to more effective environmental management decisions, behaviors, and practices. Second, EL moderates the relationship between ingenuity spirals and environmental management. When EL is high the effect of ingenuity on environmental management is likely to be high. EL propels ecological ingenuity. It also removes bottlenecks that might dampen or minimize ingenuity spirals. However, when EL is low, ecological ingenuity's effect is likely to be low because of lack of support or restrictions.

Outcomes of Ingenuity Spirals

Ingenuity spirals have proximal and distal outcomes. The proximal outcomes center on environmental management. Specifically, they influence the ecological or green decisions and behaviors of employees as well as the environmental practices of organizations. As discussed above, environmental management practices derive from the environmental strategies and systems of the organization. Green decisions refer to environment-centered decisions (Zoogah, 2011) and have panarchic implications (Holling et al., 2002a). Green behaviors are environment-centered actions of employees. They include participation in recycling, proffering of suggestions that improve sustainability, and engagement in clean-up drives. Such behaviors have generativity implications. Ingenuity spirals can also influence resource consumption and preservation as well as waste destruction and management practices of organizations by reducing the negative effects of resource overconsumption and enhancing the positive effects of resource preservation, waste destruction, and waste management.

By influencing environmental management decisions, behaviors and practices, ingenuity spirals indirectly affect environmental management effectiveness; they enable organizations to achieve ecological goals, improve climate perceptions of employees, enhance stakeholder satisfaction, and help organizations to develop sustainability-centered legitimacy. First, environmental management involves specific goals (e.g., reduction of waste by 5 per cent annually). Ingenuity spirals can facilitate, if not hasten, achievement of these goals. Employees' perception of environmental sustainability within the corporation can also be affected. Positive perceptions lead to more ecological attitudes and behaviors (Zoogah, 2013b). Another outcome is environmental stakeholder satisfaction. In this era of worthiness, organizations seeking to be good companies strive to please economic, social, and environmental stakeholders (Bassi et al., 2011). Ingenious responses to ecological problems are therefore likely to satisfy environmental stakeholders. It is also possible that ingenuity spirals will lead to legitimacy. One outcome of institutional processes is legitimacy (Scott, 1995). To the extent that ingenuity transfers externally organizations are likely to be perceived as legitimately attempting to address ecological challenges and problems. That strife enhances company reputation (Hukkinen, 1998).

Ingenuity spirals may also influence non-ecological outcomes such as increased performance, innovation, and learning within organizations. Studies show some relationship between effective environmental management and firm performance (Bassi et al., 2011), innovation (Bos-Brouwers, 2010; Florida and Davidson, 2001), and learning (von Malmborg, 2002). These are indirect outcomes of ingenuity spirals. In sum, ingenuity spirals positively influence environmental management and organizational outcomes. These effects are summarized in Figure 3.2b.

DISCUSSION

In this chapter, I introduce the concept of ingenuity spirals as a major extension of ecological ingenuity. As a dynamic concept, it builds and extends Homer-Dixon's (2000, 2006) idea of ingenuity. It is discussed in the context of environmental sustainability for two major reasons. First, it is critical for organizations that are concerned about environmental sustainability: ingenuity helps organizations to balance the opposing demands of exploiting nature for profitability and preserving nature for future generations (Homer-Dixon, 2006). Second, it is a dynamic concept like panarchy. Upward ingenuity spirals have the potential to eliminate environmental degradation traps and sustainability gaps by influencing resource consumption and preservation, waste destruction and management practices. Organizations may consider the quantity and quality of ideas that resolve environmental problems and have potential to transfer to other organizations and communities. In other words, managers may institute mechanisms that enable ingenuity spirals to spread.

Theoretically ingenuity spirals are conceptualized as mixed effects models. Ingenuity spirals at the organizational level influence organizational, practice, and individual level processes and outcomes. They can thus be examined using multilevel (single-level or cross-level) techniques. Individuals, groups, and even societies can be ingenious. So, individual and group ingenuity may also spiral in much the same way as organizational ingenuity discussed in this chapter. Research may have to untangle the differences. Further, the extent to which individual ingenuity spirals link with group ingenuity spirals may also be examined. To the extent that individuals are 'infected with the ingenuity bug' they are likely to spread it within and outside the organization through their internal and external networks. For example, ingenious ideas that are implemented at the workplace can be transferred to homes and communities to resolve environmental problems.[4]

In addition to individual level transfer, organizational level transfer could occur first through corporate social responsibility (CSR) initiatives. One of the diverse definitions refers to CSR as company activities – voluntary definition – demonstrating the inclusion of social and environmental concerns in business operations and in interactions with stakeholders (van Marrewijk, 2003). CSR also involves investment in community outreach, employee relations, creation and maintenance of employment, environmental stewardship and financial performance (Khoury et al., 1999). The second form of organizational level transfer occurs through environmental management practices (EMPs) within the industry. One example of how EMPs transfer externally is illustrated by Tom's of Maine. About 40 years ago, the company decided to become socially and environmentally friendly. The practice spread across the industry through mimetic processes because of the positive results of that decision (Spreitzer and Sonenshein, 2003). Through similar processes, ingenuity spirals can transfer to external stakeholders to affect environmental sustainability within the community or society.

Ingenuity spirals at the national level are also probable. They may influence national development (i.e., long-term performance, improvement, or a nation). National development has economic, social, political, cultural, and environmental components (Stiglitz et al., 2010). Environmental sustainability is intricately linked to social sustainability and economic sustainability (Homer-Dixon, 2006; Metzner, 1999). Ingenious spirals may therefore affect national development through environmental sustainability (Goodland and Daly, 1996; Gardner and Stern, 2002; Homer-Dixon, 2006; Metzner, 1999). As Homer-Dixon (2006) has shown, ingenuity can help resolve environmental problems. Consistent with that view, it is probable that ingenuity spirals may facilitate national development by aiding generation of ideas that lead to economic outcomes with minimal damage to the environment. Even though growth focuses on quantitative physical or material increase while development focuses on qualitative improvement (Goodland, 1995), the latter can be achieved with the aid of the former when there are specific ingenious ideas.

Ingenuity spirals may be tested in much the same way as other spirals.

First, quantitative methods can be applied to ingenuity spirals. In this age of supercomputers, nonlinear or dynamic modeling approaches can be applied. For example, Quadtree representation can be used not only to depict the spiral but also to measure the incremental changes in the upward spirals. Second, it is possible to use qualitative methods to examine ingenuity spirals. Cases of individuals, organizations, and nations can be explored. For example, by examining ingenuity spirals of individuals identified as 'ecological men', core managers, and exemplary organizations, we may understand qualitative attributes of and differences in ecological ingenuity. It is likely some cases may focus on specific components of upward spirals and other cases focus on all components.

CONCLUSION

Ingenuity spirals are conceptualized here at the organizational level as mechanisms that can enable organizations to solve ecological problems. Ingenuity spirals are proposed to positively influence environmental management effectiveness and organizational outcomes through environmental management decisions of managers, behaviors of employees, and organizational practices. The process of ecological ingenuity in general and ingenuity spirals specifically are also discussed. I hope future research explores how ingenuity spirals help advance corporate environmental sustainability research.

NOTES

1. The other two elements include social and economic sustainability.
2. The definition is also generic with regard to actors because individuals can demonstrate ecological ingenuity in much the same way that groups and organizations can be ecologically ingenious. Individual ecological ingenuity (i.e., an individual's application of ideas to solve an environmental problem) can extend to group and organizational contexts as collective phenomena. In this chapter, I limit it to the organizational level because of my focus on the organization as the unit of analysis.
3. An equivalent of this is 'simplicial complex', which in topology and combinatorics (geometry) is 'glued together' from simplex and complex. The associated combinatorial structure is called an abstract simplicial complex (http://en.wikipedia.org/wiki/Simplex).
4. Ingenious ideas can also transfer from homes and communities to the organization. However, I focus specifically on ingenuity within the workplace and how it facilitates effective corporate environmental sustainability management.

REFERENCES

Andersson, L. and Bateman, T. (2000). Individual environmental initiative: championing natural environmental issues in U.S. business organizations, *Academy of Management Journal*, **43**(4), 548–70.

Azzone, G. and Bertele, U. (1994). Exploiting green strategies for competitive advantage, *Long Range Planning*, **27**, 69–81.

Bassi, L., Frauenheim, E., McMurrer, D. and Costello, L. (2011). *Good Company. Business Success in the Worthiness Era*. San Francisco, CA: Berrett-Koehler Publishers, Inc.

Bos-Brouwers, H.E.J. (2010). Corporate sustainability and innovation in SMEs: evidence of themes and activities in practice, *Business Strategy and the Environment*, **19**(7): 417–35.

Christmann, P. and Taylor, G. (2002). Globalization and the environment: strategies for international voluntary initiatives, *Academy of Management Executive*, **16**(3), 121–35.

Clarkson, M.B.E. (1995). A stakeholder framework for analyzing and evaluating corporate social performance, *Academy of Management Review*, **20**(1), 92–117.

Collier P. (2010). *The Plundered Planet. Why We Must – And How we Can Manage Nature for Global Prosperity*. New York: Oxford University Press.

DeSerpa, A. (1971). A theory of the economics of time, *The Economic Journal*, **81**(324), 828–46.

Elkington, J. (1994). Towards the sustainable corporation: win-win-win business strategies for sustainable development, *California Management Review*, **36**(2), 90–100.

Finkelstein S. and Hambrick, D.C. (1996). *Strategic Leadership: Top Executives and Their Effects on Organizations*. St. Paul, MN: West Publishing.

Florida, R. and Davidson, D. (2001). Gaining from green management: environmental management systems inside and outside the factory, *California Management Review*, **43**(3), 64–84.

Fredrickson, B.L. (2003). Positive emotions and upward spirals in organizations. In K.S. Cameron, J.E. Dutton and R.E. Quinn (eds), *Positive Organizational Scholarship: Foundations of a New Discipline* (pp. 163–75). San Francisco, CA: Berrett-Koehler.

Gardner, G.T. and Stern, P.C. (2002). *Environmental Problems and Human Behavior*, Boston: Pearson Customer Publishing.

González-Benito, J. and González-Benito, Ó. (2006). A review of determinant factors of environmental proactivity, *Business Strategy and the Environment*, **15**, 87–102.

Goodland, R. (1995). The concept of environmental sustainability, *Annual Review of Ecology and Systematics*, **26**, 1–24.

Goodland, R. and Daly, H. (1996). Environmental sustainability: universal and non-negotiable. *Ecological Applications*, **64**(4), 1002–17.

Gunderson, L. and Holling C.S. (eds) (2002). *Panarchy: Understanding Transformations in Human and Natural Systems*. Washington DC: Island Press.

Haden, S.P., Oyler, J.D. and Humphreys, J.H. (2009). Historical, practical, and theoretical perspectives on green management, *Management Decision*, **47**(7), 1041–55.

Hart, S.L. (1995). Natural-resource-based view of the firm, *Academy of Management Review*, **20**(4), 986–1014.

Hoffman, J. and Haigh, N. (2012). Positive deviance for a sustainable world: linking sustainability and positive organizational scholarship. In K.S. Cameron and G.M. Spreitzer (eds), *The Oxford Handbook of Positive Organizational Scholarship* (pp. 953–64). New York: Oxford University Press.

Hoffmann, A.J. and Jennings, P.D. (2011). The BP Oil Spill as a Cultural Anomaly? Institutional Context, Conflict and Change. Ross School of Business Paper No. 1151. Available at SSRN: http://ssrn.com/abstract=1706096 (last accessed on 1 May 2013).

Holling, C.S. (2001). Understanding the complexity of economic, ecological, and social systems, *Ecosystems*, **4**, 390–405.

Holling, C.S., Gunderson, L.H. and Ludwig, D. (2002a). In quest of a theory of adaptive change. In L.H. Gunderson and C.S. Holling (eds), *Panarchy: Understanding Transformations in Human and Natural Systems* (pp. 3–24). Washington DC: Island Press.

Holling, C.S., Gunderson, L.H. and Peterson, G.D. (2002b). Sustainability and panarchies. In L.H. Gunderson and C.S. Holling (eds), *Panarchy: Understanding Transformations in Human and Natural Systems* (pp. 63–102). Washington DC: Island Press.

Homer-Dixon, T. (2000). *The Ingenuity Gap*. New York: Knopf.

Homer-Dixon, T. (2006). *Catastrophe, Creativity, and the Renewal of Civilization*. Washington DC: Island Press.

Hukkinen, J. (1998) Institutions, environmental management and long-term ecological sustenance, *Ambio*, **27**, 112–29.

Hunt, C. and Auster, E. (1990). Proactive environmental management: avoiding the toxic trap, *Sloan Management Review*, **31**, 7–18.

Ireland, R.D. and Hitt, M.A. (1999). Achieving and maintaining strategic competitiveness in the 21st century: the role of strategic leadership, *Academy of Management Executive*, **13**(1), 43–57.

Jackson, S.E., Renwick, D.W.S., Jabbour, C.J.C. and Muller-Camen, M. (2011). State-of-the-art and future directions for green human resource management: introduction to the special issue. *Zeitschrift für Personalforschung (German Journal of Research in Human Resource Management)*, **25**(2), 99–116.

Karliner, J. (1997). *The Corporate Planet: Ecology and Politics in the Age of Globalization*. Sierra Club Books.

Kashmanian, R., Keenan, C. and Wells, R. (2010). Corporate environmental leadership: drivers, characteristics, and examples, *Environmental Quality Management*, **19**(4), 1–20.

Khoury, G., Rostami, J. and Turnbull, J.P. (1999). *Corporate Social Responsibility: Turning Words into Action*. Ottawa: Conference Board of Canada.

Kronfeld-Schor, N. and Dayan, T. (2003). Partitioning of time as an ecological resource, *Annual Review of Ecological Systems*, **34**, 153–81.

Lepak, D.P., Smith, K.G. and Taylor, M.S. (2007). Value creation and value capture: a multilevel perspective, *Academy of Management Review*, **32**(1), 180–94.

Levy, J. and Gopalakrishnan, C. (2010). Promoting ecological sustainability and community resilience in the US Gulf Coast after the 2010 deep ocean Horizon Oil spill, *Journal of Natural Resource Policy Research*, **2**, 297–315.

Margerum, R. (1999). Integrated environmental management: the foundations for successful practice, *Environmental Management*, **24**(2), 151–66.

Metzner, R. (1999). *Green Psychology. Transforming Our Relationship to the Earth*. Rochester, VT: Park Street Press.

Nyang'oro, J. (2007). Africa's environmental problems. In A.A. Gordon and D.L. Gordon (eds), *Understanding Contemporary Africa* (4th edn) (pp. 235–64). Boulder, CO: Lynne Rienner Publishers.

Pascale, R., Sternin, J. and Sternin, M. (2010). *The Power of Positive Deviance: How Unlikely Innovators Solve the World's Toughest Problems*. Harvard Business School Press.

Pradervand, P. (1989). *Listening to Africa: Developing Africa from the Grassroots*. New York: Praeger.

Scott, R.W. (1995). *Institutions and Organizations*. Thousand Oaks, CA: Sage.

Shrivastava, P. and Berger, S. (2010). Sustainability principles: a review and directions, *Organization Management Journal*, **7**(4), 246–61.

Siegel, D.S. (2009). Green management matters only if it yields more green: an economic/strategic perspective, *The Academy of Management Perspectives*, **23**(3), 5–16.

Simon, H.A. (1982). *Models of Bounded Rationality*. Cambridge, MA: MIT Press.

Smith, R.C., Smith, L.M. and Ashcroft, P.A. (2011). Analysis of environmental and economic damages from British Petroleum's deepwater Horizon Oil spill, *Albany Law Review*, **74**(1), 563–85.

Spreitzer, G.M. and Sonenshein, S. (2003). Positive deviance and extraordinary organizing, in K. Cameron, J. Dutton and Quinn, R. (eds), *Positive Organizational Scholarship* (pp. 207–24). San Francisco: Berrett-Koehler.

Steger, U. (2000). Environmental management systems: empirical evidence and further perspectives, *European Management Journal*, **18**(1), 23–37.

Stern, P.C. (2000). Toward a coherent theory of environmentally significant behavior, *Journal of Social Issues*, **56**, 407–24.

Stevens, G.A. and Swogger, K. (2009). *Creating a Winning R&D Culture-II: How a New Five-Step Approach to Increasing R&D Group Effectiveness was Implemented More Broadly in The Dow Chemical Company*. Industrial Research Institute, Inc. Available at: http://www.winovations.com/Articles/Creating_a_Winning_RD_Culture-PartII.pdf (last accessed 20 September 2013).

Stiglitz, J., Sen, A. and Fitoussi, J. (2010). *Mis-Measuring our Lives. Why GDP Doesn't Add Up*. New York: The New Press.

Stumpf, S.A. and Mullen, T.P. (1991). Strategic leadership: concepts, skills, style and process, *Journal of Management Development*, **10**(1), 42–53.

United Nations World Commission on Environment and Development (1987). *Report of the World Commission on Environment and Development: 'Our Common Future'*.

van Marrewijk, M. (2003). Concepts and definitions of corporate sustainability, *Journal of Business Ethics*, **44**(2 and 3), 95–105.

Vlek, C. and Steg, L. (2007). Human behavior and environmental sustainability: problems, driving forces, and research topics, *Journal of Social Issues*, **63**(1), 1–9 (especially Table 1.1).

von Malmborg, F.B. (2002). Environmental management systems, communicative action and organizational learning, *Business Strategy and the Environment*, **11**(5), 312–23.

Weick, K. (1995). *Sensemaking in Organizations*. Thousand Oaks, CA: Sage Publications.

Ziegler, A. and Nogareda, J. (2009). Environmental management systems and technological environmental innovations: exploring the causal relationship, *Research Policy*, **38**(5), 885–93.

Zittrain, J. (2006). The generative internet, *Harvard Law Review*, **119**, 1974–2040.

Zoogah, D.B. (2011). The dynamics of green HRM behaviors: a cognitive social information processing approach, *Zeitschriftfür Personal forschung*, **25**(2), 117–139.

Zoogah, D.B. (2013a). Green management. In T. Lituchy, B.J. Punnett and B.B. Puplampu (eds), *Management in Africa: Macro and Micro Perspectives* (pp. 70–89). New York: Routledge.

Zoogah, D.B. (2013b). Antecedents and consequence of ecological transcendence: an organizational perspective, *Journal of Corporate Citizenship*, **46**, 103–21.

APPENDIX

Table 3A.1 Global environmental issues and problems

Issues	Problems
Pollution and toxics	Spillage from oil being transported (Exxon-Valdez) or drilled (BP in the US Gulf Coast). There is also increased toxicity leading international agencies (e.g., The Stockholm Convention of Persistent Organic Pollutants) to establish rules that would rid the world of 12 hazardous chemicals including PCBs, dioxins, and DDT.
Climate	Global warming due to concentration of carbon dioxide and atmospheric concentrations of methane. According to the WHO, climate change is also likely to lead to malaria resurgence particularly in Africa and Asia.
Energy	World energy is expected to grow by over 55% by 2030. So, there is a drive for renewable energy but the technology is lacking to satisfy the huge demand.
Biodiversity	Increases in global temperatures by about 2–5 degrees Celsius will lead to destruction of a significant proportion of world species.
Forests	The Amazon forest and other major forests in Africa and Asia are fast being destroyed partly because of global demand for beef in Brazil, wood in Asia, and energy in Africa.
Population and urbanization	World population in 2035 is expected to exceed 10 billion. There is also an increased trend toward urbanization. About 60% of the world population will be urban by 2035.
Water	Drinkable water from seas and rivers is being depleted drastically due to severe drought. It is suggested that continued reduction in drinkable water might lead to 'water wars' where countries fight to control water sources, particularly in Africa.
Fisheries	Industrial fishing has killed over 90% of the world's fish species.
Marine systems	The number of oceans and bays with 'dead zones' of water has tripled from 73 to about 219 since 1990.
Wildlife	There has been a surge in killings of endangered wildlife such as tigers, leopards, and deer, due to fur demand in the fashion industry.
Food	There is a projected food crisis, particularly in Southern Africa where about 9 million farmers have died from the AIDS epidemic.
Transportation	Traffic congestions and delays in most of the major cities in US, Europe, China, India, Brazil, and Russia that lead to billions of hours in lost time.

Sources: World Bank (www.worldbank.org), United Nations Development Program (www.undp.org), United Nations Environmental Program (www.unep.org), World Health Organization (www.who.int). See also Vlek and Steg (2007).

Table 3A.2 Corporate role in the global environmental crisis

Environment issue	Problem	Need for ingenious solution
Destruction of the ozone layer	Organizations can change the production processes to reduce chlorofluorocarbons (CFCs) the most serious ozone depletors.	DuPont Corp., for example, used to account for 25% of world production of CFCs, now develops substitutes – HCFCs and HFCs.
	Chemical corporations that produce and distribute the hazardous pesticide methyl bromide.	Eliminate or find substitutes.
Global warming	Transnational corporations account for roughly 50% of all emissions of greenhouse gases due to production of road vehicles, CFC production, as well as electricity generation and use.	Production and use of low emission (e.g., hybrid) transportation vehicles and adoption of renewal energy to replace electricity.
Persistant organic pollutants	Transnational corporations account for most of the world's toxic waste. They use chemicals to make pesticides, synthetic fibers, plastics, pulp and paper, detergents.	Finding of alternative environment-friendly chemicals that can help with production of biodegradable alternatives.
Radioactive waste	The nuclear industry, corporations that service the military-industrial complex, especially nuclear weapons contractors (e.g., Halliburton, Westinghouse and General Electric) have created worst toxic problems.	Discarding nuclear energy.
Mining	Global corporations (e.g., Rio Tinto Zinc, Kobe Steel and Broken Hill Properties) dominate the pollution-intensive mining, refining and smelting of metals such as platinum, aluminum and copper.	Development of environment-friendly mining and refining processes.

High costs of high tech	Silicon wafers and byproducts, such as sodium hydroxide, waste water and hazardous waste.	Recycling of wafers and byproducts.
Agriculture production	Agricultural chemicals from chemical giants such as Shell, Monsanto, Mitsubishi and Sandoz, Phillip Morris, United Fruit, Pepsico, Cargill, Unilever and Nestle destroy land to increase production.	Production of long lasting crops and storage that minimizes frequency of agricultural production.
Deforestation	Timber transnationals such as MacMillan Bloedel, Mitsubishi and Georgia Pacific play a major role in deforestation.	Increased reafforestation and transformation of deserts to forest.
Overfishing	Indiscriminate fishing by corporations (e.g., Spain's Pescanova, Japan's Taiyo, South Korea's Dong Won, and the United States' Arctic Alaska/Tyson Foods) create marine hazards and wipe out traditional finishing communities.	Localized fish production, standardized industry fishing schedules, and enforceable penal codes for violations, in addition to fish farms rather than ocean fishing.

Sources: United Nations Environment Program (www.unep.org), World Health Organization (www.who.int), and Corporate Watch (http://www.corporatewatch.org). See also Karliner (1997).

4 Organizational ingenuity in the commercialization of early-stage technological innovations

Larry W. Cox, Ana Cristina O. Siqueira and John G. Shearer

INTRODUCTION

Organizational ingenuity represents the ability to create innovative solutions within structural constraints using limited resources and imaginative problem solving (Lampel et al., 2011). This chapter presents a case study in which one university (University A), constrained by the fact that it possesses no intellectual property (IP) of its own, utilized organizational ingenuity and an ideation team model, in partnership with an IP-rich university (University B) to increase the commercial potential of University B's inventions. This case within a university is an instructive example of organizational ingenuity.

Understanding how and by whom entrepreneurial opportunities are discovered and created (Alvarez and Barney, 2007; Shane and Venkataraman, 2000) has persisted as a central question in the field of entrepreneurship (Wiklund et al., 2011). The subject of opportunities (Venkataraman, 1997) has received numerous conceptual and empirical studies, resulting in a critical body of knowledge in the entrepreneurship literature (Short et al., 2010).

Individuals may create business ideas through various paths, such as serendipitous discovery or fortuitous chain of events. Additionally, as a means to launch their ventures based on more promising ideas, aspiring entrepreneurs may perform a 'systematic search' to discover opportunities (Fiet, 2007). Opportunity discovery may occur via deliberate search when entrepreneurs recognize opportunities as 'problem-solution pairings' (Hsieh et al., 2007). For instance, new business ideas may emerge when individuals are unsatisfied with current products and thus develop alternatives to solve existing problems. This study develops a model that underscores the relevance of creative problem solving processes in a team setting for the discovery and creation of opportunities.

This study generates three main contributions to the entrepreneurship literature on organizational ingenuity and opportunity identification. First, we present a case study in which one university utilized organizational ingenuity and an ideation team model to increase the commercial potential of inventions. Our findings suggest that conditions such as engaging entrepreneurs with innovators earlier in the commercialization process and using principles of creativity are likely to facilitate organizational ingenuity.

Second, we extend Shane's (2000) original model of how prior knowledge influences opportunity discovery, by developing an ideation team model that emphasizes the role of individuals' prior knowledge in a team setting and the relevance of creativity processes and skills. We propose that the ideation team model helps improve the number and innovativeness of opportunities. In this way, we also extend prior research on moderators of opportunities (Short et al., 2010), by investigating how the ideation team model moderates the relationship between innovation and opportunity identification.

Third, our study contributes to the view of entrepreneurship as entailing knowledge and methods that can be learned and taught (DeTienne and Chandler, 2004; Fiet, 2007; Sarasvathy and Venkataraman, 2011), by examining how the interaction of individuals in teams can be facilitated by the use of creative problem solving processes and skills. Prior studies have offered little guidance to aspiring entrepreneurs when they describe opportunity recognition as primarily based on intuition or luck, because it is unclear how these could be taught and learned (Fiet, 2007). In contrast, aspiring entrepreneurs may use the ideation team model to organize and improve their process to identify opportunities.

THEORETICAL CONSIDERATIONS

University IP and Commercial Opportunities

Since passage of the Bayh–Dole Act in 1980, universities have increasingly been seeking ways to 'transfer' their technologies into the commercial space (Shane, 2004; Nelson, 2001). According to the Association of University Technology Managers (AUTM, 2012) the annual revenue generated by US universities from licensing their technological discoveries to commercial players has risen from around $160 million in 1991 to $2.4 billion in 2010 – a 15-fold increase over a 20-year span. Siegel et al. (2007) explain that this dramatic shift toward 'application' is a response by universities to the newfound possibility of leveraging their basic research to: a) generate additional revenue; b) create job opportunities for university-based researchers; and c) stimulate regional economic development. Yet as impressive as it sounds, this $2.4 billion in licensing revenue represents only a 4 per cent return on the $59.1 billion invested by the US government in university-based research in 2010 – a return that is substantially below what might be expected given the universities' increased motivation to garner additional revenues (Swamidass and Vulasa, 2009; Markman et al., 2005) and the over 20,000 university-based disclosures announced each year (AUTM, 2012). Further, as documented by Shane (2002), half of all university patents are never licensed and the top 20 universities account for approximately 70 per cent of those that are. Macho-Stadler et al. (2007) note that the empirical evidence 'clearly shows that significant institutional barriers to the commercialization of basic research remain' and 'smooth interactions between science and industry . . . have become a central concern of many government policies' (p. 484). Nelson (2001) adds, 'Very few university "inventions" garner significant license incomes. It is almost a sure thing that many universities are paying significantly more to run their patenting and licensing offices than they are bringing in license revenues' (p. 17).

Researchers have probed the ineffectiveness of existing efforts at 'technology transfer' from many angles. Some have looked at the university itself to see if being more 'business friendly' might result in increased IP commercialization (Franklin et al., 2001; O'Shea et al., 2005). Others have assessed the impact of university licensing strategies, policies and incentives (Markman et al., 2009), institutional prestige (Sine et al., 2003; Di Gregorio and Shane, 2003), as well as the structures, characteristics and policies of technology transfer offices (Chapple et al., 2005; Macho-Stadler et al., 2007). Still others have investigated 'the individual and social factors that shape the decisions of academics to engage in entrepreneurial behaviors' (Clarysse et al., 2011). Shane (2002), however, proposes that the best solution for commercializing university-generated technologies must center around the economic actors with a 'comparative advantage' in commercializing technologies – those individuals skilled in 'identifying customer needs, developing product concepts, designing products and processes, prototyping and manufacturing' (p. 123). That is, entrepreneurs. He argues that since university inventors rarely possess these skills, entrepreneurs are therefore better suited to the task.

Indeed, academics have long argued that entrepreneurs are essential to the commercialization of technology. Schumpeter's model of 'creative destruction', for example, posits that while novel technologies allow for the creation of new processes, new products, new markets, and new ways of organizing, it is entrepreneurs who effectuate the change in the marketplace – it is entrepreneurs who identify and then exploit the economic opportunities presented by advancements in science (Schumpeter, 1934). Recent research funded by the US Small Business Administration supports this contention. In one such study, Camp (2005) reports, 'innovation without entrepreneurship generally yields minimal local economic impact' (p. 5). It is thus reasonable to begin an investigation of the lower-than-expected rate of IP commercialization at the university level with an examination of the role of entrepreneurs in the existing process. As Wright, Birley and Mosey (2004) state, 'It is important to devote greater attention to the study of entrepreneurship in technology transfer' (p. 235).

Opportunity Identification and the Prior Knowledge of Entrepreneurs

Scholars continue to debate the extent to which commercial opportunities are 'discovered' or 'created' by entrepreneurs (Alvarez and Barney, 2007). That is, whether opportunities have an existence apart from human perception and action, or exist only as a result of path-dependent human enactment. The 'discovery' camp assumes that opportunities exist objectively and independently of human cognition and social interaction, and are only 'discovered' when individuals with certain characteristics, including 'alertness', recognize and exploit them (Kirzner, 1997; Shane, 2000). In contrast, proponents of the 'creation' view assume that opportunities do not exist apart from the perceptions and behaviors of humans, and therefore only come into being when individuals expend mental effort exploring ways to produce new products and services (Baker and Nelson, 2005; Sarasvathy, 2001). Seeking a middle ground, Zahra (2008) proposes a theory whereby certain contexts are more conducive to 'discovery' and others to 'creation'. He suggests the possibility of a 'virtuous cycle' in which the discovery of one opportunity encourages the creation of another, and vice versa. Edelman and Yli-Renko (2010) on the other hand, posit that 'opportunity perceptions' mediate between the 'objective characteristics of the environment' and 'efforts to start a new venture'. That is, opportunities exist in the environment apart from the entrepreneur, but the entrepreneur's perceptions regarding these opportunities are key to determining his or her level of entrepreneurial activity.

The authors propose an alternative view – a creative problem solving perspective which assumes that opportunity identification is a process involving first the 'discovery' of an existing (but previously unknown, unconsidered or under-considered) market problem, followed by the 'creation' of a novel and practicable market solution. As such, we assert that opportunity identification is the result of a dynamic interaction between discovery and creation. We further suggest that the identification of opportunities is not dependent on the presence of individual characteristics such as 'alertness', but rather involves 'prior knowledge' that is unique to each individual.

With regard to the entrepreneur's 'prior knowledge', the creative problem solving approach is compatible with both the discovery and creation perspectives. The discovery view (Shane, 2000) explicitly hypothesizes that 'prior information, whether developed from work experience, education, or other means, influences the entrepreneur's ability to comprehend, extrapolate, interpret, and apply new information in ways that those lacking that prior information cannot replicate' (p. 452). It further describes the domains of prior knowledge that are applicable to opportunity identification as: a) markets; b) ways to serve markets; and c) customer problems. On the other hand, the creation view, while 'silent on whether ex-ante differences are necessary and sufficient for the (identification and exploitation of opportunities) to exist' (Alvarez and Barney, 2010: 566), still proposes that one element in a set of 'means' for originating a new business is 'what an individual knows' (Sarasvathy, 2001: 253). The creative problem solving perspective builds on these approaches proposing that prior knowledge is essential for opportunity identification, but specifically for the purpose of discovering market problems.

The Role of Entrepreneurs in IP Commercialization

Figure 4.1 illustrates milestones in the process of commercializing IP (Carlsson and Fridh, 2002), as well as the steps performed by the typical university technology transfer office (UTTO; McAdam et al., 2005; Center for Genomic Regulation, 2013; Yale University Office of Cooperative Research, 2008). The first step, 'technology appraisal', involves a preliminary evaluation of university-based inventions for the purpose of eliminating from further consideration those inventions not deemed technologically viable or unique to the IP landscape. This may result in: (a) the release of a technology without further spending on commercialization or IP protection by the university; (b) passive marketing (i.e., posting the technology to an online site) for that technology; or (c) passing the technology through to the second step, 'commercial assessment'. This second step utilizes more market research and in-depth industry analysis to help university decision makers determine if commercialization and patenting should be pursued for a given technology. Technologies whose market potential is deemed insufficient, are, again, either released or posted online for passive marketing. The rest are moved forward to 'patenting writing', the third step.

Moving forward, the 'marketing' step involves the development of a

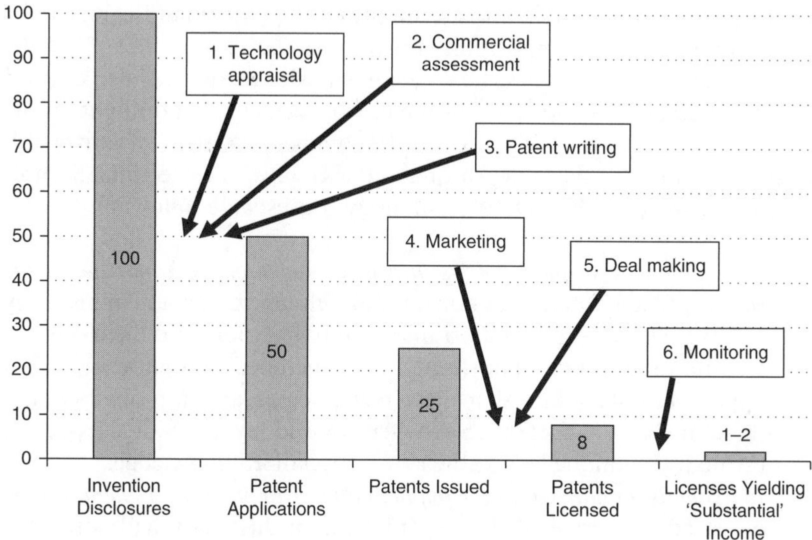

Note: Path for every 100 university disclosures.

Sources: Adapted from Carlsson and Fridh (2002) and Yale University Office of Cooperative Research (2008).

Figure 4.1 Stages, activities, and outcomes of the typical technology process

marketing campaign on behalf of those technologies that have received a patent. In this step, UTTOs seek to gain interest from targeted commercial entities in order to pass technologies into the fifth step, 'deal making', where UTTOs enter into negotiations with interested business entities over the terms of a licensing agreement. When these negotiations are complete, the licensed technologies are counted as 'successes', though it might be argued that only those deals that create substantial revenue for the university are truly 'successful'.

It is important to note that the above process and practices assume that UTTO personnel and university inventors working together are able to determine market potential for disclosed inventions apart from input from entrepreneurs. Indeed, entrepreneurs are not introduced to a given technology until it is being marketed by the UTTO to the public. It is also important to note that these practices present IP commercialization primarily as a winnowing process. That is, from the moment inventions are disclosed to the UTTO by the inventor, the goal of each subsequent step is to eliminate those technologies that will require significant additional research and development, are not unique on the patent landscape and/or do not represent great opportunity for commercialization.

However, in his seminal article, 'Prior knowledge and the discovery of entrepreneurial opportunities', Shane (2000) sheds important light on the underlying mechanisms of successful IP commercialization of university-based technology. His in-depth qualitative analysis of the eight licenses awarded around MIT's 3DP™ technology suggests the following:

a. *The commercial opportunities in a given technology do not lie on the surface.* Shane found that none of the eight entrepreneurs in his study discovered the same opportunity, and that none of them uncovered more than one opportunity. He concluded that entrepreneurial opportunities contained in new technologies are non-obvious and therefore only 'discoverable' when viewed by individuals peering through the unique lens of their own specialized knowledge.
b. *Individuals do not search for opportunities – they discover them.* Shane observed that none of the entrepreneurs in his study had contacted MIT's Technology Transfer Office to inquire about the 3DP™ technology. Rather, they had heard about it 'from someone directly involved in its development, and recognized the opportunity immediately upon hearing about it . . . almost by accident, as if they were surprised by the discovery' (p. 457). He concluded that the discovery of commercial opportunities in basic research is somewhat 'serendipitous'.
c. *Individuals exploit opportunities in markets and with methods reflecting their prior knowledge.* Each of the eight licensees in Shane's study sought to use the 3DP™ technology in a different market, with a different approach, and to solve a different customer problem. According to Shane, their approach for exploiting an opportunity was 'idiosyncratic, resulting from work experience, personal events, and education' (p. 459). In other words, in each case, the opportunity that was identified was clearly linked to a specific entrepreneur and their unique perspective.

The Ideation Team Model and IP Commercialization

Figure 4.2 shows the model for IP commercialization that was developed by the authors. The model extends Shane's (2000: 453) original conceptual model in the following ways. First, it supports his view that the relationships between innovation and opportunity identification, and opportunity identification and exploitation are moderated by prior knowledge. Yet, it incorporates two additional elements – 'team' and 'creativity processes and skills' – which, together with prior knowledge, comprise what we

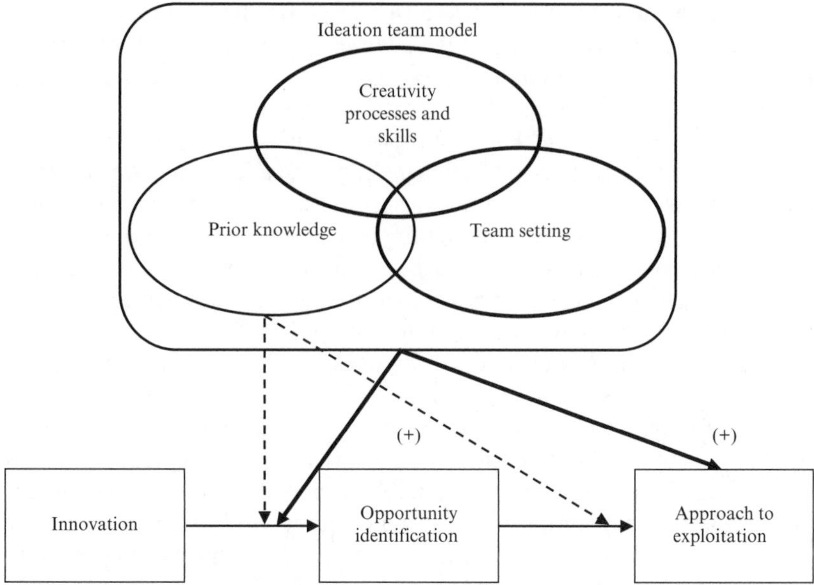

Note: The dotted arrows represent Shane's (2000) original conceptual model. The bold lines represent the authors' new elements that compose the ideation team model and indicate its effects.

Figure 4.2 Ideation team model and its effects

call 'the ideation team model'. We suggest that prior knowledge can be expanded, magnified and fostered via the ideation team model to produce superior results in identifying commercial opportunities. Finally, we assert that IP commercialization is a process of matching a given solution (i.e., technology) to multiple market problems – an application in which prior knowledge is especially important.

Enlarging the pool of prior knowledge

In general, the literature supports the notion that the identification of opportunities to commercialize technological innovations can be influenced by individuals' prior knowledge (Shepherd and DeTienne, 2005; Bingham et al., 2007). It also supports the proposition that multiple opportunities may emerge from a single technological innovation (Shane, 2000). The ideation team model emphasizes the importance of prior knowledge and posits that the utilization of multiple entrepreneurs will greatly expand the pool of prior knowledge – specifically prior knowledge of market problems – thereby substantially increasing the number of commercial opportunities that are identified.

Increasing social interaction between entrepreneurs and inventors

From a resource-based perspective (Barney, 1991), the prior knowledge of team members represents a resource (Wernerfelt, 1984) that may be utilized to leverage the development of business ideas. Specifically, 'the unique ways in which entrepreneurs think and expose themselves to a varied cross-section of social interactions allow them to accumulate the necessary and sometimes rare resources' (Alvarez and Busenitz, 2001: 768). Indeed, social interaction between serial entrepreneurs and other potential contributors – such as inventors, investors, customers, family, and friends – is essential for identifying and shaping opportunities (Sarasvathy, 2001: 249). According to Dimov (2007), the social audience with which individuals engage to discuss their ideas may affect the generation and development of those ideas. We suggest, therefore, that the pairing of entrepreneurs with inventors will enhance and strengthen the process of opportunity identification by giving the entrepreneurs access to technical facts not otherwise available. It will also increase the number of alternatives, encourage unusual connections, and assist in creatively solving problems especially with regard to exploiting the opportunity (Shalley and Perry-Smith, 2008).

Applying creativity skills

Amabile and Mueller (2008) include 'creativity-relevant processes' and 'domain-relevant skills' in their model of creativity, stating that these are two of the four components that 'should facilitate the creative process' (p. 37). According to these researchers, domain-relevant skills include 'knowledge, expertise, technical skills, intelligence and talent in the particular domain where the problem-solver is working', and creativity-relevant processes include 'a cognitive style and personality characteristics that are conducive to independence, risk-taking, and taking new perspectives on problems, as well as a disciplined work style and skills in generating ideas' (p. 35). Though not an exact match, Amabile's domain-specific skills are similar to the 'prior knowledge' element in our ideation team model, and her creativity-relevant processes are a very close match to Basadur's creative problem solving system (Basadur, 1994; Basadur and Gelade, 2006). Basadur, however, elaborates on the specific creativity skills required for idea generation, 'divergence, judgment deferral, and convergence' (Basadur, 1994).

Of the three skills included in Basadur's model, divergent thinking has received the most attention in the literature (e.g., Vincent et al., 2002). It involves non-judgmentally creating new options to increase the number and variety of ideas (Basadur and Gelade, 2006; Guilford, 1950). According to Thompson (2003), it is the toughest skill to master but is critical to the creative process. She states that, 'teams need to engage in divergent thinking in which they put aside typical assumptions' (p. 96) but adds that 'teams [actually] excel at convergent thinking [not divergent thinking]' (p. 99). Thompson continues by noting that the use of a trained facilitator can help a team diverge more effectively, a practice used by IDEO (Kelley and Littman, 2001). Cheung et al. (2006) add that a one-semester university course does in fact enhance creative abilities.

Groups practicing the creativity skills of divergence and deferral of judgment tend to generate a greater number of ideas – to accomplish what Guilford (1950) called 'fluency'. Interestingly, researchers (e.g., Silvia et al., 2008) have a difficult time empirically separating *fluency* (i.e., the number of ideas generated) from *flexibility* (i.e., the number of different types of ideas generated), and *originality* (i.e., the uniqueness of the ideas generated). Silvia et al. bemoan the fact that 'confounding [fluency, flexibility and originality] is inevitable because the likelihood of generating a unique response increases as the number of responses increases' (p. 69). Thompson (2003) adds, 'There is always a striking correlation among the three measures, such that the people who get the highest scores on originality also get high scores on flexibility and fluency. Thus, there is a strong association between quantity, diversity, and novelty of ideas' (p. 98).

Hypotheses

Figure 4.3 displays the ideation team model in a matrix. On one axis, the matrix shows social interaction between entrepreneurs and innovators. Social interaction broadens the pool of prior knowledge and metaphorically may serve as a prism. Aspiring entrepreneurs or innovators may work individually – use a single lens – to translate an innovation into a business opportunity. Or they may employ the ideation team model – use a prism – to enhance their identification of opportunities. The divergence

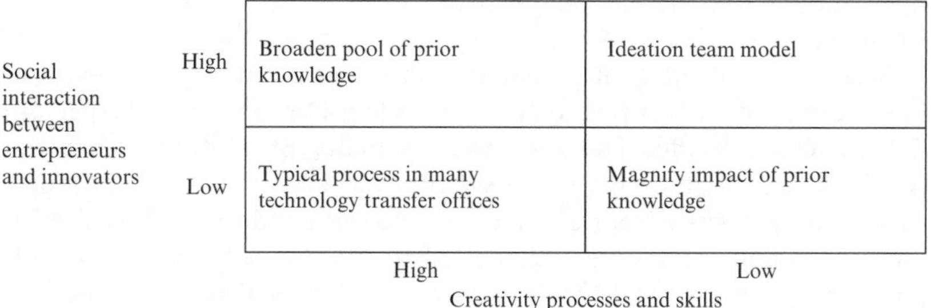

Figure 4.3 Social interaction and creativity matrix

skill implies that team members should deliberately use multiple lenses or perspectives to broaden the breadth of ideas. By changing perspectives, team members may identify a greater number of opportunities.

The other axis in Figure 4.3 shows creativity processes and skills, which can magnify the impact of prior knowledge through a better understanding of the roles of 'deferral of judgment' and 'divergence'. By providing a systematic approach that guides the interaction of team members to generate more ideas within a given time period, the creativity processes and skills above can increase the number of opportunities identified. While systematic routines aimed at innovation are not a guarantee of success, the use of such routines may help reduce the average time and cost of the search for innovation (Nelson and Winter, 1982). In this way, aspiring entrepreneurs may use the ideation team model to improve the number of opportunities identified.

Hypothesis 1: For a given innovation, the use of the ideation team model will be positively associated with the identification of a greater number of commercial opportunities.

The innovativeness of opportunities may be enhanced if team members use the creativity skills of divergence and judgment deferral during the idea finding process (Osborn, 1957; Thompson, 2003; Shepherd and DeTienne, 2005). These skills allow team members to use their imagination to a greater extent by avoiding censoring ideas in the divergence phase. They also encourage individuals to deliberately use multiple perspectives to extend the breadth of ideas. Moreover, these skills encourage individuals to generate ideas even if they seem unfeasible, untested, or apparently impossible (Basadur, 1994, 2004).

Hypothesis 2: For a given innovation, the use of the ideation team model will be positively associated with the identification of commercial opportunities that are more innovative.

A challenge of the commercial assessment phase in a typical technology transfer process is that very few people (often only a single person) determine the technical and market feasibility of a given technology (Carlsson and Fridh, 2002). Considering the fact that multiple opportunities may emerge from an invention depending on individuals' prior knowledge and experiences (Shane, 2000), the limited exposure to different people makes it difficult to uncover additional potential uses for the technology. Specifically, entrepreneurs and investors are typically invited to interact with the technology only in the final stage of licensing or commercialization. This prevents entrepreneurs from bringing a greater contribution to the process of opportunity identification.

In contrast, the ideation team model helps engage entrepreneurs with innovators earlier in the commercialization process. By using principles of creativity, the ideation team model guides the interaction between innovators and entrepreneurs during the initial commercial assessment, which may positively influence the subsequent phases toward commercialization. As a result, this model has the potential to enhance not only opportunity recognition, but also the approach to exploit opportunities.

Hypothesis 3: For a given innovation, the use of the ideation team model will be positively associated with the approach chosen for exploiting opportunities.

METHODOLOGY

Context

This research examines the performance of a team-based, creative problem solving approach (i.e., the ideation team model) to IP commercialization that was embedded in a pilot program involving two very different universities. University A is a medium-sized private university with no Engineering or Medical schools, no graduate-level Science programs and no IP of its own. University B is a large state university with a sizeable Engineering school and 'very high research activity', according to the Carnegie Foundation. The latter agreed to provide the former with access to four of its early-stage technologies and their respective scientists for the purpose of exploring the effectiveness of this unique approach. The case study specifically assesses how the ideation team model influences the number and innovativeness of identified opportunities by evaluating conditions before and after its usage with the two inventors/inventions that opted into the study.

Dependent Variables

For this study, we examine the impact of the ideation team model on the *number of opportunities identified* (Shepherd and DeTienne, 2005), as well as the *innovativeness of opportunities identified* (DeTienne and Chandler, 2007). The former is merely the total number of potential commercial applications generated by entrepreneurs, scientists and MBA students. The latter is measured with the instrument used by DeTienne and

Table 4.1 The innovativeness of opportunities

Scale	Description
1	A replication of existing products or services used in similar applications.
2	A new application for an existing product or service with little or no modification (e.g., treatment for migraines also helps with back spasms).
3	A minor modification to an existing product or service (e.g., slight modification to a knee brace).
4	A significant improvement to an existing product or service (e.g., surgical instruments that do not stick to body tissue).
5	A combination of two or more existing products into one unique product or service (e.g., camera and cell phone).
6	A product or service that is new to the world (e.g., wireless technology).

Source: DeTienne and Chandler (2007: 378).

Chandler (2007: 378) and summarized in Table 4.1. The two innovators responsible for the technologies below provided the innovativeness scores for the opportunities in Tables 4.2 and 4.3.

Technologies

This case utilized a total of four technologies. However, only three of them were released by their scientists/owners for inclusion in a research project designed to measure the impact of the ideation team model on their commercialization. Two of the released innovations for Pilot 1 came from a portfolio of technologies that had been assigned to the technology transfer office of University B. These innovations were nominated for inclusion in the program by the technology transfer office based on their potential for commercialization.

Technology 1 is a high-throughput label-free nanoparticle analyzer invented to measure nanoparticles suspended in fluids. It is capable of detection rates of about half a million particles per second, so that large sample numbers and significant population statistics can be achieved in seconds to minutes. This device has applications in the fields of pharmaceutical development, medicine, and industrial materials. It also makes it possible to produce a portable nanoparticle counting device for a range of nanoparticles for which there is no similar counting device currently in the industry.

Technology 2 includes new processes and equipment for the solution deposition of zinc oxide. Zinc oxide is a promising alternative to indium tin oxide, and might be used to replace indium tin oxide as an essential part of many semiconductor-based technologies, such as light-emitting diodes, flat-screen displays, touch-screen displays, flexible displays and thin film solar cells. The new processes can produce zinc oxide applications at a fraction of the price of indium tin oxide products. They also make it possible to significantly reduce the cost of production for the components of flat screens and solar displays.

Participants

Scientists with in-depth prior knowledge of their respective innovation

Success with the new model for IP commercialization is largely dependent on participation by scientists and inventors with extensive prior knowledge about their innovation. As creators or co-creators, their intimate familiarity with their respective technology is vital, especially during the 'fact finding' portion of the creative problem solving process, as well as the 'feasibility' and 'go-to-market' phases of the pilot program. Pilot 1 included four post-doctoral researchers and one faculty member from University B – one individual from each research team. However, only two of the post-doctoral researchers agreed to participate in the accompanying research project and signed informed consent forms.

Entrepreneurs with diverse prior knowledge of market problems

As asserted by Shane (2002), entrepreneurs are skilled in identifying customer needs. Seasoned entrepreneurs, in particular, bring vast and diverse prior knowledge to the creative problem solving process. They are especially adept at making connections between disparate pieces of knowledge for the purpose of solving market problems. This makes them extremely useful in the 'idea finding' phase of the creative problem solving process – especially after uncovering new facts from their interaction with the scientist. In Pilot 1, this group included 12 venture founders, CEOs, investors and business consultants with experience in entrepreneurship.

MBA students with prior knowledge of creativity processes and skills

MBA students are typically required to have at least five years of business experience prior to being admitted to an MBA program. As a result, they are likely to have some knowledge of market problems that will be beneficial during the 'idea finding' portion of the creative problem solving process. More importantly, however, the students invited to participate in these pilot programs had received training in creativity skills and processes, and through their coursework were also very familiar with market analysis. Both domains are key for determining the feasibility and appropriate go-to-market strategies for the commercial applications generated during 'idea finding.' For this pilot program, ten students were selected from those who applied from University A. Two students were then assigned to each of University B's four technologies. However, only four of the students agreed to participate in the research project – two from Innovation 1 and two from Innovation 2 – and signed consent forms. Students from University A all signed informed consent forms following procedures specified by the Institutional Review Board.

Timing and Process

Pre-Discovery Day

Data were collected in the spring of 2011. Only the scientists were included in this phase of the research project and 'prior knowledge' consisted solely of what they knew about their respective innovations. The scientists were specifically asked: (a) to list all of the commercial opportunities for their particular innovation which they themselves had already identified; and then (b) to rate the innovativeness of each opportunity on the 6-point scale described in Table 4.1. These questions had been pre-tested with 12 MBA students from University A, who confirmed that they were worded in a manner that was sufficiently clear and that the scale for 'innovativeness' was reasonably user-friendly with examples provided in most categories.

Discovery Day

This kickoff event was held in a single day in the spring of 2011 and attended by scientists, entrepreneurs and MBA students. As part of Discovery Day (and for each technology in turn), participants first engaged in 'fact finding'. That is, a process whereby the scientist presents their innovation and then answers questions directed to them from the other participants. This is done to help the entrepreneurs and MBA students gain familiarity with the innovation, as well as discover its boundary conditions. Next, all participants engaged in 'idea finding', a process for generating a large number of potential commercial applications for the innovation and selecting the most promising ideas from those that are generated by the group. In both 'fact finding' and 'idea finding', participants used the creativity skills of 'diverging', 'converging', and 'deferral of judgment'.

Six scientists from University B, nine MBA students from University A, twelve investors and practitioners with experience in entrepreneurship, and ten faculty and staff members from Universities A and B attended Discovery Day for Pilot 1. Once the top opportunities were selected, the

Table 4.2 Innovation 1: nanoparticle analyzer

Time	Prior knowledge	Number of opportunities identified	Innovativeness of selected opportunities	Description of selected opportunities
Before	1 Innovator	1	6	Bench-top analysis tool for nanoparticle
After	2 Ideation team	32	6	Water testing for nanoparticles
			6	Test effectiveness of other filters
			6	Medical diagnostics
			6	Check purification of viruses and cells
			6	Quality assurance and control for industrial chemicals
	3 Ideation team	NA	4	Cosmetic testing
			4	Hand-held, small, portable nanoparticle counting device
			4	Drug development screening
	4 Innovator	NA	NA	Nanoparticle analyzer

two scientists provided the innovativeness scores for these specific opportunities. The number of opportunities identified, number of opportunities selected, and description of selected opportunities for Innovations 1 and 2 are displayed in Tables 4.2 and 4.3.

Feasibility phase

During this phase, five teams each composed of two MBA students from University A, one scientist from University B and members of the Founding Faculty (acting as mentors) worked to determine the technical, financial and market feasibility of the commercial applications generated at Discovery Day. While the MBA students were able to meet on a weekly basis, the scientists were only able to meet with their respective teams twice throughout this phase. The goal of the Feasibility Phase was to narrow the list of approximately 40 potential applications for each innovation to a list of those five with the largest addressable market, least time to market, and potential to disrupt the market. This phase was conducted only as part of Pilot 1 and the teams were paid a stipend for their work.

Table 4.3 Innovation 2: solution deposition of zinc oxide

Time	Prior knowledge	Number of opportunities identified	Innovativeness of selected opportunities	Description of selected opportunities
Before	1 Innovator	4	4	Method and equipment for producing zinc films
After	2 Ideation team	27	3	Solar photovoltaic windows
			3	Medical sensors attached to the skin
			4	Alternative to vapor deposition coating
			4	Piezoelectric generator
			6	Glasses that automatically change refraction
	3 Ideation team	NA	4	Solar displays
			5	Smart windows
			6	Alternative to existing manufacturing technology for flat panels
	4 Innovator	NA	NA	Manufacturing technology for flat panels

Opportunity Day

Approximately two months after Discovery Day, the faculty founders hosted Opportunity Day, a one-day event at which student teams formally reported the results of their investigations to a large audience of business practitioners. Attendees at this event included two innovators, 11 MBA students from University A, 55 investors and practitioners with experience in entrepreneurship, and 23 faculty, staff, and students from Universities A and B. During the event, ten senior business practitioners and investors with experience in entrepreneurship were invited to complete questionnaires asking about the innovativeness of the commercial applications presented by the teams. Nine of them completed questionnaires and signed informed consent forms according to University B's Institutional Review Board procedures. These results are shown in Tables 4.2 and 4.3.

Post-Opportunity Day

The scientists were free to work with the MBA students and entrepreneurs after Opportunity Day, but there were no structured programs or planned interactions between the various groups after that day. However, approximately 4 months later, the founding faculty invited the two scientists and four MBA students from Pilot 1 to participate in interviews. Two of the MBA students accepted the invitation and each completed a 30-minute semi-structured telephone interview.

RESULTS

The Number of Opportunities

The impact of the Ideation Team Model on the number of opportunities identified for a given innovation is seen by comparing the number of opportunities considered by the scientists before Discovery Day to the number generated as a result of using the Ideation Team Model on Discovery Day (Shepherd and DeTienne, 2005). Tables 4.2 and 4.3 show the results for Innovations 1 and 2. For the nanoparticle analyzer (Innovation 1) only one commercial opportunity had been considered by the scientist prior to Discovery Day. After the use of the ideation team model, the number of potential commercial applications increased to 32. Not all of these ideas were viable at that point, but they increased the pool of possible applications. For the solution deposition of zinc oxide (Innovation 2) four opportunities had been considered by the scientist, whereas the ideation team model generated 27. These results indicate that the ideation team model is indeed associated with an increase in the number of opportunities, thus supporting Hypothesis 1.

The Innovativeness of the Opportunities

Tables 4.2 and 4.3 display the scientists' perceptions regarding the innovativeness of the opportunities. With respect to Innovation 1, the level of innovativeness remained the same at the maximum level with a score of 6 in Time 1 and Time 2. As one participant stated, 'because our technology is new and provides capabilities that were not previously available, any application of the technology is by default maximally innovative.' The results for Innovation 2 support Hypothesis 2. In this case, the variability of innovativeness increased from a consistent 4 before Discovery Day to a range of 3 to 6 on Discovery Day and Opportunity Day. This increase in the peak score for innovativeness seems to be associated with a movement from more generic business ideas – i.e., 'method and equipment' – to more specific applications – i.e., 'solar displays' and 'flat panels' – on Discovery Day and Opportunity Day. Because some opportunities for Innovation 2 are perceived to have greater innovativeness after the use of the ideation team model, this provides some support for Hypothesis 2.

The Ideation Team Model and Approach to Exploitation

Tables 4.2 and 4.3 show the path for each innovation from its initial assessment by the individual scientist to subsequent phases after the use of the ideation team model. The data in Tables 4.2 and 4.3 indicate that the opportunities have changed from more *generic* (e.g., 'bench-top analysis tool,' 'method and equipment,' and 'cleaning products') before Discovery Day to more *divergent* after Discovery Day. For instance, the variety of opportunities increased by including applications in different industries for Innovation 1 (e.g., 'medical' and 'industrial chemicals') and Innovation 3 (e.g., 'outdoor recreation equipment cleaners'). This increase in variety of opportunities seems associated with the use of such creativity skills as deferral of judgment and divergence on Discovery Day. This change from Pre- to Post-Discovery Day is consistent with Hypothesis 3.

Additionally, the description of opportunities for Innovations 1 and 2 has become even more *specific* by Opportunity Day. For instance, a focal application has become 'flat panels' for Innovation 2. For the nanoparticle analyzer, an entrepreneur provides a description of the technology that is more user-friendly for a potential customer and that describes an application of the technology as 'hand-held, small, portable sub-nanoparticle cell or particle counting device'. This entrepreneur provided the following comments to the team: 'Market potential overstated on the [team's] presentation. Cell or particle counting market is approximately $250 million.' These comments suggest that her prior knowledge of the medical industry related to the technology allows her to provide very specific guidance to the team, which implies that this type of interaction with entrepreneurs can positively influence the approach to exploit this technology. Therefore, these results based on Innovations 1 and 2 are generally congruent with Hypothesis 3.

Post-Opportunity Day, the scientists had the option of working alone. There was no facilitator to coordinate the process and there was no requirement to use the ideation team model as a systematic approach. For Innovation 1, the scientist continued his work for the most part independently from the MBA students and the entrepreneurs, one of whom reported a concern with preventing knowledge spillover when interacting with a social audience. For Innovation 2, the scientist continued to work with at least one of the MBA students to advance his innovation. Together, they searched and contacted key business practitioners who could potentially inform the project and might be interested in becoming corporate customers of the technology. Eventually, the scientist continued his work independently from the MBA students and entrepreneurs by focusing on IP issues. The findings from after Opportunity Day suggest that without continued use of the ideation team model there may be changes in the approach to exploitation, with a concentration of the work returning to the individual scientist.

DISCUSSION

In general, participants in both of the pilot projects agreed that this program clearly demonstrated the efficacy of the ideation team model for IP commercialization. Indeed, many expressed their amazement at the quantity and diversity of commercial applications generated during Discovery Day, and spoke of the importance of this process for bringing IP to market. However, in discussions with several of them following Opportunity Day, the founding faculty concluded that while the pilot programs were successful, the entire process needed to be revised in order to fully accomplish its ultimate goals. Specifically, Opportunity Day attendees expressed their disappointment that the presentations by the teams of MBA students and scientists were too 'early stage' and inconsistent. They wanted the presentations to approach what one would expect from a complete business planning process, not just from an analysis of 'feasibility'. They also noted a good deal of variability in quality and formatting from presentation to presentation. Participants also expressed frustration that for three of the five presentations, the scientist was not in attendance and that the presentations, therefore, were not 'technical' enough. For their part, the MBA students complained that the scientists were not readily accessible during the Feasibility Phase, noting that this made it very difficult for them to assess the various commercial opportunities given to them from Discovery Day. They also felt that the time between Discovery Day and Opportunity Day was far too short for students who were, for the most part, working in full-time jobs throughout the program. Most importantly, however, participants were upset by the fact that University A had no means for capitalizing on the hard work of its faculty and students. Given University A's 'in kind' and financial support for the first pilot, they were shocked that no agreement had been executed between University A and University B prior to Discovery Day. They felt that even though University A never had (and did not expect) ownership in the innovations, it should have negotiated for licensing rights or options prior to starting the project. As a result, University B was positioned to be the sole beneficiary from the increased exposure to its IP generated during the pilot, while University A was prevented from sharing in any financial gains that might result from its efforts. In addition, it was later discovered that some of the innovations were not free to be licensed – that other entities had already negotiated for exclusive rights or first rights of refusal which precluded any additional licensing arrangements.

In response to this important feedback, the founding faculty worked with University A officials to revamp the ideation team model in the following ways:

Formation of an Independent Research Foundation

Technology transfer offices are most often functionally (if not structurally) attached to a specific university (or national or commercial laboratory) with a fiduciary responsibility to exclusively promote their indigenous technologies. As a result, they rarely collaborate with one another or consider the 'bundling' of multiple technologies in an effort to create a more market ready (and valuable) 'final product'. Technology transfer offices are also exposed to the risk of lawsuits stemming from alleged patent infringement. As a result, the founding faculty formed a non-profit research foundation to both protect University A from the risk of litigation, and to provide greater managerial and financial independence for its managers. The foundation's board also approved a resolution to donate at least 50 per cent of free capital to University A.

Focus on National Laboratories

University laboratories tend to work almost exclusively on very early stage 'basic' research. University scientists are most often trying to solve a theoretical problem. National laboratories, on the other hand, are most frequently responding to a problem presented to their scientists by a government agency (e.g., the Department of Energy, the Department of Defense, etc.) seeking to solve an issue in their domain. National laboratories are certainly charged with developing basic research, but must take their discoveries to a more applied level. Their discoveries are supposed to be deployed in the field. However, while they may be adept at considering 'military' applications, for example, they have much less knowledge in developing 'commercial' applications. Yet, the technologies themselves are generally closer to being commercialized from the start. As a result, the founding faculty decided to source the foundation's technologies from national labs.

Addition of a 'Go-to-market' Phase

After completion of the first pilot, it was decided that the time from Discovery Day to Opportunity Day should be extended from 2 to 5 months. As a result, the MBA students will not only be able to assess the 'feasibility' of all of the applications generated on Discovery Day, but create go-to-market plans for those applications with the greatest market potential. The go-to-market plans should include calculations of the addressable market, results from primary market research and financial projections.

Enhanced Connection to Scientists

Encouraging entrepreneurs to interact directly with inventors in face-to-face meetings at an early stage in the development of the technology allows scientists to direct their efforts toward more defined commercial targets, and entrepreneurs to know the quality and boundaries of given technologies. It also confirms the size of addressable markets and assists technology transfer offices in shaping patents for more specific (and more protectable) fields of use. As noted by Markman et al. (2005), 'research on knowledge spillover and organizational learning suggests that continuous interactions among creators, appropriators, and consumers of technology accelerate the richness and reach of knowledge and discoveries' (p. 244).

CONCLUSION

The case of a group of faculty members seeking to implement a new pedagogical approach within a university is an instructive example of organizational ingenuity. Universities are notoriously conservative and cash-strapped. They have committees and sub-committees that are often considered 'laggards' more than 'early adopters'. In addition, universities are renowned for their politics. As Wallace Stanley Sayre quipped in 1950, 'Academic politics is the most vicious and bitter form of politics, because the stakes are so low' (Issawi, 1973: 178). The faculty founders of the ideation team model certainly encountered their share of politics and resistance, but were also able to find ardent supporters among administrators and colleagues. In the end, they were able to create an innovative program with the potential to: (a) significantly increase the quality of the entrepreneurial ventures started by their students; (b) disrupt accepted 'best practices' for IP commercialization; and (c) raise the reputation and prominence of the university through a nationally recognized program. However, future research will obviously need to examine other examples in different contexts and with different organizational actors.

Research conducted to test the efficacy of the ideation team model indicates that this approach is associated with an increase in the number of ideas generated by groups of scientists, students and entrepreneurs. Less support was found for the hypothesis that this model results in greater 'innovativeness'. However, it seems that the utilization of this model may result in more practicable solutions – possibly solutions with greater probability for solving real-world market problems. The discovery and creation of entrepreneurial opportunities have persisted as central questions in the field of entrepreneurship. Scholars have argued that discovery perspectives are incompatible with creation perspectives, while others have argued that discovery and creation are often complementary and promote one another in a virtuous cycle. By developing a model that emphasizes the relevance of creative problem solving processes for the discovery and creation of opportunities, we offer an alternative view of opportunities, namely, a creative problem solving perspective. This study extends Shane's (2000) original model of how prior knowledge influences opportunity recognition. The ideation team model adds new elements by emphasizing the role of individuals' prior knowledge in a team setting, and underscoring the relevance of creativity processes and skills. Our empirical findings generally support our hypotheses that the ideation team model positively moderates the relationship between innovation and opportunity identification, by enhancing the number and innovativeness of opportunities. In this way, we contribute to the literature on moderators of opportunities (Short et al., 2010).

This study draws attention to the implications of creativity perspectives (Basadur, 1994, 2004; Shalley and Perry-Smith, 2008) for research on entrepreneurial opportunities. Creativity entails processes that can be better understood by individuals and teams (Shalley and Perry-Smith, 2008) and facilitated through the use of well-defined tools and frameworks (Basadur, 1994, 2004). We contribute to the literature that emphasizes the relevance of creativity processes for opportunity recognition (Hsieh et al., 2007; Lumpkin and Lichtenstein, 2005), by examining creativity processes and skills as an integral part of our ideation team model.

Limitations of this study represent opportunities for future research. It investigates three technological innovations and there is a need for examining a larger number of innovations. Future research should also study innovations with different characteristics, such as non-technological innovations and innovations in service industries. Moreover, future studies should assess the application of the ideation team model in additional time periods and different settings. We hope this study inspires the use of these methodological approaches in future studies of entrepreneurial opportunities.

'Can anyone who wants to learn it be taught to do entrepreneurship well?' (Sarasvathy and Venkataraman, 2011: 117). We help answer this question in the context of opportunity identification, by arguing that knowledge about opportunity identification can be learned and taught. Prior studies have offered little guidance to aspiring entrepreneurs, because they show descriptions of opportunity recognition that appear to depend on intuition or luck, and it is unclear how these could be taught (Fiet, 2007). In contrast, aspiring entrepreneurs can use the ideation team model to organize and improve their process to identify opportunities. In this way, this study contributes to the perspective that entrepreneurship entails knowledge and methods that can be learned and taught (DeTienne and Chandler, 2004; Fiet, 2007; Sarasvathy and Venkataraman, 2011).

Organizations – such as universities and technology transfer offices – can use the ideation team model to better support their inventors in the process of matching technological innovations with commercial applications. Our findings indicate that participants tend to generate a greater number of opportunities when there is a well-defined process and facilitators guiding the interaction of innovators and entrepreneurs. These findings suggest the relevance of the ideation team model as a structured process for opportunity identification, which can be useful in a range of organizations and technology transfer offices in different universities. One key benefit from the utilization of this process with real immediate and long term monetary value, is that it results in the ability of the institution to strengthen its patents because it allows claims to be written with specific commercial applications in mind. This allows research efforts and dollars to be re-directed or re-sequenced toward the most viable opportunities. We hope this investigation inspires future research by highlighting how a case study of opportunity identification for early-stage technologies can be more fully understood from the perspective of ingenuity.

REFERENCES

Alvarez, S.A. and Barney, J.B. 2007. Discovery and creation: alternative theories of entrepreneurial action. *Strategic Entrepreneurship Journal*, **1**(1–2): 11–26.

Alvarez, S.A. and Barney, J.B. 2010. Entrepreneurship and epistemology: the philosophical underpinnings of the study of entrepreneurial opportunities. *Academy of Management Annals*, **4**(1): 557–83.

Alvarez, S.A. and Busenitz, L.W. 2001. The entrepreneurship of resource-based theory. *Journal of Management*, **27**(6): 755–75.

Amabile, T.M. and Mueller, J.S. 2008. Studying creativity, its processes, and its antecedents: an exploration of the componential theory of creativity. In Jing Zhou and Christina E. Shalley (eds), *Handbook of Organizational Creativity*. New York: Taylor & Francis Group.

AUTM. 2012. *U.S. Licensing Activity Survey Highlights: FY2012*. Association of University Technology Managers. Available at: http://www.autm.net/AM/Template.cfm?Section=FY_2010_Licensing_Survey&Template=/CM/ContentDisplay.cfm&ContentID=6874 (accessed 19 August 2012).

Baker, T. and Nelson, R. 2005. Creating something from nothing: resource construction through entrepreneurial bricolage. *Administrative Science Quarterly*, **50**(3): 329–66.

Barney, J. 1991. Firm resources and sustained competitive advantage. *Journal of Management*, **17**(1): 99–120.

Basadur, M. 1994. *Simplex: A Flight to Creativity*. Bethesda, MD: Creative Education Foundation.

Basadur, M. 2004. Leading others to think innovatively together: creative leadership. *Leadership Quarterly*, **15**: 103–21.

Basadur, M. and Gelade, G.A. 2006. The role of knowledge management in the innovation process. *Creativity and Innovation Management*, **15**(1): 45–62.

Bingham, C.B., Eisenhardt, K.M. and Furr, N.R. 2007. What makes a process a capability? Heuristics, strategy and effective capture of opportunities. *Strategic Entrepreneurship Journal*, **1**: 27–47.

Camp, S.M. 2005. The innovation-entrepreneurship nexus: a national assessment of entrepreneurship and regional economic growth and development. *SBA Small Business Research Summary*, no. 256.

Carlsson, B. and Fridh, A. 2002. Technology transfer in United States universities. *Journal of Evolutionary Economics*, **12**: 199–232.

Center for Genomic Regulation. 2013. What does a technology transfer office do? Available at: http://pasteur.crg.es/portal/page/portal/Internet/HIDE-Technolgy_Transfer/Tech%20Transfer%20Info/tt [accessed May 16, 2013].

Chapple, W., Lockett, A., Siegel, D. and Wright, M. 2005. Assessing the relative performance of U.K. university technology transfer offices: parametric and non-parametric evidence. *Research Policy*, **34**: 369–84.

Cheung, C.K., Roskams, T. and Fisher, D. 2006. Enhancement of creativity through a one-semester course in university. *Journal of Creative Behavior*, **40**(1): 1–25.

Clarysse, B., Tartari, V. and Salter, A. 2011. The impact of entrepreneurial capacity, experience and organizational support on academic entrepreneurship. *Research Policy*, **40**: 1084–93.

DeTienne, D.R. and Chandler, G.N. 2004. Opportunity identification and its role in the entrepreneurial classroom: a pedagogical approach and empirical test. *Academy of Management Learning and Education*, **3**(3), 242–57.

DeTienne, D.R. and Chandler, G.N. 2007. The role of gender in opportunity identification. *Entrepreneurship Theory and Practice*, **31**(3): 365–86.

Di Gregorio, D. and Shane, S. 2003. Why do some universities generate more start-ups than others? *Research Policy*, **32**: 209–27.

Dimov, D. 2007. Beyond the single-person, single-insight attribution in understanding entrepreneurial opportunities. *Entrepreneurship Theory and Practice*, **31**(5): 713–31.

Edelman, L. and Yli-Renko, H. 2010. The impact of environment and entrepreneurial perceptions on venture-creation efforts: bridging the discovery and creation views of entrepreneurship. *Entrepreneurship Theory and Practice*, **34**(5): 833–56.

Fiet, J.O. 2007. A prescriptive analysis of search and discovery. *Journal of Management Studies*, **44**(4): 592–611.

Franklin, S.J., Wright, M. and Lockett, A. 2001. Academic and surrogate entrepreneurs in university spin-out companies. *Journal of Technology Transfer*, **26**: 127–41.

Guilford, J.P. 1950. Creativity. *American Psychologist*, **5**: 444–54.

Hsieh, C., Nickerson, J.A. and Zenger, T.R. 2007. Opportunity discovery, problem solving and a theory of the entrepreneurial firm. *Journal of Management Studies*, **44**(7): 1255–77.
Issawi, C.P. 1973. *Issawi's Laws of Social Motion*. New York: Hawthorn Books.
Kelley, T. and Littman, J. 2001. *The Art of Innovation*. New York: Doubleday.
Kirzner, I. 1997. Entrepreneurial discovery and the competitive market process: an Austrian approach. *Journal of Economic Literature*, **35**(1): 60–85.
Lampel, J., Honig, B. and Drori, I. 2011. Discovering creativity in necessity: organizational ingenuity under institutional constraints. Call for papers for the Special Issue on 'Discovering Creativity in Necessity: Organizational Ingenuity under Institutional Constraints'. *Organization Studies*, **32**(5): 715–17.
Lumpkin, G.T. and Lichtenstein, B.B. 2005. The role of organizational learning in the opportunity-recognition process. *Entrepreneurship Theory and Practice*, **29**(4): 451–72.
Macho-Stadler, I., Perez-Castrillo, D. and Veugelers, R. 2007. Licensing of university inventions: the role of a technology transfer office. *International Journal of Industrial Organization*, **25**: 483–510.
Markman, G.D., Gianiodis, P.T. and Phan, P.H. 2009. Supply-side innovation and technology commercialization. *Journal of Management Studies*, **46**(4): 625–49.
Markman, G.D., Phan, P.H., Balkin, D.B. and Gianiodis, P.T. 2005 Entrepreneurship and university-based technology transfer. *Journal of Business Venturing*, **20**: 241–63.
McAdam, R., Keogh, W., Galbraith, B. and Laurie, D. 2005. Defining and improving technology transfer business and management processes in university innovation centres. *Technovation*, **25**: 1418–29.
Nelson, R.R. 2001. Observations on the post-Bayh–Dole rise of patenting at American universities, *Journal of Technology Transfer*, **26**: 13–19.
Nelson, R.R. and Winter, S.G. 1982. *An Evolutionary Theory of Economic Change*. Cambridge, MA: Harvard University Press.
O'Shea, R.P., Allen, T.J., Chevalier, A. and Roche, F. 2005. Entrepreneurial orientation, technology transfer and spinoff performance of U.S. universities. *Research Policy*, **34**: 994–1009.
Osborn, A.F. 1957. *Applied Imagination* (rev. ed.). New York: Scribner.
Sarasvathy, S.D. 2001. Causation and effectuation: toward a theoretical shift from economic inevitability to entrepreneurial contingency. *Academy of Management Review*, **26**(2): 243–63.
Sarasvathy, S.D. and Venkataraman, S. 2011. Entrepreneurship as method: open questions for an entrepreneurial future. *Entrepreneurship Theory and Practice*, **35**(1): 113–35.
Schumpeter, J. 1934. *Theory of Economic Development*. Cambridge, MA: Harvard University Press.
Shalley, C.E. and Perry-Smith, J.E. 2008. The emergence of team creative cognition: the role of diverse outside ties, sociocognitive network centrality, and team evolution. *Strategic Entrepreneurship Journal*, **2**: 23–41.
Shane, S. 2000. Prior knowledge and the discovery of entrepreneurial opportunities. *Organization Science*, **11**(4): 448–69.
Shane, S. 2002. Selling university technology: patterns from MIT, *Management Science*, **48**(1): 122–37.
Shane, S. 2004. Encouraging university entrepreneurship? The effect of the Bayh–Dole Act on university patenting in the United States. *Journal of Business Venturing*, **19**: 127–51.
Shane, S. and Venkataraman, S. 2000. The promise of entrepreneurship as a field of research. *Academy of Management Review*, **25**(1): 217–26.
Shepherd, D.A. and DeTienne, D.R. 2005. Prior knowledge, potential financial reward, and opportunity identification. *Entrepreneurship Theory and Practice*, **29**(1): 91–112.
Short, J.C., Ketchen Jr, D.J., Shook, C.L. and Ireland, R.D. 2010. The concept of 'opportunity' in entrepreneurship research: past accomplishments and future challenges. *Journal of Management*, **36**(1): 40–65.
Siegel, D.S., Veugelers, R. and Wright, M. 2007. Technology transfer offices and

commercialization of university intellectual property: performance and policy implications. *Oxford Review of Economic Policy*, **21**(4): 640–60.

Silvia, P.J., Winterstein, B.P., Willse, J.T., Barona, C.M., Cram, J.T., Hess, K.I., Martinez, J.L. and Richard, C.A. 2008. Assessing creativity with divergent thinking tasks: exploring the reliability and validity of new subjective scoring methods. *Psychology of Aesthetics, Creativity, and the Arts*, **2**(2): 68–85.

Sine, W.D., Shane, S. and Di Gregorio, D. 2003. The halo effect and technology licensing: the influence of institutional prestige on the licensing of university inventions. *Management Science*, **49**(4): 478–96.

Swamidass, P.M. and Vulasa, V. 2009. Why university inventions rarely produce income? Bottlenecks in university technology transfer. *Journal of Technology Transfer*, **34**: 343–63.

Thompson, L. 2003. Improving the creativity of organizational work groups. *Academy of Management Executive*, **17**(1): 96–109.

Venkataraman, S. 1997. The distinctive domain of entrepreneurship research: an editor's perspective. In J. Katz and R. Brockhaus (eds), *Advances in Entrepreneurship, Firm Emergence, and Growth*, vol. 3 (pp. 119–38). Greenwich, CT: JAI Press.

Vincent, A.S., Decker, B.P. and Mumford, M.D. 2002. Divergent thinking, intelligence, and expertise: a test of alternative models. *Creativity Research Journal*, **14**(2): 163–78.

Wernerfelt, B. 1984. A resource-based view of the firm. *Strategic Management Journal*, **5**(2): 171–80.

Wiklund, J., Davidsson, P., Audretsch, D.B. and Karlsson, C. 2011. The future of entrepreneurship research. *Entrepreneurship Theory and Practice*, **35**(1): 1–9.

Wright, M., Birley, S. and Mosey, S. 2004. Entrepreneurship and university technology transfer. *Journal of Technology Transfer*, **29**: 235–46.

Yale University Office of Cooperative Research. 2008. Technology transfer overview. Available at: http://www.yale.edu/ocr/about/documents/TECHNOLOGYTRANSFEROVERVIEW_OCRRevisions_23Sep08.pdf (accessed 16 May 2013).

Zahra, S.A. 2008. The virtuous cycle of discovery and creation of entrepreneurial opportunities. *Strategic Entrepreneurship Journal*, **2**(3): 243–58.

PART II

INGENUITY IN CONTEXT

5 Connecting regional ingenuity to firm innovation: the role of social capital

Francesca Masciarelli and Andrea Prencipe

INTRODUCTION

This chapter discusses ingenuity from a regional perspective. We propose the concept of regional ingenuity or the presence in a given region of people with 'the ability to create innovative solutions within structural constraints using limited resources and imaginative problem solving' (Lampel et al., 2011: 584). Thus, we relate regional ingenuity to firm innovation. We draw on the regional creativity literature which recognizes that creativity – considered as the presence of skilled and talented individuals in the region – affects the production of local knowledge (Florida, 2002). This chapter contributes to this literature by discussing how regional ingenuity can promote firm innovation.

The literature on creativity recognizes that skilled and talented individuals contribute to the production of local knowledge (Florida, 2002), thus, their presence in the region may favour firm innovation. This literature emphasizes that the presence of such constraints as rules or boundaries can hamper creativity (Lampel et al., 2011). Creativity is associated with the idea of freedom. However, individuals involved in creating innovative solutions have to overcome several barriers. Individuals with ingenuity are able to produce innovative solutions with the use of limited resources and by applying imaginative problem solving (Lampel et al., 2011). This chapter investigates the idea of regional ingenuity to demonstrate the relationship between ingenuity and firm innovation in a geographically bounded area.

The investigation in this chapter complements analysis of how ingenuity affects firm innovation by examining a possible enabling contingency that might help to explain how ingenuity affects innovation. Storper and Scott (2009) claim that the presence of skilled and talented individuals represents a necessary, but not sufficient, condition for innovation and suggest that other enabling conditions may be required. We suggest that these include the level of social capital in the firm's home region, that is, the level of social ties among individuals in a particular geographic location (Laursen et al., 2012b; Masciarelli, 2011; Putnam et al., 1993). This is in line with the European Commission (2003) report, which states that the knowledge embedded in people is transferred through social capital. We understand local social capital as a geographically bound phenomenon that determines the possibilities for the sharing of knowledge and resources among individuals in a given region. We argue that social capital is an important external contingency that defines how the presence of regional ingenuity can influence the firm's innovation processes. As the framework for our investigation, we use a geographical application of the concept of embeddedness (Granovetter, 1985; Polanyi, 1944), which claims that the social system exerts a strong influence on economic activities. Firm innovation is influenced by local specific factors that are tied to a given geographical area and cannot be reproduced elsewhere (Asheim and Isaksen, 2000).

We illustrate the importance of ingenuity and local social capital for the probability of firm innovation. By linking firm innovation to regional ingenuity and local social capital, this chapter integrates concepts from economic geography and innovation management.

THEORETICAL BACKGROUND

The innovation literature recognizes that firm innovation is a spatially embedded process which has been studied from the perspective of industrial districts (Becattini, 1990), regional systems of innovation (Cooke, 2001), clusters (Porter, 1990) and innovative milieux (Camagni, 1991). Work on industrial districts shows that small firms benefit from the presence of a system of active communication among the actors in their communities (Becattini, 1990; Bianchi, 1993; Brusco, 1982). The regional systems of innovation approach highlights the role of the interactions among firms, organizations and institutions within geographically bounded spaces. The literature on innovative milieux emphasizes the role of:

> the set, or the complex network of mainly informal social relationships on a limited geographical area, often determining a specific external 'image' and a specific internal 'representation' and sense of belonging, which enhance the local innovative capability through synergistic and collective learning processes. (Camagni, 1991: 3)

Work on clusters (Porter, 1990) stresses the advantages deriving from the geographic concentration of firms operating in the same or related industries. All these innovation models underline the role of proximity, which facilitates transfers of knowledge and exchanges of information among the actors. They also acknowledge the importance of social interactions to encourage innovation.

In this chapter, we discuss the role of regional ingenuity for fostering firm innovation. To discuss the concept of regional ingenuity we draw on the cultural capital (Bourdieu, 1980) and creativity (Florida, 2002) literatures. The notions of regional ingenuity and creativity recall the concept of cultural capital originally proposed by Bourdieu (1980), and its three forms of existence: (a) it describes the embodied dispositions of individuals as 'an integral part of the person' that 'cannot be transmitted instantaneously (unlike money, property rights, or even titles of nobility)' (Bourdieu, 1986: 244–5); (b) it is objectified in cultural goods such as 'pictures, books, dictionaries, instruments, machines, etc.' (Bourdieu, 1986: 243); (c) it is institutionalized in cultural institutions (Bourdieu and Passeron, 1977).

Regional ingenuity is embodied in creative individuals in the region with the ability to propose innovative solutions in a context of structural constraints. Creative individuals are not evenly distributed across geographic areas: some places attract skilled and talented people (Jacobs, 1969). Florida (2002) suggests that the presence of such individuals depends on investment in education and training, and the attributes of particular places that make them attractive to and retain talented individuals. Several studies support the idea that regional creativity increases competitiveness. Lucas (1988) shows the link between regional productivity and the clustering of human capital; Zucker et al. (1998) show that the presence of high proportions of scientists in the local population affects geographic patterns of technology start-ups; and Maskell and Malmberg (1999) suggest that regional competitiveness is influenced by the presence of human resources.

In line with Masciarelli (2011), we suggest that for regional endowments of creativity to contribute to firm innovation, efficient mechanisms are required to encourage knowledge spillovers. It has been shown that spillovers of knowledge to individuals and firms occur through social interactions, imitation and active collaboration (Belderbos et al., 2004). Knowledge spillovers are geographically bounded (Adams and Jaffe, 1996; Almeida and Kogut, 1999; Jaffe et al., 1993; Maurseth and Verspagen, 2002; Rosenthal and Strange, 2003) and require social interactions and good relational resources (Almeida and Kogut, 1999; Rosenkopf and Almeida, 2003). The presence of high levels of social capital in the region creates a social context rich in communication and social interactions. The concept of social capital is based on the idea that friends and associates are significant assets and, hence, that a large number of social ties among individuals will promote knowledge spillovers in the community (Laursen et al., 2012b).

Social capital can produce benefits for communities (a collective-good view) and individuals (a private-good view). This chapter adopts a collective-good view of social capital (Masciarelli, 2011), conceiving social capital as a geographically bounded phenomenon and a collective asset. All the firms in a given region potentially can benefit from the presence in that region of social capital. This approach to social capital is based on research conducted by Coleman (1988) and Putnam et al. (1993). Specifically, Coleman (1988) claims that the concept of social capital includes obligations and expectations, informal channels and social norms. His conceptualization of social capital (Coleman, 1990: 98–9) is exemplified by the New York diamond market:

> In the process of negotiating a sale, a merchant will hand over to another merchant a bag of stones for the latter to examine in private at his leisure, with no formal insurance that the latter will not substitute one or more inferior stones or paste replica. The merchandise may be worth thousands, or hundreds of thousands, of dollars. Such free exchange of stones for inspection is important to the functioning of this market. In its absence, the market would operate in a much more cumbersome, much less efficient fashion. Inspection shows certain attributes of the social structure. A given merchant community is ordinarily very close, both in the frequency of interaction and in ethnic and family ties. [. . .] Observation of the wholesale diamond market indicates that these close ties, through families, community, and religious affiliation, provide the insurance that is necessary to facilitate the transactions in the market. If any member of this community defected through substituting other stones or through stealing stones in his temporary possession, he would lose family, religious, and community ties. The strength of these ties makes possible transactions in which trustworthiness is taken for granted and trade can occur with ease. In the absence of these ties, elaborate and expensive bonding and insurance devices would be necessary – or else the transactions could not take place.

Putnam et al. (1993) highlight the importance of people's connections with their communities emphasizing the idea of citizens' actions contributing to the public good, via activities that encourage cooperation and foster communication. Social capital contributes to the creation of a trust-based environment characterized by respect for social norms.

In line with Laursen et al. (2012b: 784), we conceive social capital as 'a critical contextual mechanism that facilitates access to knowledge and other resources' and focus on the potential of social capital, which can be described as 'channel[ling] resources such as information, knowledge, and complementary assets within a particular geographic space'. This view of social capital is based on Nahapiet and Ghoshal's (1998: 243) definition: 'the sum of the actual and potential resources embedded within, available through, and defined from the network of relationships possessed by an individual or social unit'.

Various studies explore the relationship between social capital and firm innovation (Hauser et al., 2007; Landry et al., 2002; Laursen et al., 2012a).

Landry et al. (2002) find that perceived firm-level social capital in the form of participation and relational assets, contributes to increasing the likelihood of firm innovation. Hauser et al. (2007) studies the effect of social capital on the regional innovation capacity measured in terms of patenting activity. Laursen et al. (2012a) relate social capital to the probability that the firm will introduce an innovation, and show that being located in a region with high levels of social capital increases the effectiveness of externally acquired research and development (R&D) for product innovation in Italian regions. Italy has certain peculiarities which make it a unique case; the Italian regions have historical and cultural dissimilarities. These differences have been the motivation for several studies, one of the earliest being Banfield's (1958) *The Moral Basis of a Backward Society*, in which he relates social variables to economic performances. Banfield suggests that lack of civic engagement is one of the reasons for Southern Italy's economic backwardness.

Although these findings are important, these studies focus exclusively on the role of social capital, and do not consider the interplay with regional ingenuity: the relationship between firm innovation, regional ingenuity and social capital is generally not explored. We posit that local social capital affects the opportunities for firms to benefit from regional ingenuity.

Regional Ingenuity and Firm Innovation

Every region is different from every other region in relation to knowledge stock. According to von Tunzelmann (1995), the different distribution of knowledge contributes to explaining the different patterns of competition in regions. Studies that investigate the spatial configuration of economic activities, emphasizing the creation and diffusion of knowledge (i.e. Maskell and Lorenzen, 2004), underline the role of knowledge as increasingly important for firm competitiveness at a time when the process of globalization is making several traditional production factors available equally to all firms regardless of their location. Firms can benefit from situation in a dynamic, knowledge-producing environment created by skilled and talented individuals. The possibility to access new knowledge is crucial for firm innovation, which depends on the combination of knowledge (Fleming and Sorenson, 2001; Kogut and Zander, 1996; Schumpeter, 1912/1934). Knowledge is embodied mainly in skilled and talented individuals (Nonaka et al., 2000), who contribute to the formation of new knowledge. They produce benefits for themselves and also contribute to the overall welfare of their context (Dakhli and de Clercq, 2004). The literature on creativity shows that skilled and talented individuals can stimulate economic growth. The presence of these individuals within the region increases its employment and wages (Glaeser et al., 1992; Rauch, 1993; Simon, 1998) and positively affects the generation of new ideas (Becker, 1962). Human capital theory states that knowledge is embodied in well-educated people (Lucas, 1988). Florida (2002) and Glaeser et al. (1992) identify a wide range of amenities that attract and retain skilled and talented individuals in the region, and encourage the dynamism required for technological change and innovation. These facilities include cultural organizations, cinemas, theatres, museums, etc. To attract especially skilled and talented individuals, geographical areas need to provide opportunities, which include an environment characterized by significant cultural amenities (Florida, 2000, 2002). The concept of ingenuity contributes to the literature on creativity by considering that most creative individuals have to operate within structural constraints using limited resources and, therefore, have to develop the sets of skills, social tactics and mental orientations that define ingenuity (Lampel et al., 2011). We propose that firms located in regions with high levels of regional ingenuity are able to plug into a wider range of external sources of knowledge, and to introduce innovations based on new combinations of knowledge. In sum, we posit that:

Proposition 1: Firms operating in regions associated with high levels of ingenuity are more likely to innovate.

LOCAL SOCIAL CAPITAL, REGIONAL INGENUITY AND FIRM INNOVATION

We explore the role of local social capital in defining the relationship between regional ingenuity and innovation. The level of local social capital affects the possibilities to share knowledge and resources among the individuals in a given region. We posit that the home region's social capital influences the relationship between regional ingenuity and firm innovation. Social capital contributes to generating an environment that is conducive to collaboration and the diffusion of knowledge, and creates the conditions required for the formation of new knowledge. In line with Masciarelli (2011) and Laursen et al. (2012a), we argue that social capital operates through two different effects: the connectivity effect and the communication effect.

In relation to the connectivity effect of social capital, the home region's social capital acts to link between different groups of actors. By increasing the opportunities for face-to-face interactions involving skilled and talented individuals, the connectivity effect can produce the following benefits: (i) it encourages creative knowledge spillovers from one individual or firm to another (Knudsen et al., 2005); (ii) it defines the interactions among knowledge workers; (iii) it favours the transfer of the tacit components of knowledge; and (iv) it increases the opportunities to combine the knowledge from creative spillovers.

The communication effect of social capital affects the relationship between regional ingenuity and innovation. It contributes to creating a shared language and communication codes that foster creative knowledge spillovers. Sharing the same localized communication codes and language makes it easier for individuals to reach a mutual understanding. A shared language allows individuals to transfer ideas more easily, share information effectively, and discuss problems (Nahapiet and Ghoshal, 1998). Therefore, creative individuals are better able to overcome the limits imposed by rules or boundaries and to produce innovative solutions to existing problems. In addition, by increasing levels of mutual understanding, the communication effect of social capital can allow creative individuals to anticipate and predict the behaviours of other actors, thereby facilitating efficient coordination of activities (Bolino et al., 2002). In sum, both the connectivity and the communication effects support the idea that social capital favours creative knowledge spillovers, which, in turn, have a positive effect on firm innovation. Thus, we would argue that social capital plays a role in defining the relationship between regional ingenuity and firm innovation. Based on this argument, we posit that:

Proposition 2: Firms operating in regions with high levels of local social capital are more likely to innovate than similar firms operating in regions with low levels of local social capital.

Proposition 3: Local social capital positively mediates the effect of regional ingenuity on firm innovation by favouring the creative knowledge spillovers.

DISCUSSION AND CONCLUSIONS

This chapter discussed the idea that regional ingenuity can influence firm innovation, and that social capital emphasizes its positive effect and, therefore, that the influence of regional ingenuity on firm innovation is affected by the presence of high levels of social interaction and communication among the individuals in a region. Through the connectivity and communication effects social capital can foster creative knowledge spillovers to produce an external environment that encourages firms' acquisition of external knowledge. In line with Masciarelli (2011) and Laursen et al. (2012a), we propose that social capital is a mechanism for knowledge diffusion within regional boundaries. When levels of local social capital are low, effective flows of knowledge among the actors in the region may be reduced, resulting in less knowledge production and dissemination. This chapter has implications for theory and practice.

Implications for Theory

This chapter contributes to work showing the role of context for explaining firm innovation (Laursen et al., 2012a), by exploring the role of two environmental variables: regional ingenuity, and local social capital. Laursen et al. (2012a) find that local social capital promotes product innovation. This chapter provides further evidence of the role of social capital in determining innovation. We show also that the firm's location choice matters for its innovation processes. This is consistent with the economic geography literature which points to the importance of location for firm competitiveness in the context of industrial districts, clusters, innovative milieux, territorial innovation systems, and so on (Brusco, 1982; Romanelli and Khessina, 2005).

The present chapter also contributes to work on ingenuity by considering ingenuity as an environmental variable and showing the relationship between ingenuity, social capital and firm innovation. Regional ingenuity is represented by the presence within the region of creative actors with the ability to create innovative solutions within structural constraints; they contribute to the production of new knowledge, which, in turn, favours innovation. Local social capital facilitates the exchange of information among actors and influences the relationship between regional ingenuity and firm innovation. This supports the idea of an intrinsic link between creativity and social capital highlighted by the European Commission (2003).

Finally, by illustrating the importance of regional ingenuity for firm innovation, this chapter contributes to the resource-based view of regional economic development (e.g. Foss, 1996), which suggests that regional resource endowments can determine persistent differences in regional economic performance.

Implications for Practitioners

The ideas presented in this chapter have implications for practitioners. We argue that regional ingenuity has an impact on firm innovation. The literature on creativity (Florida, 2002; Gertler, 2003) suggests that because creative individuals often are mobile, regions (and firms) need to focus on attracting and *retaining* them. Regional policy makers should focus on these aspects.

We have shown also that local social capital is crucial for firms to benefit from regional ingenuity in their innovation processes. Social capital can be built bottom-up and public authorities should make efforts to initiate this process. The process of creation and accumulation of social capital may require deliberate public investment and coordination. Local social capital is created through a collective process that is linked to the region's historical, institutional and cultural roots (Putnam, 1995) and, therefore, is more easily established at the regional, rather than the national or international level. Location is also important; not all types of social relations attract distance costs and 'the interdependences of different types of social relations make dense combinations of them dependent upon geographical proximity' (Lorenzen, 2007: 7). Different regions show different matrixes of social relations that produce unique social codes.

We also provide some practical insights for managers and entrepreneurs. Consistent with the arguments presented in Shane and Cable (2002), Sorenson and Audia (2000), and Laursen et al. (2012a), we claim that in planning their innovation strategies, entrepreneurs and managers should take account of the regional setting and levels of ingenuity and social capital. Firms might prefer to locate in regions with high levels of regional ingenuity and local social capital.

Limitations of the Study and Directions for Future Research

We have stressed the positive net effects of social capital, but it should be remembered that there can also be negative effects. Local social capital is the product of the combination of strong and weak ties. In this context, Putnam and Goss (2002) distinguish between bonding and bridging social capital. When the networks become too dense, bonding social capital can produce negative effects by excluding outsiders, making excessive claims on group members and reducing individual freedoms (see Portes, 1998).

We discussed local social capital conceived as social interactions among individuals in a geographically bounded area. However, social capital can be international or virtual. Firms need access to local and regional networks and need to form social ties with actors located in different regions or countries. Future studies could investigate how the social capital that firms establish with actors located outside regional boundaries affects the relationship between ingenuity and firm innovation.

Finally, this chapter develops a conceptual basis for testing propositions, and to guide future empirical studies. In particular, future work might propose empirical measures of regional ingenuity although the complexity of the concept makes its measurement difficult. It might be feasible to start by using the items applied to the measurement of regional creativity. The idea of regional creativity considers aspects such as education, sector and vocation (Florida, 2005). To operationalize regional creativity, Masciarelli (2011) uses the following items: (i) level of regional human capital, which could be measured as numbers of science and technology graduates per inhabitant; (ii) regional knowledge production through R&D, which could be computed using numbers of R&D employees; and (iii) regional cultural capital, which could be measured as level of culture promotion. These items capture the basic and advanced human capital base, investment in R&D, and the presence and attraction of skilled and talented individuals. Similar efforts could be devoted to provide a measure for regional ingenuity.

REFERENCES

Adams, J.A. and Jaffe, A.B. 1996. Bounding the effects of R&D: an investigation using matched establishment-firm data. *Rand Journal of Economics*, **27**(4): 700–721.

Almeida, P. and Kogut, B. 1999. Localization of knowledge and the mobility of engineers in regional networks. *Management Science*, **45**(7): 905–17.

Asheim, B.T. and Isaksen, A. 2000. Localised knowledge, interactive learning and innovation between regional networks and global corporations. In E. Vatne and M. Taylor (eds), *The Networked Firm in a Global World. Small Firms in New Environments*. Ashgate: Aldershot.

Banfield, E. 1958. *The Moral Basis of a Backward Society*. New York: Free Press.

Becattini, G. 1990. The Marshallian industrial districts as a socio-economic notion. In F. Pyke (ed.), *Industrial Districts and Inter-Firm Co-Operation in Italy*. Geneva: International Institute for Labour Studies.

Becker, G. 1962. Investment in human capital: a theoretical analysis. *Journal of Political Economy*, **70**(5): 9–49.

Belderbos, R., Carree, M. and Lokshin, B. 2004. Cooperative R&D and firm performance. *Research Policy*, **33**(10): 1477–92.

Bianchi, P. 1993. Industrial districts and industrial policy: the new European perspective. *Journal of Industry Studies*, **1**(1): 16–29.

Bolino, M., Turnley, W. and Bloodgood, J. 2002. Citizenship behavior and the creation of social capital in organizations. *Academy of Management Review*, **27**(4): 505–22.

Bourdieu, P. 1980. Le capital social. Notes provisoires. *Actes*, **31**(2–3).

Bourdieu, P. 1986. The forms of capital. In J. Richardson (ed.), *Handbook of Theory and Research for the Sociology of Education*. Westport, CT: Greenwood Press.

Bourdieu, P. and Passeron, J.-C. 1977. *Reproduction in Education, Culture and Society* (R. Nice, trans.). London: Sage.

Brusco, S. 1982. The Emilian model: productive decentralisation and social integration. *Cambridge Journal of Economics*, **6**(2): 167–84.

Camagni, R. 1991. *Innovation Networks*. John Wiley & Sons, Inc.

Coleman, J. 1988. Social capital in the creation of human capital. *American Journal of Sociology*, **94**: 95–120.

Coleman, J. 1990. *Social Capital*. Cambridge, MA: Harvard University Press.

Cooke, P. 2001. Regional innovation systems, clusters, and the knowledge economy. *Industrial and Corporate Change*, **10**(4): 945–74.

Dakhli, M. and de Clercq, D. 2004. Human capital, social capital, and innovation: a multi-country study. *Entrepreneurship and Regional Development*, **16**(2): 107–28.

European Commision. 2003. *Building the Knowledge Society: Social and Human Capital Interactions*. Commission Staff Working Paper.

Fleming, L. and Sorenson, O. 2001. Technology as a complex adaptive system: evidence from patent data. *Research Policy*, **30**(7): 1019–39.

Florida, R. 2000. Science, reputation, and organization. Unpublished manuscript, Carnegie Mellon University at Pittsburgh, PA, pp. 2–28.

Florida, R. 2002. The economic geography of talent. *Annals of the Association of American Geographers*, **92**(4): 743–55.

Florida, R. 2005. *The Flight of the Creative Class. The New Global Competition for Talent*. New York: HarperCollins.

Foss, N.J. 1996. Higher-order industrial capabilities and competitive advantage. *Industry and Innovation*, **3**(1): 1–20.

Gertler, M.S. 2003. Tacit knowledge and the economic geography of context, or the undefinable tacitness of being (there). *Journal of Economic Geography*, **3**(1): 75–99.

Glaeser, E., Kallal, H., Scheinkman, J. and Shleifer, A. 1992. Growth in cities. *Journal of Political Economy*, **100**(6): 1126–52.

Granovetter, M. 1985. Economic action and social structure: the problem of embeddedness. *American Journal of Sociology*, **91**: 481–510.

Hauser, C., Tappenier, G. and Walde, J. 2007. The learning region: the impact of social capital and weak ties on innovation. *Regional Studies*, **41**(1): 75–88.

Jacobs, J. 1969. *The Economy of Cities*. New York: Random House.

Jaffe, A.B., Trajtenberg, M. and Henderson, R. 1993. Geographic localization of knowledge spillovers as evidenced by patent citations. *Quarterly Journal of Economics*, **108**(3): 577–98.

Knudsen, H.K., Ducharme, L.J., Roman, P.M. and Link, T. 2005. Buprenorphine diffusion: the attitudes of substance abuse treatment counselors. *Journal of Substance Abuse Treatment*, **29**: 95–106.

Kogut, B. and Zander, U. 1996. What firms do? Coordination, identity, and learning. *Organization Science*, **7**(5): 502–18.

Lampel, J., Honig, B. and Drori, I. 2011. Discovering creativity in necessity: organizational ingenuity under institutional constraints. *Organization Studies*, **32**(4): 584–86.

Landry, R., Amara, N. and Lamari, M. 2002. Does social capital determine innovation? To what extent? *Technological Forecasting and Social Change*, **69**(7): 681–701.

Laursen, K., Masciarelli, F. and Prencipe, A. 2012a. Regions matter: how localized social capital affects innovation and external knowledge acquisition. *Organization Science*, **23**(1): 177–93.

Laursen, K., Masciarelli, F. and Prencipe, A. 2012b. Trapped or spurred by the home region? The effects of potential social capital on involvement in foreign markets for goods and technology. *Journal of International Business Studies*, **43**: 783–807.

Lorenzen, M. 2007. Social capital and localised learning: proximity and place in technological and institutional dynamics. *Urban Studies*, **44**(4): 799–817.

Lucas, R. 1988. On the mechanism of economic development. *Journal of Monetary Economics*, **22**(1): 3–42.

Masciarelli, F. 2011. *The Strategic Value of Social Capital: How Firms Capitalise on Social Assets*. Cheltenham: Edward Elgar Publishing.

Maskell, P. and Lorenzen, M. 2004. The cluster as market organization. *Urban Studies*, **41**(5–6): 991–1009.

Maskell, P. and Malmberg, A. 1999. Localised learning and industrial competitiveness. *Cambridge Journal of Economics*, **23**(2): 167–85.

Maurseth, P.B. and Verspagen, B. 2002. Knowledge spillovers in Europe: a patent citations analysis. *Scandinavian Journal of Economics*, **104**(4): 531–45.

Nahapiet, J. and Ghoshal, S. 1998. Social capital, intellectual capital, and the organizational advantage. *Academy of Management Review*, **23**(2): 242–66.
Nonaka, I., Toyama, R. and Nagata, A. 2000. A firm as a knowledge-creating entity: a new perspective on the theory of the firm. *Industrial and Corporate Change*, **9**(1): 1–20.
Polanyi, K. 1944. *The Great Transformation*. New York: Holt Rinehart.
Porter, M. 1990. *The Competitive Advantage of Nations*. New York: Free Press.
Portes, A. 1998. Social capital: its origins and applications in modern sociology. *Annual Review of Sociology*, **24**: 1–24.
Putnam, R.D. 1995. Bowling alone: America's declining social capital. *Journal of Democracy*. **6**(1): 65–78.
Putnam, R.D. and Goss, K.A. 2002. Introduction. In R.D. Putnam (ed.), *Democracies in Flux: The Evolution of Social Capital in Contemporary Society* (pp. 3–21). New York: Oxford University Press.
Putnam, R.D., Leonardi, R. and Nanetti, R.Y. 1993. *Making Democracy Work*. Princeton: Princeton University Press.
Rauch, J. 1993. Productivity gains from geographic concentration of human capital: evidence from the cities. *Journal of Urban Economics*, **34**(3): 380–400.
Romanelli, E. and Khessina, O.M. 2005. Regional industrial identity: cluster configurations and economic development. *Organization Science*, **16**(4): 344–58.
Rosenkopf, L. and Almeida, P. 2003. Overcoming local search through alliances and mobility. *Management Science*, **49**(6): 751–66.
Rosenthal, S. and Strange, C. 2003. Geography, industrial organization, and agglomeration. *Review of Economics and Statistics*, **85**(2): 377–93.
Schumpeter, J.A. 1912/1934. *The Theory of Economic Development: An Inquiry into Profits, Capital, Credit, Interest and the Business Cycle* (R. Opie, trans.). London: Oxford University Press.
Shane, S. and Cable, D. 2002. Network ties, reputation, and the financing of new ventures. *Management Science*, **48**(3): 364–81.
Simon, C. 1998. Human capital and metropolitan employment growth. *Journal of Urban Economics*, **43**(2): 223–43.
Sorenson, O. and Audia, G. 2000. The social structure of entrepreneurial activity: geographic concentration of footwear production in the United States. *American Journal of Sociology*, **106**(2): 424–62.
Storper, M. and Scott, A.J. 2009. Rethinking human capital, creativity and urban growth. *Journal of Economic Geography*, **9**(2): 147–67.
von Tunzelmann, G.N. 1995. *Technology and Industrial Progress: The Foundations of Economic Growth*. Cheltenham: Edward Elgar.
Zucker, L., Darby, M. and Brewer, M. 1998. Intellectual capital and the birth of the U.S. biotechnology enterprises. *American Economic Review*, **88**: 290–306.

6 Risk perception and ingenuity in entrepreneurship in Japan*

David T. Methé

INTRODUCTION

Entrepreneurs must be ingenious in coping with the difficulties inherent in starting a new business (Carland et al., 1984). This is especially so when the new business is in technology-intensive industries where previously non-existing companies must not only legitimate themselves and their products, but also the underlying technology from which they spring (Aldrich and Fiol, 1994). Although entrepreneurship is seen as a universal human activity, many scholars recognize that different environments will have differential impacts on these activities (Thomas and Mueller, 2000; Weber and Hsee, 1998). Scholars have indicated that differences in institutional environments do affect entrepreneurial activities (Clausen, 2011; De Clerq and Arenius, 2006) and some of these differences have been examined in the context of entrepreneurship in Japan (Borton, 1992; Bracker and Methé, 1994; Methé and Bracker, 1994; Methé, 2006a; Whittaker et al., 2009).

Although the Japanese economy is dominated by small- to medium-sized businesses, both in numbers and employment (Ballon and Honda, 2000), the institutional environment for starting a business in Japan, especially a high technology business, has been consistently rated as one of the most difficult (Feigenbaum and Brunner, 2002; Bosma and Harding, 2007; Kelly et al., 2011). The fostering of high technology start-ups is particularly important in that these types of organizational forms tend to be more likely to generate radical innovations (Methé et al., 1996) and are also seen as a primary engine of growth and renewal in economies.

Risk has been characterized as an inherent part of the managerial function, whether that manager is in a larger established company or a newly founded start-up (Foss et al., 2007; Boone and de Brabander, 1997). How those risks are perceived, how they are handled and the consequences of those risks, however, are still a matter of contention in the academic literature (March and Shapira, 1987; Sitkin and Pablo, 1992; Miller, 2009). Hence the starting point of this study was a straightforward empirical question of how managers perceive risk and what they did to cope with that perceived risk.

An area of continued fascination for both scholars and practitioners alike is the study of entrepreneurship and entrepreneurs (Sarasvathy, 2001; Ireland and Webb, 2007; Busenitz et al., 2003). This study also examined how managers in various types of organizations confront the risks inherent in their activities as business managers. Entrepreneurs are often seen as engaging in unconventional and imaginative behaviors in order to overcome the challenges they confront (Sarasvathy, 2001; Murmann and Sardana, 2013). In doing so, entrepreneurs can be seen as engaging in a type of 'organizational ingenuity: the ability to create innovative solutions within structural constraints using limited resources and imaginative problem solving' (Lampel et al., 2012) which we term entrepreneurial[1] ingenuity.

Our primary focus was on managers of venture businesses and those managers involved in new business activities in established companies, in particular venture businesses in high technology areas such as semiconductors and other related electronics fields. These managers are involved in a specialized aspect of management, which is entrepreneurship. Although we centered our research on understanding how these entrepreneurial managers perceived and coped with risk, we were also interested in what role these perceptions and coping mechanisms played in the development of new high technology businesses in Japan. As such, this research borrows from and touches upon many different disciplines and areas of interest, such as resource-based theory (Alverez and Busenitz, 2001), regime theory and institutional theory (Wijen and Ansari, 2007; Phillips and Tracey, 2007).

These interviews were conducted over a three-year period from 2005 to 2007 and, in this multiple case study, involved about 21 companies and managers. In summary, eight venture businesses (VB), seven established small-/medium-sized companies (SME/EC), two foreign subsidiaries of non-Japanese venture businesses (FS) and four venture Angel (VA) companies, which were also newly established ventures in their own right, were examined for this study. These VAs are self-classified as Angel investors. This type of investor is involved in the very early funding of a venture business, and usually is involved in providing funds that range from several hundreds of thousands of dollars up to a million dollars. All these companies were involved in the semiconductor industry, an industry characterized by successive waves of technological innovation in both products and processes (Methé, 1991, 1992, 2006a). Hence the opportunity for innovative activity by either start-up or established companies remains high.

In some of the companies' cases, especially with the SME/EC companies, two or three managers were present during the interview, but in these companies' cases the main person interviewed was the CEO. In the case of the venture businessess, the founder, president or CEO of the company was interviewed as were the founders of the VA companies. The two foreign subsidiaries' managers were the general manager in charge of the subsidiary. These foreign subsidiaries were formed from venture businesses that had started up overseas (in either the US or Europe) and then set up a subsidiary in Japan.

This paper uses the qualitative data collected during the interviews, supplemented where possible with data collected from other archival sources. We believe that the qualitative approach taken here has value given the exploratory nature of both the field of entrepreneurship, risk and especially how these concepts are activated in the Japanese context (Gephart, 2004; Pitt, 1998). The interviews themselves provide a rich set of data since these ranged from one and a half hours to three hours in length. The data presented below is exploratory and should be seen only as a starting point for further research.

ON DEFINITIONS OF RISK: CLASSICAL AND EMPIRICAL

How risk has been conceptualized over the years has changed and evolved as researchers have examined the topic (Knight, 1921; Shackle, 1955; March and Shapira, 1987; Weber and Milliman, 1997; Faro and Rottenstreich, 2006; Miller, 2009; Basili and Zappia, 2010). In our study we examined how the various managers involved with new business activities defined risk. As such we are looking at the simple empirical question of 'how do managers view risk?' Most of these comments came in relation to the question concerning how each manager defined high risk, but often these comments came in the general discussion that took place during the interview. In this section we will examine how managers involved in new business activities defined risk and how their definition differs from the classical definition of risk often used by researchers who are studying decision making under uncertainty.

One of the comments was as follows:

> Risk is not having enough money to cover your costs and going bankrupt.
> (President, but not founder, of a VB)

Another manager noted as follows:

> There is no risk here at [ABC] Japan, because the HQ has a long history and good profits, so it is unlikely that they will go bankrupt, but there is a risk that they could be acquired by a bigger company and I will lose control.
> (General manager of a subsidiary of a foreign VB)

Another entrepreneur noted the following:

> Risk is higher when you start without a product and reputation because it is harder to get money, but once you have a product and reputation your risk goes down.
>
> (Founder of a VB)

Another comment was as follows:

> Real risk is not at the starting point, but gradually risk gets higher and higher. In the seed phase if bankruptcy occurs there is very little money and people involved, but as we grow the level of risk increases. The level of risk is now higher than at founding.
>
> (Founder of a VB)

Finally we have this comment:

> Generally people say investing in ventures holds high risk, but I do not think so. I think the risk is not as high as that, rather there might be only a little risk. The reason is the scale for the investment is not that big since we invest in venture companies at a very early stage. Even if they fail, the loss does not damage us badly.
>
> (Founder of a VA company)

In these comments the concept of risk as being related to bankruptcy, failure or damage to the project or company is quite evident. This perception was common among the respondents in this study. Unlike the classical definition of risk in decision theory, where risk is concerned with the variance of the probabilities of various gains and losses generated from an alternative (see for example Pratt, 1964; Arrow, 1971), in our study the interviewed managers defined risk as the chance of loss. This is consistent with findings from other studies. Other studies have noted that managers involved in new business activity tend to consider the negative consequences of the activities as risk (March and Shapira, 1987). There is a difference then, between 'risk' as a measure of the distribution of possible outcomes from a choice as is often the case in the classical literature on decision making under uncertainty and 'risk' as a problem or hazard as is often described by practicing managers.

Second, these comments highlight how managers consider the chance or uncertainties that are an important part of how managers define risk change over time. These comments are illustrative of how managers perceive the issues concerning the sources of risk and will become an important part of our discussion of the managerial assumptions about controlling risk. We introduce them at this time, because they are indicative of another element of managerial definitions of risk, that is, that risk is perceived as dynamic and changes over time, as the conditions surrounding the project change (Conrath, 1967; Miller, 2009).

In essence, risk can be adjusted, but it can only be done by influences placed upon a variety of processes that in turn generate some saleable product and develop some kind of reputation for the company, which then make it easier to acquire the resources needed to prevent going bankrupt. As indicated by these managers, these processes and products might range from being 'known' to 'unknown but knowable' to 'as yet undiscovered and thus not knowable'. Hence, the type of 'risk' discussed by these entrepreneurs appears to share characteristics closer to 'ambiguity' and 'uncertainty' (Knight, 1921; Mosakowski, 1997; Meyer and Shi, 1995; Murmann and Sardana, 2013). In any case, 'risk' as perceived by the managers in this study was not reducable to some simple formula, let alone a single quantifiable concept.

Third, the last two comments also hold risk as derived from the negative outcome of bankruptcy or failure and as generated by a complex and dynamic set of processes, but illustrate the importance of the magnitude of the consequences on the perceptions of risk, both for those who are starting a venture business and for those who are involved with funding the venture business. As the size of the investment, both in terms of people and money increases, the level of risk is seen as increasing. At first reading the third comment may appear to contradict the following two comments; however, we interpret it as a complement. This is because the focus of the last two comments are more on the magnitude of the consequences of failure; that is, the amount of the potential loss. Again, this is consistent with other research that has found that the amount at stake often has a greater impact on the decision, than simply the probabilities related to the loss. This has been called loss aversion (Tversky and Kahneman, 1991) or regret aversion (Bell, 1983).

RISK AS AN INTEGRAL PART OF MANAGEMENT AND THE MANAGER'S DUTIES TO BRING ABOUT GAINS

In this section we will examine how managers engaged in new business activity feel about risk and in particular what their attitude is towards the risks they take in carrying out their business tasks and also to explore how the thoughts about risk relate to the process of generating positive gains.

> Initiating a project or not initiating one, or taking an action or not taking an action in business involves risk.
>
> (Founder of a VB)

> In any situation, how we carry our business we should always keep in mind the risk. For example, if an earthquake occurred, how quick we would come back to our business and start supplying our products to our customers. We also need to employ good workers and expect them to master skills to cope with customers' needs. Financially, we need to build up our money in case we want to invest the money in a new business. In terms of risk, having workers is also a part of risk. Raising money from the market, and performing well for our shareholders is also pressure on me. Management or running a company always involves risk. I want our employees' families to be happy too. Taking all responsibilities of this company is risk and this is my responsibility.
>
> (CEO of an SME/EC)

These two comments are important because they illustrate a common impression that the managers interviewed all seemed to share. That impression was that they were not seeking out risk, or thrills just for the sake of the risk itself, or simply to have some excitement in their lives (Xu and Ruef, 2004). They saw risk as built into the very nature of their jobs, especially in those tasks that related to decision making and guiding the company.

Managers involved in new business development activities saw their ability to handle the risks involved in guiding their companies as a key part of who they were as managers. Their ability to deal with the ambiguities and challenges of high-risk decisions and projects, were seen as the key element in the success or failure of the project. As one senior manager noted about his CEO:

> Although the market share of our product has risen to about 70 per cent in Japan and now is the main source of profit for our company, this business was a decision with high risk at first. Mr [ABC], the president, took strong initiative as a leader. Without his leadership this business would not have succeeded. Tough times for this business lasted quite a long time, and in this period many objections were raised, since the business was making deficits. It was Mr [ABC]'s strong consistency and support for the business that enable it to be where it is today.
>
> (Senior manager of an SME/EC)

Risk was something that was inescapable for these managers. Even if they did not take on a new project or activities within an ongoing project there was risk involved. For these managers it was simply accepted that what they did entailed risk, and in particular the risk of failure of the business. However, they also had an attitude that this was part of their job and that they could and must handle these risky decisions. Although no systematic analysis was done in the survey concerning how these managers viewed themselves compared to their colleagues on issues of risk tolerance and such, an informal review of the answers to these questions would indicate that managers in this study see themselves as better able to tolerate, handle and, therefore, take risks.

ON RETURNS AND GAINS: THE STARTING POINT OF ENTREPRENEURIAL INGENUITY

These managers were also concerned about the returns that they would receive from successfully carrying out their new business development activities. They usually phrased their ideas about returns either in technological or in market share terms, and only occasionally in terms of profits or returns on investment. These following comments were typical of many discussed during the interviews.

> I really have to stress the potential of this technology. In 10–20 years this will be make up 60 per cent of the market.
>
> (Founder of a VB)

> Products based on our technology have become the standard in the industry.
>
> (General manager of a foreign subsidiary of a VB)

> We are pioneering a new materials technology and are the only company in Japan actively working on commercializing this technology.
>
> (Founder of a VB)

> This company is a high-risk high-return business. There is risk, but return is if you can make the products you can make billions so it is also high return.
>
> (Founder of a VB)

There is an asymmetry between the risk and the returns. The entrepreneurs and mangers intuitively differentiate between the domain of loss and the domain of gain (Weber and Milliman, 1997; Sadler-Smith, 2004). The domain of gain appears to come from an underlying faith in the product or technology to conquer the market. In the entrepreneurs' view, returns will flow from the technology or product and this flow was not always seen as directly connected to the sources of risk. The risks are related to the potential failure of the business, which one can minimize, but doing so does not guarantee the success of the business.

It appears that in the mind of these entrepreneurs, overcoming the risks is necessary for the survival of the business, but not sufficient for success in business. It has been indicated that this bifurcation between the domain of loss/risk and the domain of gain/return may play an important role in the sense of confidence, perhaps overconfidence either in the decisions to enter in the first place (Hayward et al., 2006) or to continue even when most of the results are negative (Lowe and Ziedonis, 2006), that some entrepreneurs appear to exhibit. It is less clear how this confidence plays out in the behavior of entrepreneurs pursuing their dream.

The entrepreneurs in this study imagined something clear and distinct in the technologies and products they were developing. They had to make choices to pursue those technologies and choices about how to overcome the risks that inevitably followed. As noted by Basili and Zappia (2010), in their discussion of G.L.S. Shackle's work:

> . . . Shackle's point is that choice is always choice among thoughts since the future is yet to be created: when choosing, the future necessarily is a product of thought (Lachmann, 1976, p.57). Therefore, in a 'kaleidic', open-ended world both the nature of the outcomes and the process that gives rise to them are conceptualized in terms of novelty and surprise (Dequech, 2000).
>
> (Basili and Zappia, 2010: 471)

It appears that the starting point of entrepreneurial ingenuity is in the imagining of a future based on the potential perceived in the yet to be developed technology. This is the energizing wellspring from which the entrepreneurs draw in order to confront and cope with the risks inherent in their endeavors.

It is important to understand the link between this starting point for entrepreneurial ingenuity and overcoming entrepreneurial risk. Although the two domains of risk and gain may be separated in thought, in the minds of the entrepreneurs, there is a connection between these two domains in the actions of the entrepreneurs in overcoming the challenges posed by the sources of risk in pursuit of gains from the new technology or products. Overcoming the risks in order to pursue the gains requires an openness and flexibility of mind that leads to a process or path to problem solving that uses imaginative and unconventional ways of securing and using resources; in short, entrepreneurial ingenuity. How these entrepreneurs embarked on this path is beyond the scope of this chapter to consider in detail; however, it is important to recognize that the sweep of this path has been characterized as following a continuum of strategic processes from what has been termed mechanistic, synoptic, rational, causal or planned at one end of the continuum to organic, effectual, incremental or intuitive at the other end (Burns and Stalker, 1961; Sarasvathy, 2001; Methé et al., 2000).

ENTREPRENEURIAL INGENUITY AS EXHIBITED THROUGH THE INTER-PLAY BETWEEN THE SOURCES OF RISK AND COPING MECHANISMS FOR RISK

Irrespective of whether these entrepreneurs began by rational plan or intuitive insight along the path to developing their technology and business, they soon encountered unforeseen difficulties that would require imaginative and ingenious solutions to cope with the risk that could end their business. In this section we will examine how the managers involved with new business activities saw various factors as sources of risk and how these managers coped with these various risk generators. From our study we observed that ingenuity came about as the entrepreneurs confronted some source of risk by utilizing available resources and often by combining these resources in a unique and unconventional way to enable them to provide a solution that was imaginative. From the study it became apparent that the sources of risk and the sources for coping with the risk were often different sides of the same coin. Some challenge emanating from a source of risk would plunge the entrepreneur into a headlong struggle to cope, resulting in a solution that would recombine elements of the business, allowing it to go forward. Consequently, entrepreneurial ingenuity and entrepreneurial risk often operated in an obverse relationship to each other. Without a source of risk to challenge the survival of the business there was no opportunity for ingenuity, and without an ingenious solution the project or business ceased. The learning from this process of overcoming or not overcoming a source of risk was carried forward by the entrepreneur into the next project or business and became part of the plan or intuitive source of inspiration for that next project or business. This relationship between risk and ingenuity also exhibited a helical or dynamic duality as well, in that the ingeniously recombined resources that provided a solution to the initial source of risk sometimes would become the base from which the next sources of risk would arise. However, no matter how the relationship between risk and ingenuity spun, the initial trigger for action was the source of risk.

We now turn our attention to how the entrepreneurs in this study considered the sources of risk. As one founder of a venture business that had been in business for about 15 years noted:

> I have three major problems to contend with, especially at the initial founding stage. The first was money. The second was customer relationships and the third was hiring good engineers. All three are important and the third one is very important now, because this is a high technology business.
>
> (Founder of a VB)

We will examine these three major resource acquisition problems more or less in order in this section, because each is cited as a major source of risk[2] for managers involved in new business development activities (Alverez and Busenitz, 2001). These three problems are critical because each plays an important part in the development of the value chain of the venture business. We will also examine how the managers interviewed used various strategies and developed coping mechanisms for minimizing the risk involved in their new business activities. Most of the managers felt that they could control or manipulate the risks involved, and that these risks were not given to them but could be changed through their actions.

Funding, Cash Flow, Money

As noted above, getting funding for the venture business is a major element of ensuring the success of the venture. In the early stages of venture business development, as was noted, the company may not have a product or reputation to offer as a concrete track record to entice investors to provide funds for the company. The chances are very high that the company will not have enough money to fund the development activities needed to move the technology out of the research laboratory and into a saleable product.

A fundamental dilemma for a venture business founder is how do you cover the costs of developing a product when you do not have any revenue? This is why early stage funding in the United States and other market-based models of capitalism (Clausen, 2011), through Angels and venture capitalists, is so vital to the survival of the venture business. These funds can cover the costs until a product is developed and begins to generate sales revenue. However, these funds rarely come without conditions, and are not a form of charity. The Angels and venture capitalists are investing their funds with the expectation of a return on that investment. So for the venture business manager one source of risk, lack of enough funds to cover current costs, is reduced through the use of a venture capitalist, but another form of risk, loss of control over the venture business resulting from miscommunication or incompatible expectations between the venture capitalist and the venture business manager, can result.

The following comments from the founder/CEO of a venture business help to illustrate how the risk is perceived and how it is minimized.

> So the former president of this company wanted to get some money from venture capital companies, but I disagreed with him. I feel that venture capital companies in Japan do not use proper criteria for evaluating high technology venture businesses. They (the venture capital companies) only want to get profits from the venture business as soon as possible, even if only a small profit, so they do not really evaluate the technology potential. Venture capital

> companies in Japan have been destroying venture businesses, especially the high technology/high risk/high return types. Angel investors do evaluate the technology potential and are useful.
>
> (Founder/CEO of a VB)

Now it may seem that the obvious way to minimize the risk of getting money from venture capital companies who do not understand your technology would be to go to Angel investors who do. However, in Japan there are further problems as noted by this same founder/CEO.

> Funding risk in Japan is high because there are not many Angels.
>
> (Founder/CEO of a VB)

This founder/CEO did have a mechanism for dealing with the need for early stage funding that understood the technological potential of his venture business. This particular founder/CEO had been working in a large established Japanese electronics company for about 30 years before taking on his current new business development project of starting his company. As he stated:

> I have my own network and I use it. They (the people he knows) understand technology. Even though they are still employed in established companies and do not want to start their own venture business, they have money that they would be willing to invest. I talk with them and get them to invest in my company. These people, in my network, are technology pros and understand how to evaluate the potential of the technology. They are also risk tolerant because they have steady incomes from their jobs and can wait for the technology to develop.
>
> (Founder/CEO of a VB)

The use of a personal network of friends or business acquaintances as a source of initial and continued early stage funding is an example of recombining available resources in an ingenious way to fill an institutional void in the Japanese economy. An institutional void is a gap in the arrangement of an economy that makes it difficult to bring the buyers and sellers together in an efficiently functioning market (Khanna, 2002). Institutional voids can lead to market failure, but also to market opportunities if someone can imaginatively fill the void. The ingenuity of this entrepreneur's solution arises from his using a network of business colleagues who have money and understand the technology, even though these colleagues are not and have no intention of becoming Angel investors as a business. Consequently, even though there are few Angel investors in Japan, this lack of early technology evaluation capability in the Japanese new business development activity has been bridged by the ingenious use of personal networks.

One reason there are few Angels in Japan is that Angels are usually people who have been successful entrepreneurs themselves and have made enough money that they move away from founding their own start-ups to helping other people found start-ups. Usually Angels will stick close to the technology/product/market combinations they understand best. So if you founded a successful venture in integrated circuits or biotechnology you will most likely end up being an Angel for other integrated circuits or biotechnology companies respectively. In the United States there are many Angels, while in Japan there are few. This institutional void in the early funding stages of venture business development also presented a business opportunity for one of the companies interviewed for our study. This company developed a system for evaluating other start-ups technology and providing that information to various early stage funding organizations. As the founder of the company noted:

> We are the only one doing this type of rating in Japan, maybe even in the world. It is limited because evaluating technological potential is difficult. We have 500 people working for us. These 500 are retired from technology companies about 2 to 3 years ago. They do onsite evaluations of venture businesses that take from two weeks to one month to complete. This evaluation can be used by a venture capital company or bank to determine the credit worthiness of the venture business.
>
> (Founder of a VA business)

This company is one of the few filling the institutional void that exists in the early stage funding market in Japan. Interestingly enough, this founder of the company also has no plan to use the venture capital companies that he has as clients as sources of funding for his VA start-up.

In order to gain access to revenue, some of the managers of the venture businesses interviewed did contract research with larger companies. This strategy was used but more by venture businesses that were developing new technologies, which were still closer to their science knowledge base and less likely to immediately yield a commercial product. This provided a solution to the funding source risk, but opened up the possible risk of capture/acquisition or loss of the intellectual property right to the larger firm. Some of these relationships did linger, however, even when products had been commercialized. As one venture business founder noted:

> About 20 per cent to 30 per cent of our revenue comes from co-research and development with large companies.
>
> (Founder of a VB)

Other sources of money to cover expenses were used as well. These included grants from the government and public research institutes and universities. While the work with these various organizations were seen as useful, both in terms of the technological knowledge the activities yielded and for the funds they provided, the ultimate outcomes of the government/public research institute/university and the venture business are not the same. As one venture business founder put it succinctly:

> Getting a patent or solving some technological problem in itself is not doing a business. This is where our company comes in, evaluating if the patent has potential for being a business and how much potential.
>
> (Founder of an VA company)

Thus solving the funding source risk through government or university affiliation could lead to being trapped with technology and patents but no commercial products. Another issue with public funding in Japan is that 'one size does not fit all'. As one venture business founder noted:

> Compared to bioventures that require very little money to invest this business (IC related ventures) requires much more and so these are very different types of ventures. Part of the problem is that the government funding is limited and using the wrong criteria to evaluate the venture business. . . . The government is using the wrong criteria and is too tight with the money once it is invested. . . . The government system right now is focused too much on research . . . they are really strict on documentation and that increases the length of time for decision making and this is really hurting the development of venture businesses, especially high technology venture businesses in Japan.
>
> (Founder of a VB)

Finding Customers and Developing Customer Relationships

As noted, many of the venture businesses may have used some form of public funding or university connection in their earliest stages of development, but the need to have a commercially viable product and to find customers is compelling. The ability to develop a steady and stable source of sales revenue is critical for moving the venture out of its earliest stage of development and away from the need to find continued external funding and into a stage where it can develop into a fully functioning company through funding generated by internal activities. Finding that first customer, especially if it is a large established company, does a lot to legitimize the venture business and allow it to overcome its 'liability of newness' (Aldrich and Fiol, 1994). Sometimes the first major customer is also a major investor. As one venture business president noted:

> We visited many large companies asking for funding but could not get anything. Venture capitalists were also not putting money into our company. [FJH] company (a large established electronics company) originally came in as

> a co-research partner and then about six months later invested a lot of money in us. After that we could get money easily because we had the chip and a strategic partner and we had gotten good press.
>
> (President of a VB)

This strategy of using a major company as both a customer and an investor, however, has its own pitfalls, which can in turn increase risk for the venture business manager. The dynamic duality or the helical movement of the risk/ingenuity relationship is illustrated in the following comment from a venture business founder:

> Our first partnership was with [RST] company. They were looking for technology our company was working on. The managing director of that company convinced his company to not only tie up with us as suppliers, but also to invest in us through a capital tie up. We became a production subsidiary. This relationship lasted three years and led to our finding [MNO] as a customer. We had to terminate the capital tie up with [RST] company because staying with them makes it difficult to develop other customers. At that time a venture capitalist offered us an investment that would allow us to buy out [RST] and become independent.

The pull of the customers' needs is considerable even when discussing new product development in established businesses. When asked about the sources for new ideas for products and whether these were technology push ideas that originated in the lab from engineers or market pull ideas that came from customer needs, one established company CEO noted:

> Of course yes. But we are a manufacturer, so covering customers' problems is the easiest way for making profits in terms of business strategically speaking. We have ideas of new products and techniques that originate in the lab, but we keep them until the time they will help our customers. When we believe this product or technique is extraordinary, then we approach the best sponsor, paying close attention to the trend of the market. From the point of a manufacturer, we think the trend of the market is very important . . .
>
> (CEO of an SME/EC)

For venture businesses this strategy of using customers to help in the co-research and development of a product, and also as a means for coping with the lack of funds source of risk, also had its drawbacks, however. While it does generate revenue to cover current costs, it can divert important human resources away from the main research needed to bring the venture business's product to market. Further, there was always the chance that the larger companies could, because of their already established position in the industry and access to customers, develop the product before the venture business did. Here again the solution of finding a customer as a source of risk sets the conditions for further risk to develop.

The use of patents as a form of intellectual property rights protection (IPRP) was seen as helpful, but not in all cases. As one founder/CEO of venture business noted:

> Having patents is not the best way to protect the intellectual property, since [KLM] company knows the technology and if we get a patent we make it public knowledge and they can copy it. So we try to keep it as a trade secret. Also we are trying to get into a niche market that [STU] and [SVW] companies are not likely to enter. [STU] and [SVW] cross-license and enter markets through that cooperation without having to pay license fees, but for a small company with few patents even when the patent is important, it is difficult to enter with big companies since they have lots of patents and can use these to get fees from the small company since the small company cannot do a comprehensive cross-license agreement.
>
> (Founder/CEO of a VB)

Venture business managers noted that they had to be very careful in selecting research partners. Usually they would select a company that was a potential customer, but was less likely to be an immediate direct competitor, because that company lacked some of the capabilities to do work in the technological area that the venture company is pioneering. This required careful attention to each potential customer as to its capabilities and intent. As one venture business manager noted:

> One of my main customers is [STU] company. I do not sell to [MNO] company even though they are in the same business [as STU]. [STU] is not as comprehensive as [MNO] so it cannot or at least does not copy my technology, but [MNO] can.
>
> (Founder of a VB)

> We sell to all the major electronics companies in Japan. Of course we are concerned about losing our intellectual property, but this technology requires a lot of capability in analog, and these major companies like [RST] and [TUV] got out of that area a while ago. They do not have the capabilities now to copy our technology.
>
> (General manager of a subsidiary of a foreign VB)

Most of the managers noted that they found their first customers for the new product from their own existing network of contacts. It was noted that serendipity or coincidence often played a role in these connections. Serendipity has been noted as being an underemphasized element in our understanding of the innovative and entrepreneurial process (Dew, 2009). Its role in fostering, or inspiring ingenuity is also not well understood. Serendipity did play an important role in a number of the successes of the venture businesses in our study. In discussing the move into a totally new business area that was to become the major source of sales revenue for his company, the CEO of an established company noted:

> It was really good timing. I mean, [DEF], which is now our customer in the semiconductor equipment business, was setting up a committee aiming for improving this area of technology. The committee was going to test products from different companies. We were not in the business at the time, but had products we thought could fit, so we asked them to test our products too. This chance might be the most key point in making our products the success they are today. [DEF] was looking for good products since they were not happy with what they had at that time.
>
> (CEO of an SME/EC)

There are limits, however, to the serendipity generated by extensiveness of personal contacts and the luck of good timing in finding customers that have needs that can be solved by your products. For many companies another important risk occurs after the company has developed a saleable product from its technology; the risk of not being able to find customers who are willing to pay for it.

Again, the existence of an institutional void that has some unique Japanese features to it is the primary generator of this risk. As noted by the founder of the VA company:

> One big advantage we offer to venture businesses is market access. We have a big OB [Old Boy] network. In Japan, the big companies, if they trust the people and companies they are buying from they will buy the product. The price and quality does not matter as much as the trust of the people and companies. Even more to the point, selling to the big companies requires that you know who specifically to sell to and to have an introduction to that person. The 500 OB working for our company all worked for the big companies and know who is doing what and can provide an introduction.
>
> (Founder of a VA business)

Again, this institutional void generated a market opportunity for this particular VA company. Those venture businesses that could gain access to this particular network may be able to reduce the risk involved in finding customers. Most of the venture business managers noted that they had to fall back on their own personal networks here as well. There was a clear distinction made with the perceived situation in the United States, where venture companies were able to find customers by simply having a product with unique technological features or capabilities. Many of the managers at venture businesses did not go to the United States or overseas, because of concern about their lack of English language skills and their lack of knowledge about the market situation and customer relations' etiquette. However, those that did go were able to get some results in spite of the difficulties.

> I made a decision to go to the United States [to find potential customers] and after finding a company, within four days I had a contract. It took up to one year for me to do the same thing here in Japan.
>
> (Founder of a VB)

Even when a customer is found, who is willing to take a chance on an unknown company selling a product that has not been tested for a long time in the market, using some novel technology, there are difficulties in the way large companies treat smaller companies here in Japan. Again, it must be emphasized that one of the fundamental issues that venture companies have to contend with is maintaining a positive cash balance, or they go bankrupt. Any delays in the payment for a product can be catastrophic for a fledgling company starting out. One venture business manager related how he had difficulty even after he had confirmed orders for his product, which was a specialized piece of semiconductor manufacturing equipment. After they had developed the prototype and had demonstrated it, they received an order for five units.

> We received an order for five sets from a large Japanese semiconductor materials company. We had the factory building, but we did not have enough money to start production. Because we had the order we could get money from a venture capitalist and we began our production and sent the five completed sets to the company. But the custom in this business is to give an oral order for equipment, so nothing is written down. The big company did not pay us for the equipment we delivered.
>
> (Founder/CEO of a VB)

This venture business went bankrupt and part of it was sold to another more successful venture business, which eventually did get paid by the large company. Some of the original venture business was also merged with another start-up working in a different line of business and is now part of a venture business holding company, which has five venture businesses in its portfolio. So, for this entrepreneur, he was able to re-configure the original company, and through his personal contacts work the situation out so that he has been able to continue to grow his new company. He learned from these past experiences to be very careful to structure contracts for his products so that he receives payment as soon as possible. As he noted:

> The biggest risk for me is to get into an expensive contract (one requiring a lot of investment upfront) without knowing I will get a return. I only make a

> contract with a certain cash inflow. I like to structure my contracts so that I get paid within two months.
>
> (Founder/CEO of a VB)

Another venture business manager stated the following:

> A main difference in the States is that if the technology is good and highly efficient then it is accepted in the market no matter the company's history, but in Japan it is not that way and it depends on the contact history of the company. In the US the new technology replaces the old, and in a sense it is the Anglo-Saxon culture that is 'dry' in that we move to the new and better without concern for the relationship.
>
> In Japan there is a more close relationship between the main company and component company so it is hard to get into the relationship but once you get into it the customer company will tell you that some competitor has a better product and can you make it better and they will support you to make it better, rather than leave you for the company with the better component right away, as in the States.
>
> (Founder of a VB)

This manager expressed a view that was echoed by several of the other managers interviewed for this study. The knowledge needed to cultivate relationships with customers was fine grained in that it was very specific to the technology, products and customers involved. Further, some of this could be taught in training sessions but because of the idiosyncratic nature of the knowledge it was not easily learned except through direct experience.

This human relations form of contracting is particular to the business and even the geographical region in Japan. This same manager went on to say that:

> It is strict to the local culture. Chiiki micchaku gatta or 'the local way of doing things', are important to entering and maintaining the relationship.

He went on to contrast the business style in several countries noting that:

> It is this way in Korea also. I have a business relationship in Taiwan, but in Taiwan it is business based. The personal blends with the business in Taiwan, whereas in Japan and Korea the personal is business and business is personal. In Taiwan the personal can become business, but the two are not the same.
>
> (Founder of a VB)

These relationships can become institutionalized to such a degree that they affect all aspects of customer development from knowing what the future needs of customers will be to how the buyer and seller manage the financial aspects of the contract to what happens with issues of quality control and after sales service. This is a common problem for both the indigenous Japanese venture business company and for the foreign subsidiary setting up in Japan. Neither company has a history or reputation that is well known. Knowledge of who is doing what kind of work in the customer companies and would be the specific person to pitch your new product to is unknown and access is also guarded unless an introduction is granted. Persistence is required, but time and lack of cash are working against any new entrant. The one advantage that the foreign subsidiary has in this is that it may not be under the same cash flow pressure as the Japanese venture start-up, assuming the mother company has adequate financial resources from its operations in its home and other markets. Working against the foreign subsidiary, however, is an almost total lack of understanding of the local business culture.

> Finance is not important since that is handled mostly by the HQ in the home country. In conducting sales we have a problem because of the Koza or bank account. In other countries the seller checks the buyer's financial information, but in Japan the customer checks the seller's financial situation and reputation. This is the Koza system. If my company wants to sell to Panasonic, for example, but if we have no Koza then one has to be set up and Panasonic has to pay money for this. So if the business is very big (in terms of amount of sales) then they will, but if it is small, then it is too much trouble to set up the Koza.
>
> (General manager of a subsidiary of a foreign VB)

He went on to point out that finishing a project is critical for survival in Japan, because trust and credibility is an important factor in doing business here. He noted that:

> Once you fail with a customer, even one or two years later it is almost impossible to come back. In the United States even if you fail once, if you have a good product and price you can come back, but in Japan there is less of a second chance so if you fail once your reputation is hurt and it is almost impossible to come back.

He offered the following example:

> One Japanese company that did have relationships with other big Japanese companies began to shift its business over to Nokia because it was growing and the Japanese companies were not. After a couple of years this company tried to re-establish its relationships with the big Japanese companies and these companies said 'no, because a couple of years ago you did not support us so you are untrustworthy'.

He offered another example as follows:

> I once sold one million units of our product to a Japanese customer and one unit was defective. The Japanese customer asked to check the entire production process and find out where the problem was. Doing so costs a lot of money, and the HQ said they did not care about the one in a million and why doesn't he just give them another one as a replacement. The Japanese customer did not like this but the HQ did not understand.

For this manager of a foreign VB, one of the largest sources of risk came in the form of having to have to balance the needs to compete in Japan and meet the expectations from his headquarters company. He has two ways of coping with this set of risks. One way he is able to cope with this risk is to pay very close attention to who is hired. He was very careful to make sure that those who were hired here in Japan truly understood how to do business in Japan. In this case it meant hiring only people who had had experience working in Japanese companies. In order to reduce the risk involved in hiring, he is personally involved in the interviewing process and he uses only selected headhunting firms, not open recruitment. He tries not to use the foreign affiliated headhunter companies, however, because as he said:

> These guys get a lot of people who are good at English and have worked only in foreign affiliated companies, so they do not know what it is like to work in a Japanese company, even when they are Japanese. They do not know the proper business etiquette (reigi).

The careful attention to the recruitment and hiring of new personnel ensures that he has a staff that understands not only the technology, but also has had experience with the local business culture (local to the type of integrated circuit products he sells) and so he will minimize any misunderstanding with the customers.

The second way he copes with this risk is to acculturate the HQ people to the differences here in Japan. Since he has had more than ten years' work experience in the United States in selling integrated circuits, he has credibility with his own HQ. As he puts it:

> I try to incrementally talk to the HQ. You cannot push all at once, so you push little by little. With the Japanese customers, they understand that there are these issues and so I ask them to be patient and I work on changing the HQ view little by little.
>
> (General manager of a subsidiary of a foreign VB)

FINDING GOOD PEOPLE, HUMAN RESOURCES AND ENGINEERS

As noted by the above manager, one of the sources of risk is the people who are brought into the company. Having people who understand the technology and can interact with customers is critical for the survival of the company. Being able to hire the right people for the company is difficult.

> Ten years ago when I graduated from university, people did not really think being in a venture business was good. It was better to enter an established big company. People do not respect the entrepreneur and do not see the importance of ventures. The United States has a different view which puts more value on entrepreneurs and ventures. Parents, families, friends will support and give money to a venture business in the United States, but to do this in Japan is very difficult. You need to change these cultural beliefs and norms.
>
> (Founder/CEO of a VB)

> Also in the university with the professors at the labs and such, the professor knows about how many students get into big name companies, but not how many start their own business. So the prestige and reputation of working in a big company is evident and supported and the feeling is that the not so smart students cannot get in and start their own business. This ends up being a self-fulfilling prophecy because smart students are encouraged to go into big name companies and the not so smart ones go into venture businesses so the failure rate of ventures is high. This has to change and professors have to support smart people going into venture businesses.
>
> (Founder/CEO of a VB)

> Other things vary, but human resources are important all the way through.
>
> (Founder of a VB)

Even when an established company has been in business for many years, it has to be concerned with attracting good people to work for it. One company which had been listed for many years on one regional stock exchange decided to go through the effort to get listed on the Tokyo Stock Exchange. When asked why, the CEO responded as follows:

> The most important advantage we think is to acquire good quality human resources. We think that human resources support a company, so we need good quality ones.
>
> (CEO of an SME/EC)

Founder and Top Management Founding Team as a Human Resource

One of the key human resources for any new business development project is the person who is in charge of managing the project. This is because the entrepreneur is the first source for 'legitimating' the venture business as a real company. The characteristics, especially the attitude, experience and skills of the manager are critical both in terms of that person's ability to cope with risk and to inspire others to join what are objectively risky and speculative endeavors. Without this legitimating capability the venture business manager would be unable to attract the funds necessary to start the business, the human resources necessary to develop the business or the customers necessary to grow the business.

From our study, we also found that these managers rarely operated as single individuals. They worked with someone else, most frequently two or three other people, who made up the 'founding team'. Because of the limitation of time and resources we focused on the founder or CEO of the company. In the discussion that follows, the characteristics of the founder or CEO are described.

When asked what did the various managers in charge of the Angel capital companies look for in the people who were running the venture businesses, the following comment was typical:

> The character of the entrepreneur is important. Having a positive outlook and clear thinking and having a 'hot atmosphere'. After that the person's specialty and human network. So in evaluating a venture we look to see if they have a good idea, the entrepreneur has a good character and good support. Those are the keys to success.
>
> (Founder of a VA)

This Angel capital manager went on to say that he would try to make adjustments if any of the above were lacking from the entrepreneur running the venture business. He noted that:

> If the idea is good but the entrepreneur's character is not good then we try to replace him with someone who has a good character; that is, someone who has a positive outlook and an outgoing personality and good management and personal skills and the person with the good ideas should be put into another area, say research and development. Ideally we like if we have a pair of people, one who is outgoing and one with technical creativity.
>
> (Founder of a VA)

From our study, especially during the interviews, it was evident that the managers we talked with were all articulate and outgoing. They clearly did appear to have good people skills, at least during the interviews. Most of the managers we interviewed also had engineering and technological knowledge about the fields they were engaging in business. All the managers interviewed had previous experience working in companies; again, most were working in electronics companies before starting up their venture business.

On the surface the level of experience, which ranged from about 10 to 30 years, should have been adequate to prepare these managers to take on the role of entrepreneur. However, many of the managers interviewed in this study indicated that running your own business was an experience in and of itself.

First, many noted that failure of the entrepreneurial enterprise often meant not only bankruptcy of the venture business, but also personal bankruptcy as well. The process of early stage funding often required that the entrepreneur put his personal assets up as collateral to secure funding. This was not something that was encountered in working for an already established company.

As one venture business founder noted:

> The image of risk is different. In Japan you lose everything if the business fails. You lose the business and you also lose your personal assets if the business fails.

Second, the training and experiences that were encountered in an established company prepared these managers for handling some of the strategic elements of their jobs as entrepreneurs, such as developing a business plan, and for many of the subtle elements of personnel or human resource management; however, most noted that they felt ill-prepared for issues related to the management of money. As one Angel capitalist noted:

> Especially in the high technology companies the managers are engineers and have no sense of finance and other business matters and often fail.

As one venture business manager with many years of experience in a large established electronics company put it:

> Through my experiences at the company I feel I learned a lot and that gave me the initial confidence that I could manage a start-up business. In addition to my engineering experiences I spent one and a half years in the Strategic Planning department of the company and learned about the importance of the developing business strategy. However, I did not anticipate that there would be so many problems in getting the company going and I almost went bankrupt. I am in the red now, at least in terms of my personal financial return, but since my children are grown and my wife is OK with my taking this venture, I am sure it will work out.
>
> (Founder of a VB)

Another entrepreneur noted that he had a lot of experience in a foreign subsidiary of an electronics company in Japan. This foreign electronics company expected that its senior managers would be able to develop business models for the new business areas they were moving into. As he noted:

> My experience helped me with dealing with the venture capitalists, since I had to develop business models and work with many of the other management and business issues, but I still was not prepared enough with the financial aspects of running a business. Nowadays, I have hired a financial expert to help me with this critical skill.
>
> (Founder of a VB)

This is why many of the venture business mangers and the Angel managers interviewed for this study noted that the actual experience of running a venture business is vital to understanding how to evaluate and guide potential entrepreneurs in founding their venture businesses.

SOME CONCLUDING THOUGHTS AND MANAGERIAL IMPLICATIONS

Our study is preliminary and should be seen only as a starting point for further research. Our study focuses on the perception of risk and the mechanism that managers and entrepreneurs in Japan use to compensate for risk. Entrepreneurial risk is seen as primarily related to factors such as the acquisition of funding, customers and employees that can cause the project or business to fail or go bankrupt. Risk is perceived as inherent in the management process and all the managers believed that they had to deal with risk. Risk was also perceived as manageable and could be manipulated in ways that would minimize it. Minimizing risk required entrepreneurial ingenuity and took the form of re-configuring the sources of risk to provide a solution. This solution would often form the base for new sources of risk that would again require ingenuity to be countered. We observed a dual dynamic between entrepreneurial risk and entrepreneurial ingenuity that spun helically over time. Our research suggests that while the spark of imagination concerning the potential contributions from the technology served as a source of motivation for the entrepreneur to draw upon in sustaining the efforts necessary to overcome risk, the entrepreneurial risk was the trigger for those efforts.

The entrepreneurs in our study commented that acquiring cash, customers and employees posed the greatest sources of business related entrepreneurial risk. The fundamental dilemma of not having the product to show to get the resources necessary to develop the product was seen as independent of the purely technical difficulties in developing the product. This fundamental entrepreneurial dilemma was most prominently illustrated in the acquisition of adequate funding. Cash flow problems existed throughout the experience of the companies, even managers in the well established companies, some with over 100 years of existence, were concerned with maintaining positive cash flows. The concerns for the venture businesses were particularly acute, however, since they had no internal source of funding once the initial investment of the founder himself had run out. From an institutional point of view, in creating the value chain for the venture business, entrepreneurs all had to turn to external sources for the funding and other resources necessary, either Angels, venture capitalists, government programs, other large companies, banks, or universities. Each of these sources of funds and other resources, while useful in dampening the first risk of bankruptcy, generated other sources of risk for the venture business manager. Consequently an important implication for managers about to embark on an entrepreneurial venture is to become aware of this helical dynamic and expect that the solution to one set of problems will often generate the next set of problems. Learning how to manage these relationships is critical to both the short-term and long-term survival of the venture.

The knowledge and capabilities of the entrepreneur are critical to the survival of the business. The entrepreneur acts as the first 'legitimator' of the venture business and the entrepreneur and the team assembled to run the business are often the only sources of human and financial resources available to found the business and move it through the initial stages of technology proof and prototype development. However, as Stanworth and Curran (1976: 100) pointed out, entrepreneurial endeavors are an 'ongoing social entity constructed out of the meanings and actions of those who participate in the firm or who are "outsiders" in relation to the firm as social grouping but nonetheless interact with the participants' and consequently, 'definitions and meanings attached to situations are *socially generated, socially sustained* and . . . *socially changed*'. As a result, the particular institutional environment that the entrepreneur finds him or herself in can affect greatly the identity of the entrepreneur. Although this was not studied directly, among the three entrepreneurial identities suggested by Stanworth and Curran (1976) the 'artisan' or perhaps the 'manager' types appear to be the most prevalent in Japan. Each of these types has different aspirations for the growth of their businesses, and as such different motivations when engaging in entrepreneurial ingenuity. While the necessity to engage entrepreneurial ingenuity to overcome risk is there for all three types, the desired end state will differ among the artisan, manager and classical entrepreneur. An important implication for both the entrepreneur and those attempting to foster a business ecology more supportive of entrepreneurs would be to allow the current set of venture business managers to have a greater role in developing and implementing the various funding, and other resource generating activities so necessary for healthy business development. Listening more carefully and taking more seriously the views and opinions of the venture business managers could put better criteria for evaluating and funding technological proposals put into place. This is something that government and private organizations such as venture capital companies and banks need to consider whether in Japan or elsewhere.

Further, by lowering the level of consequences that can result from the failure of a business, and allowing for second chances, the experience, attitudes and skill sets of entrepreneurs can be more easily developed and passed on. It is important to recognize that the perception of failure from a private business point of view is actually just an experiment when viewed at the level of the society (Knott and Posen, 2005). That experiment embodied in the entrepreneur contains much useful information that can be used by other business managers in incumbent firms and entrepreneurs about to launch new ventures. In most cases, a 'learning by doing' process characterized by a lot of trial and error dominates the creation of new businesses. Each entrepreneur learns from his/her own previous experiences, which if not catastrophic at the outset can enhance the chances of not only survival but also success in a second or third attempt. Entrepreneurial ingenuity, it appears, does not always spring fully formed from the mind of the entrepreneur, but at times emerges through an incremental process that may be quite disjointed.

Although not discussed in detail in this report, many of the venture business managers alluded to previous 'failures' that turned out to be valuable learning experiences that have enhanced the ingenuity of their current actions that allowed them to survive in their current venture business endeavors. These previous failures were often 'private'; that is, out of sight of other actors (government, banks, venture capital companies, potential customers and such). However, the learning effects from failure are not just the direct effect of learning by doing for each entrepreneur, but also can emanate from vicarious learning (Denrell, 2003). If greater allowances for trial and error could be made, it would be possible to provide an environment that does not 'throw the baby out with the bath water'. Change in the Japanese business environment is neither straightforward nor evenly paced, but it does occur (Methé, 2005, 2006a, 2006b). Overcoming the draconian effects on reputation that follow from a public business failure will not occur in one jump, but if Japan is to truly revitalize its economy, this aspect of its business ecology must fade away.

This has very important managerial implications not only for how an established business treats its intrapreneurs who fail but also for financial, educational, governmental and other businesses that encounter an entrepreneur from a failed start-up. The experience gained from confronting risk and ingeniously overcoming it is obvious. Less so the experience gained from failing to overcome. But these failures also hold the seed of ingenuity in them, if the entrepreneurs themselves are given opportunities to use the lessons learned and if others are made aware of the pitfalls through the sharing of those experiences.

NOTES

* This paper is based on research conducted for a project on Risk Cognition, Technological Innovation, Institutional and Managerial Environments: Entrepreneurs and Intrapreneurs in Japanese High-Tech Industries Grant-in Aid for Scientific Research (B) by the Society for the Promotion of Science, MEXT (Kaken Grant) (2005–2007).

1. In Japan, a start-up is called a venture business and those who start a new business are called venture business managers. In this chapter, the term entrepreneur refers to anyone that engages in developing a new product, process or technology, not necessarily only those who start up a new business. Thus, the term entrepreneur or intrapreneur is used interchangeably throughout to refer to managers who are engaged in new product/process/technology development, whether in an established company or a new venture business.
2. The participants in our study recognized that there were technical difficulties that needed to be overcome in order to develop the product process or technology, and so there was a technological risk component. This technical or technological risk component, however, was not seen as posing the threat to the business endeavor that finding funding, customers and employees posed. It appears that the technological risk component was integral to the new technology venture and formed a base line of risk. The 'business resource' related risk then was the risk that appeared above the base line and caught the entrepreneur's attention.

REFERENCES

Aldrich, Howard and C. Maelene Fiol (1994) 'Fools rush in? The institutional context of industry creation', *Academy of Management Review*, **19**(4), 645–70.

Alverez, Sharon A. and Lowell W. Busenitz (2001) 'The entrepreneurship of resource-based theory', *Journal of Management*, **27**, 755–75.

Arrow, Kenneth J. (1971) *Essays in the Theory of Risk Bearing*. Chicago: Markham.

Ballon, Robert and K. Honda (2000) *Stakeholding: The Japanese Bottom Line*. Tokyo: The Japan Times Ltd.

Basili, Marcello and Carlo Zappia (2010) 'Ambiguity and uncertainty in Ellsberg and Schackle', *Cambridge Journal of Economics*, **34**, 449–74.

Bracker, Jeffrey and David T. Methé (1994) 'A cross-national study of planning characteristics of U.S. and Japanese entrepreneurs', *International Journal of Management*, **11**(2), 634–40.

Bell, D.E. (1983) 'Risk premiums for decision regret', *Management Science*, **29**, 1156–66.
Boone, Christophe and Bert de Brabander (1997) 'Self-reports and CEO locus of control: a note', *Organization Studies*, **18**(6), 949–71.
Borton, James W. (1992) *Venture Japan: How Growing Companies Worldwide Can Tap Into The Japanese Venture Capital Markets*. Chicago, IL: Probus Publishing Company.
Bosma, Neils and Rebecca Harding (2007) *Global Entrepreneurship: GEM 2006 Summary Results*. Babson College and London Business School. Available at: http://gemconsortium.org.
Burns, T. and G.M. Stalker (1961) *The Management of Innovation*. London: Tavistock.
Busenitz, Lowell W., G. Page West III, Dean Shepherd, Teresa Nelson, Gaylen N. Chandler and Andrew Zacharakis (2003) 'Entrepreneurship research in emergence: past trends and future directions', *Journal of Management*, **29**(3), 285–308.
Carland, James W., Frank Hoy, William Boulton and Jo Ann Carland (1984) 'Differentiating entrepreneurs from small business owners: a conceptualization', *Academy of Management Review*, **9**(2), 354–59.
Clausen, Tommy Hoyvarde (2011) 'Comparing start-up activity across capitalist economies', *Acta Sociologica*, **54**(2), 119–38.
Conrath, David W. (1967) 'Organizational decision making behavior under varying conditions of uncertainty', *Management Science*, **13**, B487–B500.
De Clercq, Dirk and Pia Arenius (2006) 'The role of knowledge in business start-up activity', *International Small Business Journal*, **24**(4), 339–58.
Denrell, Jerker (2003) 'Vicarious learning, undersampling of failure and the myths of management', *Organization Science*, **14**(3), 227–43.
Dequech, D. (2000) 'Fundamental uncertainty and ambiguity', *Eastern Economic Journal*, **26**, 41–60.
Dew, Nicholas (2009) 'Serendipity and entrepreneurship', *Organization Studies*, **30**(7), 735–53.
Faro, David and Yuval Rottenstreich (2006) 'Affect, empathy, and regressive mispredictions of others' preferences under risk', *Management Science*, **52**(4), 529–41.
Feigenbaum, Edward A. and David J. Brunner (2002) 'The Japanese entrepreneur: making the desert bloom', Stanford University.
Foss, Kirsten, Nicolai J. Foss and Peter G. Klein (2007) 'Original and derived judgment: an entrepreneurial theory of economic organization', *Organization Studies*, **28**(12), 1893–1912.
Gephart, Robert P. (2004) 'Qualitative research and the *Academy of Management Journal*', *Academy of Management Journal*, **47**(4), 454–62.
Hayward, Mathew L.A., Dean A. Shepherd and Dale Griffin (2006) 'A hubris theory of entrepreneurship', *Management Science*, **52**(2), 160–72.
Ireland, R. Duane, and Justin W. Webb (2007) 'A cross-disciplinary exploration of entrepreneurship research', *Journal of Management*, **33**(6), 891–927.
Kelly, Donna J., Neils Bosma and Jose Ernesto Amoros (2011) *Global Entrepreneurship Monitor: 2010 Global Report*. Global Entrepreneurship Research Association. Available at http://gemconsortium.org.
Khanna, T. (2002) *Local Institutions and Global Strategy*. Boston, MA: Harvard Business School Publishing.
Knight, F. H. (1921) *Risk, Uncertainty and Profit*. New York: Houghton Mifflin.
Knott, Anne Marie and Hart E. Posen (2005) 'Is failure good?', *Strategic Management Journal*, **26**, 617–41.
Lachmann, L. (1976) 'From Mises to Shackle: an essay on Austrian economics and the kaleidic society', *Journal of Economic Literature*, **14**, 54–62.
Lampel, Joseph, Benson Honig and Israel Drori (2012) 'Discovering creativity in necessity: organizational ingenuity under institutional constraints', *Organization Studies*, **33**(2), 279–81.
Lowe, Robert A. and Arvids A. Ziedonis (2006) 'Overoptimism and the performance of entrepreneurial firms', *Management Science*, **52**(2), 173–86.

March, James C. and Zur Shapira (1987) 'Managerial perspectives on risk and risk taking', *Management Science*, **33**(11), 1404–18.

Methé, David T. (1991) *Technological Competition in Global Industries: Marketing and Planning Strategies for American Business*. Westport, CT: Quorum Books.

Methé, David T. (1992) 'The influence of technology and demand factors on firm size and industrial structure in the DRAM market – 1973–1988', *Research Policy*, **21**(1), 13–25.

Methé, David T. (2005) 'Continuity through change in Japanese management: institutional and strategic influences', in R. Haak and M. Pudelko (eds) *Japanese Management: The Search for a New Balance between Continuity and Change*. New York: Palgrave Macmillian.

Methé, David T. (2006a) 'Institutional, technological and strategic factors in the global integrated-circuit industry: the persistence of organizational forms', in Yoshitaka Okada (ed.) *Struggles for Survival: Institutional and Organizational Changes in Japan's High-Tech Industries*. New York: Springer-Verlag.

Methé, David T. (2006b) 'Employment practices in Japanese firms: can Mikoshi management survive?', in R. Haak (ed.) *The Changing Structure of Labour in Japan: Japanese Human Resource Management: Between Continuity and Innovation*. New York: Palgrave Macmillian.

Methé, David T. and Jeffrey Bracker (1994) 'A cross-national study of perceptions of the business environment by U.S. and Japanese entrepreneurs', *International Journal of Management*, **11**(1), 599–603.

Methé, David T., William Mitchell and Anand Swaminathan (1996) 'The underemphasized role of established firms as the source of major innovations', Second Special Issue on Telecommunication Policy and Strategy, *Industrial and Corporate Change*, **5**(4), 1108–203.

Methé, David T., D. Wilson and J.L. Perry (2000) 'A review of research on incremental approaches to strategy', in J. Rabin, G.J. Miller and W.B. Hildreth (eds) *Handbook of Strategic Management: Second Edition, Revised and Expanded*. New York: Marcel Dekker, Inc.

Meyer, Robert J. and Yong Shi (1995) 'Sequential choice under ambiguity: intuitive solutions to the armed-bandit problem', *Management Science*, **41**(5), 817–34.

Miller, Kent D. (2009) 'Organizational risk after modernization', *Organization Studies*, **30**(2/3), 157–80.

Mosakowski, Elaine (1997) 'Strategy making under causal ambiguity: conceptual issues and empirical evidence', *Organization Science*, **8**(4), 414–42.

Murmann, Johann Peter and Deepak Sardana (2013) 'Successful entrepreneurs minimize risk', *Australian Journal of Management*, **38**(1), 191–215.

Phillips, Nelson and Paul Tracey (2007) 'Opportunity recognition, entrepreneurial capabilities and bricolage: connecting institutional theory and entrepreneurship in strategic organization', *Strategic Organization*, **5**(3), 313–20.

Pitt, Martyn (1998) 'A tale of two gladiators: 'reading' entrepreneurs as text'. *Organization Studies*, **19**(3), 387–414.

Pratt, J.W. (1964) 'Risk aversion in the small and in the large', *Econometrica*, **32**, 122–36.

Sadler-Smith, Eugene (2004) 'Cognitive style and the management of small and medium-sized enterprises', *Organization Studies*, **25**(2), 155–81.

Sarasvathy, Saras D. (2001) 'Causation and effectuation: towards a theoretical shift from economic inevitability to entrepreneurial contingency', *Academy of Management Review*, **26**(2), 243–63.

Shackle, G.L.S. (1955) *Uncertainty in Economics and Other Reflections*. Cambridge: Cambridge University Press.

Sitkin, Sim B. and Amy L. Pablo (1992) 'Reconceptualizing the determinants of risk behavior', *The Academy of Management Review*, **17**(1), 9–38.

Stanworth, M. John and James Curran (1976) 'Growth and the small firm-an alternative view', *Journal of Management Studies*, **13**(2), 95–110.

Thomas, Anisya S. and Stephen L. Mueller (2000) 'A case for comparative entrepreneurship: assessing the relevance of culture', *Journal of International Business Studies*, **31**(2), 287–301.

Tversky, A. and D. Kahneman (1991) 'Loss aversion in riskless choice: a reference-dependent model', *The Quarterly Journal of Economics*, **106**(4), 1039–61.
Weber, Elke U. and Richard A. Milliman (1997) 'Perceived risk attitudes: relating risk perception to risky choice', *Management Science*, **43**(2), 123–44.
Weber, Elke U. and Christopher Hsee (1998) 'Cross-cultural differences in risk perception, but cross-cultural similarities in attitudes towards risk', *Management Science*, **44**(9), 1205–17.
Whittaker, D. Hugh with P. Byosiere, S. Momose, T. Morishita, T. Quince and J. Higuchi (2009) *Comparative Entrepreneurship. The UK, Japan and the Shadow of Silicon Valley*. Oxford and New York: Oxford University Press.
Wijen, Frank and Shahzad Ansari (2007) 'Overcoming inaction through collective institutional entrepreneurship: insights from regime theory', *Organization Studies*, **28**(7), 1079–100.
Xu, Hongwei and Martin Ruef (2004) 'The myth of the risk-tolerant entrepreneur', *Strategic Organization*, **24**(4), 331–55.

7 Reinvesting dance with meaning: authenticity and ingenuity in the artistic dance field*

Tamar Sagiv

INTRODUCTION

Creative industries represent a rapidly growing economic sector that exerts an extraordinary influence on our values, attitudes and lifestyles (Caves, 2000; Flew and Cunningham, 2010; Lampel et al., 2000). Cultural goods are 'nonmaterial' goods serving aesthetic and expressive needs rather than a clearly utilitarian function (Becker, 1984; Hirsch, 1972; Hirsch, 2000). Cultural goods derive their value from subjective experiences that rely heavily on the use of symbols or abstract ideals to manipulate perception and emotion (Lampel et al., 2000).

After the establishment of the Creative Industries Task Force (CITF) in the 'post-industrial' UK in the late 1990s, an opportunity has emerged for the sector of cultural goods to gain status as a sector that contributes to national income, overall employment and wealth creation (Flew and Cunningham, 2010). Since creativity is one of the resources crucial to the success of a cultural product (Jones and DeFillippi, 1996; Miller and Shamsie, 1996), the shift in conceptualization of cultural products as providing both economic value and symbolic value has promoted new discourses on the role of creativity in the generation of economic success (Amabile, 1998; Caves, 2000). The potential contribution of cultural products to economic growth has inspired researchers to study how the dilemmas experienced by culture entrepreneurs (Swedberg, 2006) have much in common with those of other industries, where knowledge and creativity are central in sustaining competitive advantage, such as IT, biotech, consulting, and even higher education (DeFillippi et al., 2007; Lampel et al., 2000).

Competition in cultural industries is driven by the search for novelty. A major dilemma concerns to what extent new artistic genres emerge *de novo* or whether artistic innovations derive from pre-existing types or categories (Bornstein, 1989; Hekkert et al., 2003; Lampel et al., 2000; Lounsbury and Rao, 2004). For example, cultural entrepreneurs can differentiate products without making them fundamentally different from others in the same category. This kind of novelty represents a recombination of existing elements and styles that differentiates, but does not break existing artistic and aesthetic conventions. On the other hand, cultural entrepreneurs can pursue innovation beyond existing limits. This type of novelty often results in new types of cultural products that may expand or fundamentally change the market (Lampel et al., 2000). Evidence from empirical studies is mixed. A study on film teams (Perretti and Negro, 2007) revealed that teams innovate both through crafting new practices with new members and by combining standardized or established practices with more experienced members. A study that examines the balance between product differentiation and market innovation in the early motion picture industry (Mezias and Mezias, 2000) finds that genre creation could be regarded as a form of product differentiation rather than market innovation. Other research has shown that theater managers avoided standardized solutions to recurring problems (Eikhof and Haunschild, 2007). The tension between differentiation and innovation has also been addressed in the context of the rise of Impressionism in France toward the end of the nineteenth century (Wijnberg and Gemser, 2000). Impressionism makes a decisive break with established styles and traditions in the visual arts and at the same time transforms the innovativeness of a work of art into the dominant criterion by which its value is judged.

Novelty in cultural industries has also been addressed by two lines of scholarship, which differ in their level of analysis. The first, the production of culture perspective (Bourdieu, 1983; Peterson and Anand, 2004) emphasizes that symbolic elements are shaped by the systems of production in which they are embedded. The second approach departs from societal-level analysis and provides a micro-foundation perspective that acknowledges the contribution of the agency of exceptional individuals pursuing creative careers in facilitating creative processes (Jones and Thornton, 2005; Jones, 2010; Svejenova, 2005). Important attempts have been made to complement the production of culture perspective with a micro-foundation approach and bring together action and structure as the engines that drive novelty in cultural industries through constant interactions between collective action and individual agency (Bourdieu, 1990; Dougherty, 1992; Feldman and Orlikowski, 2011; Svejenova, 2005).

The tension between product differentiation and market innovation in the creative industries has been addressed also in scholarly debates on defining the relationship between authenticity and creativity. The cultural good must be new and valuable according to some external criteria, and creativity is the production of novel and useful ideas in any domain (Amabile et al., 1996; Amabile, 1998; Sternberg, 1999). Cultural goods that are regarded 'authentic' are accorded distinctive significance, are perceived to carry special meaning and are perceived as more valuable than other non-authentic objects with the same characteristics (Carroll and Wheaton, 2009; Jones et al., 2005). Some scholars see creativity and authenticity as opposing terms (Peterson, 2005) while others see them as complementary (Jones et al., 2005; White and White, 1965). A cultural entrepreneur is said to have a 'creative voice' if he or she is judged to have an interpretation that makes his or her presentation distinctive and clearly recognizable (Perry-Smith and Shalley, 2003; Peterson, 1997). Authenticity hints at reference to an existing implicit template (Peterson, 2005). The creation of 'authentic' schema requires the adoption of claims that involve representation of a distinguished style or tradition (Beverland, 2005; Khaire and Wadhwani, 2010; Spooner, 1986), belonging to a specific place (Jones and Smith, 2005; Lindholm, 2008), ascription of group membership or kinship (Urquia, 2004) or commitment to a specific lifestyle (Fine, 2004; Grazian, 2005). Putting aside the scholarly debates on whether authenticity and creativity are opposing or complementary terms, current literature on cultural industries stresses that the construction of authenticity and creativity in organizations are processes that enhance one another and that claims of authenticity are essential to support claims of originality and creativity. In other words, to gain creativity it is essential to claim authenticity (Dewey, 2005; Jones et al., 2005; Peterson, 1997; Ryan and Peterson, 1982). However, the relative neglect of the role of authenticity during the creative process has limited our understanding of the nature of the interaction between authenticity and creativity within the artistic realm. Moreover, neither authenticity nor creativity are 'real' things or attributes that can be objectively determined. Rather, both authenticity and creativity are socially constructed phenomena determined by audiences in a particular social context (Carroll and Wheaton, 2009; Peterson, 1997; Peterson, 2005). This 'social construction' perspective on authenticity and creativity implies that assessment of the relationship between the two constructs is much more dynamic and complicated than interactions of facilitation or contradiction. The relationship might be subjected to changes and redefinitions along the creative process itself and over time. Moreover, the literature on the construction of authenticity and creativity has paid less attention to the potential influence of the limited resources and structural constraints inherent in cultural industries (Caves, 2000; Fine, 2004; Hirsch, 2000). Despite the challenges raised by limited resources and structural constraints on the operation of firms, the ways in which constraints force cultural entrepreneurs to enhance organizational ingenuity and compete for novelty are relatively underexamined. More empirical studies are needed to examine the various constraints that trigger creativity as a free exercise of ideas in response to coerciveness. For example, a potential scenario that has never been tested is whether actors reject external constraints but are strongly motivated by various inner constraints in their thinking and artistic development. The purpose of this chapter is to shed new light on our understanding of constraints in cultural industries, specifically the tendency to see constraints in these industries as essentially external, without due regard for the artist's own sense of creative integrity, which also acts as a constraint. I investigate how organizational ingenuity serves to overcome artists' inner constraints and enhance creativity and novelty. Further, I show how the development of adequate structure and institutional infrastructure is crucial for gaining novelty at the field level, far beyond the artist's own creative voice.

To encompass the dynamic interaction between the agency of cultural entrepreneurs and the social structures within which they operate, I adopt a conception of the field of cultural production (Bourdieu, 1983), a structured space with its own laws of functioning and its own power relations. One of the salient features of the field of cultural production is that it is structured by an opposition between two sub-fields: the field of restricted production and the field of large-scale production (Bourdieu, 1983). The field of restricted production concerns what we think of as 'high' art (e.g., fine arts, non-commercial theater, classical music, belles-lettres and artistic dance). In this sub-field, cultural entrepreneurs seek consecration and symbolic, prestigious and artistic reputation instead of economic profit. The field of large-scale production involves 'mass' or 'popular' culture: commercial television, cinema production, radio and mass-produced literature. Its cultural entrepreneurs seek economic profit. The restricted field of production is conducive to constant innovation whereas the dependence of the large-scale production field on the broadest possible audience makes it less susceptible to formal experimentation (Bourdieu, 1983).

I use the field of artistic dance as the context to examine the relationship between processes of organizational ingenuity and authentication. The context of the field of artistic dance is well suited for my study for several reasons. First, the artistic dance field belongs to the restricted field of production (Bourdieu, 1983); therefore, it offers an environment in which constant experimentation and innovation are particularly crucial. Dance production represents not just the creativity of individual artistic directors, but rather each work is an expression of the field as a whole (Baumann, 2007; Becker, 1984; Bourdieu, 1983). Second, the field of artistic dance is characterized by the dominance of traditional 'high culture' genres like classical ballet and modern dance (DiMaggio, 1987; DiMaggio, 1992) and very conservative classification systems that provide strong institutional constraints for new dance genres to evolve. Third, most dance companies constantly struggle for scarce resources, including financial support from governmental subsidies and private sponsorship, allocation of adequate space for rehearsals and performances, and competition for audiences. Dance companies are similar in many respects to other small- or medium-sized private organizations. They have products that must be marketed and sold and they have administrative operations and boards of directors. These similarities enable at least partial generalization of my research findings across other organizational populations.

My study is based on the analysis of data from in-depth interviews with the choreographers[1] of 15 Israeli contemporary dance companies and a content analysis of visual material from 50 video-recorded dance performances. All the companies have a well-established international reputation and operate as not-for-profit organizations. Financial support comes from governmental and municipal sources as well as revenues from the box office. The companies chosen for the sample have been active for at least 20 years, putting on at least one new production on an annual basis. I have interviewed the choreographers at least twice: once at their studios alone and another time during rehearsals or improvisations. In this way, I obtain data immediate to the creative process itself. I benefit from the artists' willingness to share with me the key dynamics of authenticity and ingenuity in their work process as it evolves. This data collection process also allows me to formulate questions related to the meaning of authenticity and creativity, to witness how artists pursue authenticity and ingenuity in their work and not to be confined to retrospective evidence or conceptual constructs alone.

The focus on the relationship between authenticity and creativity enables me to develop a conceptual framework that contributes to existing studies on the nature of authenticity (Baugh, 1988; Peterson, 1997; Peterson, 2005) and creativity (Amabile et al., 1996; Amabile, 1998; DeFillippi et al., 2007; Jones and DeFillippi, 1996). Drawing on my analysis, I outline how artistic innovations are shaped by a strong need to maintain or recapture authenticity, which acts as a constraint. Choreographers reject external constraints, but are strongly motivated by authenticity as an inner constraint on their thinking and artistic development. This case sheds new light on our understanding of constraints in creative industries, specifically the tendency to see constraints in these industries (as elsewhere) as essentially external, without due regard to the artist's own sense of creative integrity, which also acts as a constraint.

The remainder of the chapter is organized as follows. The next section presents the theoretical framework for assessing the importance of authenticity to cultural goods and provides definitions of key concepts. I introduce the artistic dance field as the context to examine the relationship between authenticity and ingenuity. The third section provides background for the research context by tracing the professional dance industry in Israel from its foundation in the 1920s until today. In my review I highlight the specific structural constraints and limited resources that the Israeli dance companies have had to confront. The fourth section describes the research design and methodology. In the fifth section, I discuss the relationship between authenticity and ingenuity manifested in the artists' work and discourse. The final section addresses the question of whether processes of organizational ingenuity in response to artists' own sense of creative integrity, which acts as a constraint, can lead not only to aesthetic but also institutional novelty.

THEORETICAL FRAMEWORK

Authenticity: Classification and Social Construction

Cultural goods regarded as 'authentic' are accorded distinctive significance and are perceived to carry special meaning and embody more value than other non-authentic objects with the same characteristics (Carroll and Wheaton, 2009; Jones et al., 2005). Especially today, as technology permits the multiplication of prestige objects, there is an increasing appeal of authentic producers or products in modern markets despite their disadvantage in terms of economy of scale (Appadurai, 1988). What might cause cultural consumers to be attracted to perceived authenticity is a reaction against the perceived loss of a personalized self in contemporary mass society. We speak of authentic art, authentic music, authentic food, authentic dance, authentic people and authentic products (Lindholm, 2008). A sociological definition stresses that authenticity is not a 'real' thing or attribute that can be objectively determined. Rather, authenticity is a socially constructed phenomenon determined by audiences in a particular social context (Carroll and Wheaton, 2009; Peterson, 1997; Peterson, 2005).

Scholars often apply either of two general meanings of the term authenticity with respect to cultural objects (Baugh, 1988; Grazian, 2005). The first posits that an object clearly fits some particular classification that it has been assigned or someone has claimed for it. Carroll and Wheaton (2009) term this meaning *genre* or *type authenticity*. In type authenticity, something is an authentic X if it is an instance or member of the class of Xs (Davies, 2001). The focus is on whether the object meets the criteria for inclusion in the genre or category, presupposing the existence of such classificatory criteria.

A second general meaning of authenticity conveys moral meanings about the values and choices embedded in an object. Baugh (1988) utilizes the term *moral authenticity*. A person, for instance, is said to be authentic if he or she is sincere, assumes responsibility for his or her actions and makes explicit value-based choices concerning those actions and appearances rather than accepting pre-programmed or socially imposed values and actions. For example, the values of an 'authentic' cultural entrepreneur in the field of restricted production include creating something appreciated over something profitable (Becker, 1984; Caves, 2000; Swedberg, 2006). Performances or other art works are perceived as morally authentic to the extent that they resist simply being consumed, and the end they express is not solely that of entertaining but something indeterminate and singular (Baugh, 1988): 'Authentic objects, persons and collectives are original, real, and pure; they are what they purport to be, their roots are known and verified, their essence and appearance are one' (Lindholm, 2008, p. 2).

Authenticity and the Search for Meaningful Careers in Creative Industries

Agency and authenticity are especially valuable for the shaping of meaningful careers in art (Svejenova, 2005; White and White, 1965). Two distinct strategies may be used for claiming authenticity. One is to subject one's creative voice to the perpetuation of tradition and to copy canonical works as closely as possible, for example, symphony performances of classical music. A second route to authenticity entails being original and offering a distinctive approach (Jones et al., 2005). The first strategy is compatible with the classification of 'type authenticity' (Carroll and Wheaton, 2009), while the second strategy is compatible with the classification of 'moral authenticity' (Baugh, 1988; Carroll and Wheaton, 2009). According to Peterson's insight (Peterson, 1979; Peterson, 1997; Ryan and Peterson, 1982), managing impressions of authenticity, through claims to tradition or a distinctive personae and approach, is the central task of the creative industries.

Four stages through which authenticity plays a role in career creation have been identified (Svejenova, 2005): (1) exploring aspects of multifaceted identity and image; (2) narrowing down and focusing identity expression and image manufacturing; (3) enhancing one's control over the creative and business aspects of the artwork; and (4) a quest for professionalism. These stages are bridged by authenticity work, which is manifested in the duality of identity expression and image manufacturing (Jones et al., 2005). Individuals seeking authentic careers are willing to take initiative and responsibility. They are able to achieve congruence not only between past and future but also between the private and public domains of one's self (Svejenova, 2005). In a culture where things are valued according to their perceived authenticity, people must establish their identities, know their roots and verify their origins. They try to bring together essence and appearance. Most of all, they seek their own authentic selves. However, authentic individuality is often pursued through conformity, because the only proof of authenticity is what others think. The value of authenticity lends itself to constructing, asserting and ranking groups or individuals. Authenticity is a modular value, which can be configured to fit different social and cultural needs (Lindholm, 2008; Parish, 2009).

Perceived Authenticity in the Dance Field

Little is known about the search for authenticity within the social context of artistic dance. The academic literature on artistic authenticity comes mainly from scholars of aesthetics and art criticism, who are primarily interested in literature, classical music and fine arts (Baugh, 1988; Davies, 2001; Glynn and Lounsbury, 2005; Sartre and Elkaïm-Sartre, 1948). Works that connect interpretive cultural studies of authenticity and organizational studies tend to focus on commercial cultural products such as country music (Peterson, 1997; Peterson, 2005), blues music tourism (Grazian, 2005), motion pictures, book publishing and records (Hirsch, 1972), oriental rugs (Spooner, 1986), advertising (Moeran, 2005) and self-taught artists (Fine, 2004). Studies that examine the construction of authenticity within different dance genres have focused on types of dance perceived to have an association with distinct ethnic groups, for example, salsa (Urquia, 2004; Waxer, 2002), flamenco (Aoyama, 2007) and tango (Garibaldi, 2010). Dancers' claims of authentic group affiliation are essential to become legitimate practitioners of these ostensibly national genres. Thus, authenticity is an element not of the dance itself but rather an attribute of the performers. Moreover, although the relationship between authenticity and creativity has been examined in some commercial artistic genres like country music (Peterson, 2005), the exact relationship between authenticity and creativity has not, to my knowledge, been studied within the context of 'high culture', or the sub-field of restricted cultural production (Bourdieu, 1983; DiMaggio, 1987; DiMaggio, 1992), in particular, the artistic dance field. In this chapter I examine the idea that in certain cases the artist's own sense of creative integrity might also act as an inner constraint that might risk his or her artistic expression. This is where ingenuity, the ability to create innovative solutions within constraints using imaginative problem solving, might eventually lead to artistic innovation.

Type Authenticity in the Dance Field

The evolution of types is dynamic; new forms are established, become institutionalized or become extinct (Hannan and Freeman, 1986; Ruef, 2000; Zuckerman, 1999). The field of artistic dance is by no means different. Classical ballet and modern dance are two traditional dance types that have become accepted and institutionalized among knowledgeable audiences over time and have set the criteria for inclusion or exclusion from the high culture model. Ballet and modern dance have culturally defined schema that set the expectations for what they should look like and what kind of gestures, aesthetic shapes and story lines are to be manifested in each performance. Later in the chapter, I examine to what extent works of contemporary choreographers actually draw on the classical ballet and modern dance types.

Classical ballet is a style of expressive dance that represents a variety of dramatic and emotional situations. The highly formalized technique of classical ballet is based on precise conventional steps and gestures. The foundation of classical ballet consists of five basic positions of the legs (all of them in the mode of 'turn out') coupled with five basic positions of the arms (Adshead-Lansdale, 1988; McFee, 2012).

The terminology of *modern dance* is similar to that of classical ballet; however, its vocabulary is much more diverse and focuses on the transformation of basic movements into heroic and theatrical expressive gestures. Some elements of the modern dance technique were developed deliberately in opposition to the balletic forms (e.g., fall to the floor, contraction and release, bent limbs and flat feet of the dancers). Modern dance deals with contemporary concerns in contrast to the narrative aspects of ballet (Adshead-Lansdale, 1988; McFee, 2012).

Moral Authenticity in the Dance Field

The essence of moral authenticity in the artistic dance field comes to the fore when the artistic director represents his or her work as primarily reflecting intrinsically motivated choices rather than extrinsic ones. Conceptualizing the artistic dance field as a 'sub-field of restricted production' (Bourdieu, 1983) implies that moral authenticity results from a systematic inversion of the fundamental principles of various economies:

1. *The economy of business*. Moral authenticity excludes the pursuit of profit and does not guarantee any sort of correspondence between investments and monetary gains. Categories that invoke this moral basis are, for example, artists that question the accepted aesthetics

and forms that pervade both classical ballet and modern dance and import to their works motives from distinctive cultural societies (e.g., the Noh, the Balinese and the Butoh dances). Another representation of moral authenticity is the self-funding of new projects that aims to cover lengthy research and development periods in order to maintain rigorous artistic integrity and quality.

2. *The economy of power*. Moral authenticity condemns honors and temporal greatness. Categories that invoke this moral basis are, for example, the relinquishing or refusal of choreographers and dancers to work in established, financially supported dance companies despite the fact that these dance companies provide the fastest and safest way to wide recognition and exposure.
3. *The economy of institutionalized cultural authority*. Categories that appeal to this moral foundation include, for example, the absence of formal training or consecration, the rejection of the necessity for traditional spaces, designing productions for theaters in which the proscenium arch is disposed of, such as 'alternative theaters' in old warehouses, train stations, shelters and even residential apartments, and the focus on the direct relationship between performers and audiences.

To provide a better understanding of the links between ingenuity processes, authentication and creativity, it is important to understand the specific background in which the field of artistic dance in Israel has evolved. The following section reviews briefly important chronological landmarks that anchor the evolution of the dance field to the Israeli context.

THE ARTISTIC DANCE FIELD IN ISRAEL

The story of the beginning of artistic dance in Israel in the 1920s is a story of individuals who immigrated to Israel, brought their own rich cultural background and tried, without any resources, to create something out of nothing (Eshel, 1991; Manor, 1986; Roginsky, 2009; Rottenberg and Roginsky, 2009). Between the 1930s and the mid-1960s the dominant school was the German expressionist; it was easier for expressionist dancers to get a job in a low-budget country (relative to classical ballet dancers) because the performances were mainly solo dances, and the primary focus was on expression and less on formal technique mastery, which is much more time and resource consuming (Eshel, 1991; Manor, 1978; Manor, 1980; Rottenberg and Roginsky, 2009). In 1949, Sara Levi-Tanai founded the Inbal Dance Theatre Company. It was the first dance company that received formal funding from the state and won international recognition (Manor, 1975; Manor, 2002).

By the mid-1960s, the influence of the European expressionist school had started to fade and the American school became dominant. In 1964, the Bat-Sheva Dance Company was founded and the dances and technique of the American modern dancer Martha Graham took control of the accepted canon. This marked a turning point in the history of artistic dance in Israel. Graham's influence was so strong that her artistic language and formal technique became the leading and, actually, the only style of dance in Israel for about 20 years.

In 1971, the Kibbutz Contemporary Dance Company was founded. Both Bat-Sheva and the Kibbutz Contemporary Dance Company eventually became public organizations with a permanent budget from the Ministry of Education and Culture. These accomplishments contributed to the development of modern dance in Israel. The two companies' works combined elements of the two strongest schools of modern dance – the personal and/or political expression of the German expressionist school and the virtuoso technique of the American school (Rottenberg and Roginsky, 2009).

The establishment of classical ballet dance companies in Israel took a different path. Israel in the 1920s, a land of swamps and malaria, was not an ideal place for satin point shoes and tutu dresses. Moreover, the Jewish pioneers who immigrated to Israel during those years had developed a hostile ideology toward ballet. For them, classical ballet represented the aristocratic world of Eastern Europe that they wished to abandon. With neither a legacy nor an audience, it seemed impossible that professional ballet would take hold in Israel. In 1967, however, the Israeli-Ballet Dance Company was founded. The Israeli-Ballet is an institutional dance company that performs works from the pre-eminent world classical and neo-classical repertoires. In contrast to modern dance, new generations of choreographers have not been nurtured (Eshel, 2001).

Despite the constant creative activity, artistic dance in Israel has been conceptualized as a threat. In the art of dance, the corporeal body experiences sensual manifestation. This stands in contrast to the perception of the body in the Jewish religion. The human body in Jewish religion is a taboo; therefore, it cannot accept the presence of an artistic discipline that constantly aims to free the body from physical and institutional chains. In a country where religion and political power are tightly coupled, the field of artistic dance had to struggle for recognition.

In the last 20 years, there has been a tremendous flowering of dance in Israel. The Suzanne Dellal Centre was founded in 1989 in Tel Aviv by the Dellal family of London, England, in honor of their daughter Suzanne, as well as by the Municipality of Tel Aviv Yaffo, the Tel Aviv Foundation and the Israeli Ministry of Culture and Education. The mission of the Suzanne Dellal Centre has been to cultivate, support and promote the art of contemporary dance in Israel. The Centre pursues this mission by offering a variety of events, festivals and workshops across artistic disciplines, ensuring that dance stays in close relationship with all art forms. The Suzanne Dellal Centre has two primary goals: the first is to support and facilitate top-quality presentation of Israeli and international choreographers; the second is to create world-class dance productions and engaging educational activities. As a result the Centre has launched dozens of successful and innovative programs to nurture and support new work and emerging artists, providing visible platforms to share new creations with large audiences. Some facts on the Centre include:

1. The Suzanne Dellal Centre is the most visited tourist site in Tel Aviv.
2. Since it was founded the Centre has established over 100 initiatives for artistic creation.
3. On average the Centre welcomes over 0.5 million visitors per year.
4. Since it was founded the Centre has featured over 850 premieres.
5. On average, the Centre hosts over 750 cultural events and performances per year.
6. The Centre has become a model of cultural and urban revitalization, hosting several international delegations each year that come to study its physical layout and dynamic programming.

Another important milestone in the development of Israeli contemporary dance was the foundation of the Israeli Choreographers Association in 1998, by a number of independent choreographers sharing the same goal of exposing and promoting their personal work. The choreographers I interviewed expressed a desire to expand the audience of dance lovers in Israel and to have their works reach all of Israel, through collaborations and special programs. Together with this diverse, ongoing local creation, the artists also strive to achieve international exposure. The Association includes among its members most of the independent dance artists working in Israel. The dozens of artists making up the Association have rich professional experience in the art of dance, and all manage companies or create new projects every year. When an artist joins the Association, he or she receives all the different services and manages all dance activities in Israel and abroad through the Association's account. In addition, the Association provides various administrative services, such as accounting, legal consultation, insurance, studio room subsidies, press advertising and public relations. The Association coordinates information from external and internal sources and assists in all matters related to filing financial assistance applications and implementing budgets. Every year the Association is joined by more and more choreographers whose activities are established, who meet the criteria set by the Association and who desire to advance in the scope of their activities in terms of art and budget. The Israeli Choreographers Association is supported by the Ministry of Culture and Sports – Administration of Culture – Department of Dance and by the Tel Aviv municipality. Other supporters of the artists are the Mifaal Hapais Lottery Company and public and private foundations. Encouraging original quality artistic creation and peaceful artistic lives, expressing the spirit of the place, holding an up-to-date mirror up to dance – these are the values guiding the Association's work.

The Association produces special projects held annually, which aim to provide its artists exposure:

1. *'Naim'*. The Association's flagship project enables moving shows and workshops to, and producing projects in, the periphery. The Association markets all artists' shows fully and also creates integrated Association artist shows, as well as weekends in which the best dance pieces are shown. Coming to the periphery is intended to enhance the dialog between dance and the audience, and for this reason, shows in this project are integrated with discussions with the audience and artist workshops. The Naim project is made possible by the support of the Ministry of Culture and Sport.
2. *Exposure to dance buyers*. These events are held throughout the year and are intended for potential buyers in the Israeli dance industry. The purpose of the events is to provide exposure for directors of cultural institutes, community culture coordinators, dance teachers and all potential consumers of shows and workshops, and to promote the sale of shows.
3. *Curtain Up Workshops*. The Association markets Curtain Up events held in the Suzanne Dellal Centre to associations and assists in bringing audiences to Curtain Up events in the Jerusalem Theater. As part of its marketing activities, the Association proposes to communities throughout Israel to host artists for workshops and master classes throughout the year and to create a preliminary acquaintance with artists, which enables a larger audience to connect with the pieces.
4. *Marketing and production of courses and seminars for artists in various fields of management*. The Association produces courses intended to educate marketing people and producers in the field of dance. Most artists need production and marketing people, but they find it difficult to locate professionals specializing in this field. The courses integrate

lectures by professionals in all fields of marketing and production, who specialize in matters related to these professions in the world of dance.

The establishment of the Suzanne Dellal Centre and the foundation of the Choreographers Association have facilitated the development of contemporary dance in Israel by providing a home and international exposure to young choreographers and independent ensembles that need financial support as well as marketing and public relations services. Most of the new dance companies are contemporary dance ensembles with a wide diversity of artistic production.

METHODOLOGY

I collected data from interviews with 15 choreographers of contemporary dance companies and from video recordings of their dance productions over the years. I sampled interviewees based on artistic preeminence – all of the choreographers in my sample have been significantly active in the dance field for at least the last 20 years, and some of them have been artistically active for more than 40 years. Many of their works have won international recognition. My sample ranged from the artistic directors of well-established and institutionalized companies to independent choreographers who either manage small companies or create new projects every year.

In order to take full advantage of the richness of the data I collected and to ground my theoretical formulations within the work situations of the choreographers, I utilized the narrative approach to in-depth interviews (Gubrium and Holstein, 2002; Rubin and Rubin, 2005). In-depth interviews proved excellent for gaining insights about individuals' mindsets and perceptions. As 'conversations with purpose', they enable one to simultaneously capture objective and subjective nuances (Thorpe and Holt, 2008). Using the narrative approach, I looked at the actions taken in particular situations in order to understand how they reflect a larger system of meaning (Geertz, 1973; Langley, 1999; Lieblich et al., 1998). In this chapter, sections of in-depth interviews with 15 choreographers are utilized to elaborate my findings about the link between authenticity and creativity in the field of dance.

The interviews with the choreographers lasted from one to three hours. They were conducted at the artists' studios and were aimed at unveiling how they saw the development of their careers, how they defined their work and how they achieved creativity. I was careful to take into consideration both their artistic inclinations and the shifting environmental influences on their work. Most of the studios were located in Tel Aviv, the Israeli metropolis known for its rich cultural life. The high cost of living and the need for large spaces and quiet led some of the choreographers, however, to leave the city and relocate their studios to the countryside. In most cases, they settled in peripheral kibbutzim that were going through bankruptcy. To gain some revenue, the kibbutzim charged an upfront fee for permission to reconstruct old and unused buildings originally slated for agricultural purposes. After the initial investment, the choreographers could open huge studios to house their companies and to operate dance schools.

Since the purpose of my study was to analyze how (if at all) processes of ingenuity and authentication are manifested in dance itself, a major part of the data analysis was a content analysis of video-recordings of dance performances. I utilized a comprehensive collection of 50 dance pieces that spanned the period between 1985 and 2010. The videos enabled me to examine artistic production as such, from a non-verbal perspective that concentrated on its physical aspects. I analyzed not only specific movements, gestures and patterns but also looked at the choreography holistically. By utilizing multiple sources, I sought to build trust in the congruence of my data and reduce the risk of bias. The content analysis comprised successive stages. First I developed narratives that described the ingenuity process associated with: (1) three categories of external constraints: financial, institutional and physical constraints that were discerned as part of the larger Israeli context; and (2) claims for authenticity. I made sure that the categories were discernible in all cases. Second, I analyzed the forms of movement for each dance. I then grouped the movements into distinct patterns of physical expression. I differentiated between patterns consistent across the different dance companies and patterns consistent within the works of an individual dance company over the years. In this way I could follow patterns shared among the various dance companies, patterns unique to individual choreographers but consistent over time, patterns both shared by all dance companies and consistent over time and movements unique to a specific dance production. Third, I identified the physical and kinetic representations of type and moral authenticity. To achieve a set of robust descriptions for each category of authenticity, I moved back and forth from looking at the movement separately to looking at the context of the dance as a whole. In this way I could also compare movements across dance companies, across works of individual artistic directors along the years and within each dance performance as a unit of analysis. I used ATLAS.ti Qualitative Data Analysis and Research Software (Version 6.2.25) for both the narrative analysis of the in-depth interviews and the content analysis of the video-recordings of dance performances.

Finally, I came to this research with my own experiences as a professional dancer and choreographer. Years of training as a dancer, followed by years of performing on stage and working and touring with ensembles, has provided me the opportunity to be part of this world from within and to develop an in-depth understanding of the field. My deep familiarity with the dance field as a professional practitioner, coupled with my extensive training as a social scientific researcher of organizations, has enabled me to apply the theoretical framework of organizational studies to the field of dance with minimum risk of reduction and/or oversimplification.

FINDINGS

Invoking Intrinsic Worth: In Search of 'Real Art'

Despite the differences between their artistic expressions, the choreographers I interviewed exhibited some similarities in their search for 'real art' and creative voice. Some of them described a turning point in their careers when they had decided to leave behind established dance companies in favor of developing independent careers. In a few cases, this turning point occurred at the peak of their success in the conventional dance field. Several of the choreographers mentioned the advantages of keeping an organization small enough to allow control over the schedule and over artistic choices, despite the enormous economic risk. Some chose to move to the periphery, where they could afford to have their own studios and earn their living from community-based activities. As one interviewee put it:

> I always say that we are working in a tomato factory. It is exactly the same . . . I worked for the [dance] company for 17 years, doing all the big solos. I was very famous. But when the company became public and got its funds from the government things changed . . . when we started we had a very high vision, but with all these committees, decisions became mediocre and had nothing to do with our dreams. I felt that I had to leave to have my own company and do my own thing . . .
>
> (Interview with author, December 2011)

This case exemplifies the essence of moral authenticity; the artist stresses how after a certain point her career choices were primarily intrinsically motivated rather than extrinsically motivated. The artist left behind a career as a solo dancer in a large, well-established and reputable dance company in favor of an independent career. In this case external constraints such as limited resources or lack of reputation did not play a role in the artist's career choices. Rather, leaving behind an established dance company exemplified the agency of cultural entrepreneurs that seek ingenious solutions to break through (inner) barriers.

Another choreographer invoked moral authenticity in a completely different manner:

> [F]rom the performing art center's perspective – my raison d'être is to disseminate the work to the public. This means putting on a new production very often within a limited time. [But] I am slow. Deciding where is left and where is right takes me time . . . From the beginning, we were very successful and were offered attractive mainstream projects. But I knew that this would burn my creativity. I had to stay faithful to my art.

Later in the interview, she says:

> [T]he path to success kills your free soul. You do all the right training, get into the best schools, get your certificates, [and yet] the audience wants more of what they liked in your previous works. And then you look back. You are no more than a clerk. But now you have a lot to lose if you change. When you have nothing to lose, then over time your greatest disadvantage becomes your greatest benefit. When you have nothing to lose, you can challenge yourself and look at life from interesting perspectives. You keep your free soul.
>
> (Interview with author, January 2012)

Here too, the interviewee described how she made moral choices based on intrinsic motivations instead of economic or reputational considerations. In contrast to the first interviewee, this choreographer made a choice, from the beginning of her career, to take an alternative route and not to go through the well-established training institutions and dance companies. Her message was clear: the way to constantly replenish her creative resources is to nourish a 'free soul', even if it impedes progress along the prescribed path to success.

Sometimes, the construction of moral authenticity challenged the authority of gatekeepers:

> [F]irst of all, it is important for me always to be involved in creative processes. This gives legitimacy to my existence. However, this process takes a long time, a lifetime that is dedicated to research. I always make sure that I leave something open. There are people who hate my work for that reason. They say that my work is too dark, too secret. They need closure. Most journalists work this way. They know how much power they have over you. They can determine your whole future. A journalist writes a bad review about you and then one of your sponsors sees it and revokes your budget. This has a tremendous influence on your work.
>
> (Interview with author, January 2012)

Moral authenticity was invoked here through the inversion of the fundamental principles of the economy of institutionalized cultural authority, in this case the moral choice to risk rejection by critics. The choreographer's comments reflected on the reality of the Israeli dance field; breaking institutional constraints by nonconforming to prevailing aesthetic norms and searching for artistic autonomy might lead to economic penalties and the loss of financial resources.

Giving the Audience a 'Backstage Pass': Stripping Dance Down to the Bare Bones of What Performance Is

Companies that moved their studios to the geographical periphery took two successive steps: first, they moved their studios from the metropolis of Tel Aviv but kept their headquarters in the city. They lived in the periphery during rehearsal periods but maintained their performances in well-known established theaters. This strategy helped to keep their audience despite the change in geographic location. Over time, more and more operations were transferred to the periphery. The choreographers explained how moving to the periphery was a part of their effort to strip dance down to the bare bones of what performance is. They rejected traditional spaces and instead started to perform in alternative spaces far from the center. Here too, the motivation behind their agency was not to find creative solutions to overcome structural constraints or limited resources, but rather to invoke ingenuity that enhanced artistic creation in response to internal moral constraints. Most of the choreographers conceptualized their work as a mode of performance that breaks down the barriers between the dancer and the audience. To achieve such a goal, the alternative spaces challenged the traditional proscenium arch and served to foster a direct relationship between performer and audience by utilizing the imagination of the audience in conjunction with the imagination of the performer. Relating the relocation to the periphery directly to artistic inspirations brings to the fore the importance of ingenuity as means to overcome authenticity that acts as an internal constraint. Many of the new spaces were designed in such a way that provided the audience a direct look at the backstage or even totally eliminated the backstage, allowing transparency of the production system.

One of the pioneers of the trend toward performing in alternative spaces talked about her motivation:

> There was always something beyond myself or the company that has guided me. Suddenly, I understood that I want to perform outside, from a very deep human-environment connection, not just 'let's get ecological' because it is in vogue today. It was not in vogue then. It was unconscious, maybe because I

> had become a mother, and in the theater, people told my kids to be quiet . . . I told them that this is my own show on stage . . . I realized that what I want is to have my performances outdoors, where kids can come and have lots of space to make noise and play . . . I felt that I needed to go against this individualistic separation of modern society. After years of art and creativity, I understood that as a female artist I want to change lifestyles . . . people that come to our artists' village get much more than what I can give in an hour-long show at the art center in Tel Aviv. They come, stay for a day or two, see our studio, see the performance, see how we live and ask questions . . . the experience is less formal and audience gets the whole picture of what art is. And for me, life and art go together . . . we are the kind of artists that the 'craziness' doesn't leave. You have to keep on doing what you feel is creating reality – to give meaning.
>
> (Interview with author, January 2012)

In fact, claims of moral authenticity ran through virtually all of the interviewees' narratives. In addition to describing the need for a work environment in which they were fundamentally autonomous, the artistic directors expressed similar notions about the core of the creative approach. They spoke about reinvesting dance with meaning, particularly when they felt it had been stripped away through decades of strict adherence to the formalized techniques of classical and modern ballet. As one of the most experienced artistic directors explained:

> It took a long time until I dared to say, 'That's it! I can't stand it anymore!' I made the move and founded my own dance company and started with a new language. I rebelled against Martha Graham. I had had enough of that style. I thought that there were other things that I wanted to express. First of all, I wanted to get rid of all these big narratives from the Bible and Greek mythology on stage – always big stories . . . For a while, I tried to imitate Martha, but I felt that I needed to clean up and move to something completely different, not to tell stories, not to use grandiose sets, to avoid all these hairstyles and this makeup. I needed something more modest that is related to the place where I belong. I come from a different family; do you understand? I had to express my own world, so I left the modern techniques and moved to practice Yoga, Feldenkrais, everything that was [going on at] that time . . . one of the big rebellions of the 1960s was to go out and dance in the streets, dance in different spaces . . . not everything so theatrical. I started to use objects to help me find a more abstract language . . .
>
> (Interview with author, December 2011)

In this passage, the pure types of classical ballet and modern dance represent a reference point from which the artist could move toward developing a creative dialog. The conscious revision of ongoing forms and routines (Hannan and Freeman, 1986; McKendrick et al., 2003) and the experimentation with new types of dance are facilitated by moral claims targeted at the strict requirements of classical ballet and modern dance and their formal techniques that constrain self-expression.

The utilization of classical ballet and modern dance as a basic ground upon which other dance techniques could be layered was evident throughout several of the interviews:

> . . . I always believe in continuity. Look, classical ballet was not invented for nothing. Great people invented it. It took a lot of time to create classical ballet, and I use this technique a lot. Classical ballet is a language; it is knowledge. [But] I am not stuck in their stage performances. I don't deal with Gisele, with this sweetness and the heaviness of this style. Although sometimes I do . . . But this technique is a language. It is like saying these are the musical notes and I continue this line without being stuck in it. When I teach I also use the technique as body language. Language is very important: if you stand on one leg or two legs. If you cross legs or stay in parallel. These are very important things and I develop and add everything that I have learned all these years on movement and on human beings. In the end, this is a humanistic and a very intimate language. I have to understand this magic of life, to bring in my dreams. That is why for more than 20 years I practice the same routines every day, ballet or modern dance. This is my 'mantra'. The routine helps me to fly without the need to trust the mind or the memory, to free my soul . . .
>
> (Interview with author, November 2011)

Another choreographer, when asked about her attitude toward classical ballet and modern dance, replied:

> I have in my body a lot of movement techniques from all the years of studying. From artistic gymnastics to the classical ballet of the American Dance Company – I was an outstanding student on point shoes, modern dance, Balinese, Mexican dance, many forms, many techniques that I have learned. I think that what has saved me over the years is this small island of individual artistic work. Through my creative work, I produce alienation, oblivion and then recombination through my own anatomy . . . I am less connected now to the worlds of ballet and modern dance. Personally, they are not within the zone of my capabilities in such a way that I could have continued to investigate what is in them. I am sure that every technique has the essence of the original. In ballet, I think that the dominance of the formal morphology is too overwhelming, so it prevents me from dealing with other things that today look to me more important and essential to my process. It is too much information to deal with when you have to count all the time to keep in pace with the rhythm . . . I feel that I have to give up and not to deal with the things that ballet and modern dance deal with so I can be devoted to what I think is dance.
>
> (Interview with author, February 2012)

The two citations above demonstrate the utilization of type authenticity (Carroll and Wheaton, 2009). The choreographers try to rebel against the institutionalized types of classical ballet or modern dance but at the same time, they retain aspects of the artistry and mastery. Audiences still recognize that the knowledge, skills and techniques of the choreographers and the dancers are beyond the normal person's reach, requiring special training and apprenticeships, and a range of specialized experiences. This recognition manifests the ingenious way in which choreographers provide cognitive pleasure and enjoyment to audiences who need a familiar template for their aesthetic pleasure but at the same time allow novelty by seeking new meanings of their art.

Interestingly, the efforts to disconnect from established formal dance types such as ballet and modern dance were accompanied by mixed feelings. Musing retrospectively, one of the choreographers recalled:

> Only by looking backward can I tell about all these changes. Life and maturity change dance and dance changes you, helps you to grow up. After 15 years of intensive and rich work, I realized that actually – and the critics also wrote about it – I have never really moved completely away from Martha Graham; I actually continue her in a way but with slight differences. For example, I have never used this dramatic facial expression, but I was never alienated. The drama was always inside of me; it was always there. Also the rites and rituals are there . . . attenuated but there . . . [S]he had huge decorations and I use small objects . . . so her blood still runs in my own blood and her voice is still in me.
>
> (Interview with author, November 2011)

Her words illustrate how claims of authenticity were coupled with the re-use of past resources (in this case the use of the well-established formal technique and physical vocabulary of the classical canon) in imaginative ways that require ingenuity.

Another choreographer described the dynamics of the institutionalization of dance as a constant struggle over definition:

> I have broken through institutions many times – also at school. And I have never worked for an established dance company. We had to invent a dance company because we could not fit into well-established institutions. And then, in the end, you become the institution, and there are people inside your institution and they want to break it and change it . . .
>
> (Interview with author, April 2012)

Here too, the findings show how the artist rejected external constraints such as institutional barriers on the path to success and instead uses creative solutions to overcome intrinsic barriers.

ANALYSIS OF THE VISUAL MATERIALS

Considering Content: Uncovering Idiosyncrasy

The categorization of movements across dance companies and within an individual dance company over time revealed some salient common features:

1. Most of the works represented a cross between dance and theater, consisting of short units of dialog and action, often of a surreal nature.
2. The use of everyday movements was a valid performing art.
3. In some dances, the invisible division between the real person and the stage character seemed to collapse so that one often has the sense of watching an exaggeration of real life events. At the other extreme, there were dance pieces that approached complete surrealism and abstraction.
4. Unlike performers in classical ballet, the dancers were physically pushed to the limit and exhibited their exhaustion and pain quite openly on stage.
5. Repetitions of movements were used to push familiar interactions to their extremes, so that they swayed on the edge between humor and anguish. The artistic directors never completely relinquished the formal techniques of classical ballet and modern dance. To break from formal conventions, however, the vocabulary of classical ballet was sometime charged with energy completely different than the energy of classical ballet (i.e., 'high-speed' pirouettes, the use of point shoes for men instead of women, deliberate repetitions of classical bounce), which ultimately transformed classical phrases into newly contextualized dances. The deconstruction and reconstruction of elements from the classical canon of dance built an independent mosaic that breaks away from accepted artistic conventions.

The analysis of the visual materials pointed at how the virtuoso mastering of formal techniques facilitated both an adherence to a widely accepted aesthetic language as well as a departure toward novel physical experimentation. This use of physical expression was more prevalent among a group of choreographers who shared a common career path: all of them had trained in and performed with established classical or modern dance companies, and, at a certain point in their career, they sought to follow their own intuitive way.

Different ingenious solutions were invoked by another group of choreographers, on a different career path, who were not trained from early childhood as ballet or modern dancers and who were instead introduced to dance at a relatively 'old' age. Most of them had always remained far from the center of artistic creation, whether geographically or by refusing to work for the more established dance companies. In their performances, the use of everyday movements such as getting dressed and undressed, washing the hands and face, hugging, drying sweat off the forehead, walking and running were dominant. In some of the pieces, it was clear that the performers were not 'real' dancers; for example, in some cases the dancers were disabled. In one of the works, a pregnant woman was thrown again and again into the air by the other dancers, and in another work, the artistic director's mother performed with the ensemble. The exposed vulnerability of the performers contributed to the authenticity of the dance piece by evoking strong feelings from the audience about the quirky and idiosyncratic occurrences they saw on stage.

DISCUSSION AND CONCLUSIONS

Discovering Creativity in Necessity: Overcoming a Strong Need to Recapture Authenticity that Acts as a Constraint

The leitmotif that repeated throughout the interviews was the need to find an authentic expression that would reflect the choreographers' own sets of values and beliefs and their unique physical and psychophysical 'histories'. The choreographers described their actions in terms of explicit value-based choices they had made that went against pre-programmed or socially imposed values and actions. Thus, as cultural entrepreneurs who struggle against financial, physical and institutional constraints, the artistic directors represented themselves, first of all, as committed to creating something appreciated over something profitable or widely accepted, making a commitment to 'morally authentic dance'.

As the content analysis of the video-recordings revealed, the dance performances conveyed a dual message. On the one hand, the retention of formal technical mastery served as an essential platform for gradual (in contrast to breakthrough) innovative work. On the other, however, a vast use of challenging elements and blending the boundaries between dance and other artistic disciplines forced the audience to look at dance in a completely novel way.

My findings point to a complex ingenious process that invokes the ability of contemporary dance choreographers to create innovative solutions and imaginative problem solving to overcome high demand for creative integrity. In their discourse the choreographers in my sample presented their career path in terms of rejecting external constraints and looking for creative solutions to reinvest their dance with meaning. However, my discussion would not be complete without taking into consideration not only the micro-processes of the career developments of the choreographers but also the macro-level transformations that support the agency of exceptional individuals pursuing creative careers. In the Israeli case, the establishment of the Suzanne Dellal Centre and the foundation of the Choreographers Association provided the artists with a suitable institutional environment as well as financial support that facilitated creative processes at the macro level. Novelty thus was gained not only for the artistic expression itself but also at the field level. The Suzanne Dellal Centre provided an ingenious solution that served the need of contemporary dance choreographers, offering them the home they abandoned by turning their backs on the major established dance companies, as well as exposure to audiences and other stakeholders through organized festivals and dance events. These festival and dance events and competitions served to gather together many choreographers working in small anonymous groups, creating a 'critical mass' of performances that could attract audiences and culture purchasers. The founding of the Choreographers Association exemplified the production of culture as a collective action. Choreographers did not really reject external constraints. Instead, they found an ingenious solution that afforded them the ability to fulfill their financial and other external and material needs through collective action while at the same time communicating moral integrity and high aesthetic demands from their art on an individual level. This combination had the potential to foster novelty in artistic dance. In other words, ingenious processes that unfolded as means to cope with inner constraints were coupled with ingenious processes of collective action. Novelty could be gained through acts of agency supported by structural transformation.

My findings have contributed to our understanding of novelty in creative industries. The use of innovative solutions and imaginative problem solving derive from: (1) challenging preconceptions of what high art should address and the role of agency among exceptional individuals; and (2) emphasizing the importance of the process through which collective action shapes the institutional environment.

The integration of macro and micro processes that support each other have the potential to facilitate novelty in creative industries.

NOTES

* I would like to thank the Henry Crown Institute of Business Research in Israel, The Faculty of Management, Tel Aviv University for supporting this research.

1. Large dance companies also hire outside choreographers for special projects for limited periods of time. In our sample, when the choreographers were affiliated with a large company, they were also the artistic directors of these companies, who work on a permanent basis, which means that they are responsible for the company's artistic direction and plans in addition to their active role as choreographers.

REFERENCES

Adshead-Lansdale, J. (1988). *Dance Analysis: Theory and Practice*. London: Dance Books Ltd.

Amabile, T.M. (1998). How to kill creativity. *Harvard Business Review*, September–October, 77–87.

Amabile, T.M., Conti, R., Coon, H., Lazenby, J. and Herron, M. (1996). Assessing the work environment for creativity. *Academy of Management Journal*, **39**(5), 1154–84.

Aoyama, Y. (2007). The role of consumption and globalization in a cultural industry: the case of flamenco. *Geoforum*, **38**(1), 103–13.

Appadurai, A. (1988). *The Social Life of Things: Commodities in Cultural Perspective*. Cambridge: Cambridge University Press.

Baugh, B. (1988). Authenticity revisited. *The Journal of Aesthetics and Art Criticism*, **46**(4), 477–87.

Baumann, S. (2007). A general theory of artistic legitimation: how art worlds are like social movements. *Poetics*, **35**(1), 47–65.

Becker, H.S. (1984). *Art Worlds*. Berkeley, CA: University of California Press.

Beverland, M.B. (2005). Crafting brand authenticity: the case of luxury wines. *Journal of Management Studies*, **42**(5), 1003–29.

Bornstein, R.F. (1989). Exposure and affect: overview and meta-analysis of research, 1968–1987. *Psychological Bulletin*, **106**(2), 265.

Bourdieu, P. (1983). The field of cultural production, or: the economic world reversed. *Poetics*, **12**(4–5), 311–56.

Bourdieu, P. (1990). *The Logic of Practice*. Stanford, CA: Stanford University Press.

Carroll, G.R., and Wheaton, D.R. (2009). The organizational construction of authenticity: an examination of contemporary food and dining in the US. *Research in Organizational Behavior*, **29**, 255–82.

Caves, R.E. (2000). *Creative Industries: Contracts between Art and Commerce*. Cambridge, MA: Harvard University Press.

Davies, S. (2001). *Musical Works and Performances: A Philosophical Exploration*. Oxford: Clarendon Press.

DeFillippi, R., Grabher, G. and Jones, C. (2007). Introduction to paradoxes of creativity: managerial and organizational challenges in the cultural economy. *Journal of Organizational Behavior*, **28**(5), 511–21.

Dewey, J. (2005). *Art as Experience*. New York: Perigee.

DiMaggio, P. (1987). Classification in art. *American Sociological Review*, **52**(4), 440–55.

DiMaggio, P. (1992). Cultural boundaries and structural change: the extension of the high culture model to theater, opera, and the dance, 1900–1940. In M. Lamont and M. Fournier (eds) *Cultivating Differences: Symbolic Boundaries and the Making of Inequality*, pp. 21–54. Chicago: University of Chicago Press.

Dougherty, D. (1992). A practice centered model of organizational renewal through product innovation. *Strategic Management Journal*, **13**(S1), 77–92.

Eikhof, D.R. and Haunschild, A. (2007). For art's sake! Artistic and economic logics in creative production. *Journal of Organizational Behavior*, **28**(5), 523–38.

Eshel, R. (1991). *Dancing with the Dream: The Development of Artistic Dance in Israel 1920–1964*. Tel Aviv: Sifriyat po-alim.

Eshel, R. (2001). Movement-theatre in Israel 1976–1991. PhD thesis. Arts Faculty of Tel Aviv University.

Feldman, M.S. and Orlikowski, W.J. (2011). Theorizing practice and practicing theory. *Organization Science*, **22**, 1240–53.

Fine, G.A. (2004). *Everyday Genius: Self-Taught Art and the Culture of Authenticity*. Chicago: University of Chicago Press.

Flew, T. and Cunningham, S. (2010). Creative industries after the first decade of debate. *The Information Society*, **26**(2), 113–23.

Garibaldi, D. (2010). El Tango Extranjero. Unpublished Honors thesis. Available at: http://hdl.handle.net/10161/2253 (last accessed 25 September 2013).

Geertz, C. (1973). *The Interpretation of Cultures: Selected Essays*. New York: Basic Books.

Glynn, M.A. and Lounsbury, M. (2005). From the critics' corner: logic blending, discursive change and authenticity in a cultural production system. *Journal of Management Studies*, **42**(5), 1031–55.

Grazian, D. (2005). *Blue Chicago: The Search for Authenticity in Urban Blues Clubs*. Chicago: University of Chicago Press.

Gubrium, J.F. and Holstein, J.A. (2002). *Handbook of Interview Research: Context and Method*. Thousand Oaks, CA: Sage Publications.

Hannan, M.T. and Freeman, J. (1986). Where do organizational forms come from? *Sociological Forum*, **1**(1), 50–72.

Hekkert, P., Snelders, D. and Wieringen, P.C.W. (2003). 'Most advanced, yet acceptable': typicality and novelty as joint predictors of aesthetic preference in industrial design. *British Journal of Psychology*, **94**(1), 111–24.

Hirsch, P.M. (1972). Processing fads and fashions: an organization-set analysis of cultural industry systems. *American Journal of Sociology*, **77**(4), 639–59.

Hirsch, P.M. (2000). Cultural industries revisited. *Organization Science*, **11**(3), 356–61.

Jones, C. (2010). Finding a place in history: symbolic and social networks in creative careers and collective memory. *Journal of Organizational Behavior*, **31**(5), 726–48.

Jones, C. and DeFillippi, R.J. (1996). Back to the future in film: combining industry and self-knowledge to meet the career challenges of the 21st century. *The Academy of Management Executive*, **10**(4), 89–103.

Jones, C. and Thornton, P.H. (2005). *Transformation in Cultural Industries*. Amsterdam and Oxford: Elsevier JAI.

Jones, C., Anand, N. and Alvarez, J.L. (2005). Manufactured authenticity and creative voice in cultural industries. *Journal of Management Studies*, **42**(5), 893–899.

Jones, D. and Smith, K. (2005). Middle Earth meets New Zealand: authenticity and location in the making of *The Lord of the Rings*. *Journal of Management Studies*, **42**(5), 923–45.

Khaire, M. and Wadhwani, R.D. (2010). Changing landscapes: the construction of meaning and value in a new market category – modern Indian art. *The Academy of Management Journal*, **53**(6), 1281–304.

Lampel, J., Lant, T. and Shamsie, J. (2000). Balancing act: learning from organizing practices in cultural industries. *Organization Science*, **11**(3), 263–9.

Langley, A. (1999). Strategies for theorizing from process data. *The Academy of Management Review*, **24**(4), 691–710. Retrieved from http://www.jstor.org/stable/259349 (last accessed 25 September 2013).

Lieblich, A., Tuval-Mashiach, R. and Zilber, T. (1998). *Narrative Research: Reading, Analysis and Interpretation*. Thousand Oaks, CA and London: Sage Publications.

Lindholm, C. (2008). *Culture and Authenticity*. Oxford: Blackwell Publishing.

Lounsbury, M. and Rao, H. (2004). Sources of durability and change in market classifications: a study of the reconstitution of product categories in the American mutual fund industry, 1944–1985. *Social Forces*, **82**(3), 969–99.

Manor, G. (1975). *Inbal: Quest for a Movement Language*. Tel Aviv: Tavnit Press.

Manor, G. (1978). *The Life and Dance of Gertrud Kraus*. Tel Aviv: Hakibbutz Hameuchad Publishing House.

Manor, G. (1980). *The Gospel According to Dance: Choreography and the Bible: From Ballet to Modern*. New York: St. Martin's Press.

Manor, G. (1986). *Agadati, the Pioneer of Modern Dance in Israel*. Tel Aviv: Emmett Publishing Co.

Manor, G. (2002). *Sara's Way: Sara Levi-Tanai and her Choreography*. Tel Aviv: Inbal Dance Theatre, and the Ethnic Multicultural Center.

McFee, G. (2012). *The Philosophical Aesthetics of Dance: Identity, Performance and Understanding*. Alton, UK: Dance Books Ltd.

McKendrick, D.G., Jaffee, J., Carroll, G.R. and Khessina, O.M. (2003). In the bud? Disk array producers as a (possibly) emergent organizational form. *Administrative Science Quarterly*, **48**(1), 60–93.

Mezias, J.M. and Mezias, S.J. (2000). Resource partitioning, the founding of specialist firms, and innovation: the American feature film industry, 1912–1929. *Organization Science*, **11**(3), 306–22.

Miller, D. and Shamsie, J. (1996). The resource-based view of the firm in two environments: the Hollywood film studios from 1936 to 1965. *Academy of Management Journal*, **39**(3), 519–43.

Moeran, B. (2005). Tricks of the trade: the performance and interpretation of authenticity. *Journal of Management Studies*, **42**(5), 901–22.

Parish, S.M. (2009). Review essay: are we condemned to authenticity? *Ethos*, **37**(1), 139–48.

Perretti, F. and Negro, G. (2007). Mixing genres and matching people: a study in innovation and team composition in Hollywood. *Journal of Organizational Behavior*, **28**(5), 563–86.

Perry-Smith, J.E. and Shalley, C.E. (2003). The social side of creativity: a static and dynamic social network perspective. *The Academy of Management Review*, **28**(1), 89–106.

Peterson, R.A. (1979). Revitalizing the culture concept. *Annual Review of Sociology*, 5, 137–66.

Peterson, R.A. (1997). *Creating Country Music: Fabricating Authenticity*. Chicago: University of Chicago Press.

Peterson, R.A. (2005). In search of authenticity. *Journal of Management Studies*, **42**(5), 1083–98.

Peterson, R.A. and Anand, N. (2004). The production of culture perspective. *Annual Review of Sociology*, **30**, 311–34.

Roginsky, D. (2009). The national, the ethnic and in-between: sociological analysis of the interrelations between folk, ethnic, and minority dances in Israel. In H. Rottenberg and D. Roginsky (eds), *Dance Discourse in Israel*, pp. 95–125. Tel Aviv: Resling Publishing.

Rottenberg, H. and Roginsky, D. (eds) (2009). *Dance Discourse in Israel* (1st edn). Tel Aviv: Resling Publishing.

Rubin, H.J. and Rubin, I. (2005). *Qualitative Interviewing: The Art of Hearing Data*. Thousand Oaks, CA and London: Sage Publications.

Ruef, M. (2000). The emergence of organizational forms: a community ecology approach. *American Journal of Sociology*, **106**(3), 658–714.

Ryan, J. and Peterson, R.A. (1982). The product image: the fate of creativity in country music songwriting. *Sage Annual Reviews of Communication Research*, **10**, 11–32.

Sartre, J.P. and Elkaïm-Sartre, A. (1948). *Qu'est-ce que la littérature?* Paris: Gallimard.

Spooner, B. (ed.) (1986). *Weavers and Dealers: The Authenticity of an Oriental Carpet*. Cambridge: Cambridge University Press.

Sternberg, R.J. (1999). *Handbook of Creativity*. Cambridge: Cambridge University Press.

Svejenova, S. (2005). 'The path with the heart': creating the authentic career. *Journal of Management Studies*, **42**(5), 947–74.

Swedberg, R. (2006). The cultural entrepreneur and the creative industries: beginning in Vienna. *Journal of Cultural Economics*, **30**(4), 243–61.

Thorpe, R. and Holt, R. (2008). *The Sage Dictionary of Qualitative Management Research*. London: Sage Publications.

Urquia, N. (2004). 'Doin' it right': contested authenticity in London's salsa scene. In A. Bennett and R.A. Peterson (eds), *Music Scenes: Local, Translocal, and Virtual*, pp. 96–112. Nashville, TN: Vanderbilt University Press.

Waxer, L. (2002). *Situating Salsa: Global Markets and Local Meanings in Latin Popular Music*. New York and London: Psychology Press.

White, H.C. and White, C.A. (1965). *Canvases and Careers: Institutional Change in the French Painting World*. New York: John Wiley & Sons.

Wijnberg, N.M. and Gemser, G. (2000). Adding value to innovation: Impressionism and the transformation of the selection system in visual arts. *Organization Science*, **11**(3), 323–9.

Zuckerman, E.W. (1999). The categorical imperative: securities analysts and the illegitimacy discount. *American Journal of Sociology*, **104**(5), 1398–438.

8 Ingenuity as creative unfolding: framing the frame in haute cuisine

Ninja Natalie Senf, Jochen Koch and Wasko Rothmann

1. INTRODUCTION

Organizations are increasingly forced to be creative and innovative to ensure sustained competitive advantage (Lampel et al., 2000). However, it can be assumed that the space for creativity is often limited by institutional forces. In this chapter we explore the institutional context and its impact on and interplay with the creative freedom of organizations embedded in that frame. For that purpose we focus on the case of haute cuisine, i.e. high-end gourmet restaurants. This setting is instructive due to two main reasons: first, the field of haute cuisine can be considered a highly institutionalized context with apparently strict standards and codes with which the restaurants have to comply (Durand et al., 2007; Fauchart and von Hippel, 2008; Ottenbacher and Harrington, 2007a; Rao et al., 2003). Second, creativity and innovation have evolved into a major evaluation criterion to determine the gastronomic quality and hence the success of the organizations (Ferguson, 1998; Rao et al., 2003; Fauchart and von Hippel, 2008). Therefore, creativity and change are vital for long-term competitive advantage. Hence, the institutional frame on the one hand encourages and even requires creativity and innovation, while on the other hand induces clear limits and standards within which 'innovative' actions have to take place (Durand et al., 2007; Svejenova et al., 2007). This reciprocal, but surprisingly neglected (Stierand and Sandt, 2007), relationship between the institutional frame and the space for culinary creativity of the chefs will be analyzed in the following.

The aim of this chapter is to unfold the apparent paradox of – so to say – 'enabling limits'. We focus on two questions in particular: How is the space for creative freedom impacted by the institutional frame? And in what way can the embedded organizations frame the frame in order to enlarge their scope of acceptable innovative action? For this purpose we will explore the institutional forces in place in haute cuisine and draw implications as to how organizations can deal with them ingeniously. In doing so we propose a procedural and interactive perspective on the evolution of institutional frames within which organizations operate and which constrain and enable their strategic actions.

2. CREATIVITY AND INSTITUTIONAL CONSTRAINTS

The quest for organizational creativity is ubiquitous, as it has proven to be an important source of sustained competitive advantage through fostering innovation and renewal. This has long been true for cultural industries, which are dependent upon constantly creating a unique and lasting experience to the consumers, but has increasingly spread to other fields as well (Lampel et al., 2000). However, innovation does not take place within an open space, but is rather embedded within normative, regulative and cultural-cognitive dimensions of the institutional context (Scott, 2001). This implies that the creative freedom of an organization is impacted by demands and restrictions of the field, as they determine the acceptance of innovations (Stierand and Lynch, 2008; Stierand et al., 2009). Adopting such a 'limiting' perspective implies to understand creative work within an already given institutional framework. Whereas institutional theory has basically emphasized such a constraining impact of the context as in 'ruling the game' (Barley and Tolbert, 1997; Powell and DiMaggio, 1991; Meyer and Rowan, 1977; Scott, 2001; Zucker, 1977), institutional entrepreneurship calls into question this one-sided influence of institutions on organizational actors and hence on the space for creativity. Further on it refrains from the assumption that institutional frames are taken-for-granted, clear, consistent, self-evident, stable and equal for all organizations competing in the 'same' institutional context. Thus, frames have to be considered plastic, making the same (frame) different. In this vein, creativity has to be considered not only an innovative activity within an existing framework but also transcending such frameworks by 'gaming the rules'. Consequently this stream of thought acknowledges the importance of the actor's participation and agency in the process of shaping and changing the institutional frames in which they are embedded (Battilana et al., 2009; Dacin et al., 2002; Drori and Landau, 2011; Sewell, 2005). It is just this conception of 'creative adaption or reformulation of institutional constraints' that Lampel et al. (2011: 1305) have termed 'organizational ingenuity'. This also implies that the creative freedom is neither given nor constrained by the context per se, but can be actively shaped by the embedded organizations. Consequently, it is of great importance to gain a deeper understanding of the respective institutional frame (i.e. its rules, expectancy structures and key players), in order to determine opportunities for organizations to unfold creativity within it – be it through creative adaptation to given constraints or be it through actively reshaping or reframing the frame.

3. CREATIVITY AS NECESSITY IN HAUTE CUISINE

As opposed to other cultural industries, which often lack a clear standard of quality (Lampel et al., 2000), haute cuisine appears to be a highly institutionalized field (Durand et al., 2007; Fauchart and von Hippel, 2008; Ottenbacher and Harrington, 2007a; Rao et al., 2003). There exist strict guidelines and rules, which are continually authorized by gourmet guides and food critics (Bouty et al., 2013; Johnson et al., 2005; Svejenova et al., 2007). Those institutional players shape and maintain the institutional frame through continuously evaluating the actions of the respective organizations (Ferguson, 1998; Karpik, 2000; Bouty et al., 2013). Organizations competing in that field thus assumingly face equal and relatively stable evaluation criteria to which they have to adhere (Durand et al., 2007). Among the existing guidebooks, the *Guide Michelin* can be regarded as the most influential one (at least in the European field), having established a widely known ranking system of none to three stars (Christensen and Strandgaard Pedersen, 2011; Karpik, 2000). The central unit of external observation is the chef/restaurant dyad, whereby those with one to three stars (according to the *Michelin* ranking) can be considered haute cuisine (Durand et al., 2007). Over time, the star system has remained rather stable, probably due to its relatively conservative awarding of stars (Johnson et al., 2005) and its fixed standards (Christensen and Strandgaard Pedersen, 2011). Besides technical excellence, as well as outstanding products and services, creativity and innovativeness of the chef have become central (Bouty and Gomez, 2010). Thus, in order to become and remain competitive, restaurants have to continuously work on their creative output (Bouty and Gomez, 2013; Harrington, 2005). It does not suffice for organizations to continually deliver excellent quality: 'Successful chefs must be able to adapt and evolve if they want to be successful in the short- and long-term. Thus, to succeed in this competitive environment, fine dining restaurants must systematically develop innovations' (Ottenbacher and Harrington, 2007b: 444f.).

However, as in cultural industries in general, the institutional frame of haute cuisine forces chefs to balance contradictory expectations (see Lampel et al., 2000). While on the one hand they are required to be creative, on the other hand innovations should neither break with existing standards nor appear random. Instead, chefs are supposed to develop their own recognizable and distinctive culinary style and identity over time (Svejenova et al., 2010; see Durand et al., 2007; Ferguson, 1998; Rao et al., 2003). This can be achieved either through the use of occupational rhetoric (Fine, 1996), or through innovative action. As Durand et al. (2007) point out in their study, there exist different forms of the latter – some maintaining the profile built-up in the past, and some breaking with the past. According to their results, external reactions to both categories of change vary depending on the organization's status, age, profile and relative degree of change compared to their competitors. Thus it seems as if there exist certain organization-related and history-dependent aspects, which influence external judgment of creativity and restrict the space of innovative action. Hence, the 'quest for authenticity' logically precedes the 'quest for creative freedom' (Svejenova et al., 2010: 419). Exploring whether and why this is the case will be the subject of the empirical investigation described subsequently.

4. METHOD

4.1 Empirical Background

For gaining a deep understanding of the institutional context and its impact on the creative freedom of the embedded organizations, we chose to investigate the field of haute cuisine in Berlin, Germany. This is justified by two main reasons. First, this institutional context can be considered an instructive case. Besides being highly institutionalized with evident and clear evaluation criteria, there is also a rich database of magazine articles, press reports and restaurant reviews on the area which allow us to observe the tentative process over time by which the institutional frame reacts to and judges actions taken by the embedded organizations. Second, the chosen regional focus is helpful for capturing the context as a circumscribed case. 'Cuisine is a regional phenomenon' (Durand et al., 2007: 461). Accordingly, geography and cuisine are connected through, for example, the availability of ingredients, gourmet restaurants in one region can be regarded as competitors for the same customers and – most importantly for our investigation – they presumably face the same institutional context and thus rules and expectations to which they have to adapt. Furthermore, Berlin has undergone a major and drastic development with regards to haute cuisine and the gourmet scene since the fall of the Berlin Wall, which can be considered unique. Within the timespan of only 20 years, it has evolved from a gastronomical desert into the area with the highest number of starred restaurants in Germany (Berlin currently has

12 starred restaurants holding 16 stars total, based on the *Michelin Guide 2013*) – with amount of restaurants and number of stars awarded continually increasing. This 'start from scratch' enticed many chefs to come to Berlin in search of creative freedom and is still attracting new internationally renowned stars – such as Pierre Gagnaire, who opened the restaurant 'Les Solistes' in the new Waldorf Astoria Hotel in January 2013 – as well as newcomers.

4.2 Data Collection

For reconstructing the institutional context and exploring its plasticity and impact on the degree of creative flexibility the actors have, we chose a qualitative approach in the form of an exploratory case study. According to Yin (2009), this method is recommended in cases where the focus is on exploring a complex contemporary phenomenon within a real-life context, over which the researcher has no control and which is studied with the goal to develop propositions for further inquiries. In order to grasp the richness of the phenomenon we have drawn on multiple sources of evidence. First, we started out by extensively researching secondary data on haute cuisine in Berlin over the past 20 years, in order to understand the development that has taken place. More specifically, we first looked for press articles on *Berlin haute cuisine*, *Berlin star chefs* and *Berlin star restaurants* in relevant databases (wiso, lexis nexis, local newspaper databases). Additionally, we scanned the relevant and leading gourmet guidebooks, namely *Michelin*, *Gault Millau* and *Feinschmecker* with regards to their evaluation of the restaurants and chefs over time. After identifying the relevant restaurants and chefs, which have had and still have a high evaluation in the guidebooks, we continued by researching articles on them specifically. Furthermore, we researched haute cuisine in general and looked for case studies by other researchers that gave insight into relevant characteristics of haute cuisine, the institutional context and relevant players. Second, we collected primary data in the form of two rounds of narrative interviews with the most relevant Berlin gourmet critics, culinary journalists and experts of Berlin haute cuisine. Those were identified and selected through a multistep method. All relevant authors of articles found in the first round of research on haute cuisine, restaurants and chefs in Berlin were collected, which resulted in a group of experts. Then, through the research we came across an annual award ('Berlin Master Chefs'[1]), aiming at honoring Berlin's master chefs and promoting the culinary excellence in the city, which has been in place since 1997 and has had a great impact of the rise of Berlin as a culinary capital. The award is given by the city of Berlin to the Berlin Master Chef of the Year, the Berlin Shooting Star of the Year, the Berlin Maître of the Year and the Berlin Sommelier of the Year, as well as Gastronomic Innovator and Brandenburg Master Chef. The winners in the six categories are selected by a jury, which consists of the most relevant experts in this field in Berlin. Through merging the jury members with the group of experts we identified before, we were able to create a list of 15 experts that we contacted for an interview. Out of those, we were able to arrange interviews with 13 of them in the first round, which took place in October 2011. The interviews were held in an open-ended, conversational manner and focused mostly on understanding the institutional forces of haute cuisine in general and in Berlin; the most relevant external evaluators, the evaluation criteria and their subjectivity/objectivity as well as their comprehensibility and predictability; the evaluation process of the critics, the (perceived) influence of the critique on the development of the chefs and the style of the cuisine in general; the development of the Berlin market; the success factors of star chefs and restaurants; their (perceived) reactions to feedback from the environment; the relationship between chefs and critics as well as among chefs themselves; and predictions as to how the market will evolve. The first 13 interviews had an average length of 75 minutes and were recorded and fully transcribed. The conducted interviews yielded a data saturation, which is why we refrained from talking to more experts at that time. We completed the data collection with short follow-up interviews (of approximately 16 minutes on average via telephone) with seven of the experts whom we regarded as most representative after the first round. Those took place in March 2012 and were aimed at evaluating, commenting and explaining the *Michelin* and *Gault Millau* ratings, which were announced after the first round of interviews was completed and which showed unexpected results with regards to forecasts made during the first round.[2]

4.3 Data Analysis

All interviews were fully transcribed and subsequently coded, categorized and analyzed with the help of the computer assisted QDA software Atlas. ti. As a result of our research design and data collection method, the codes were derived in an exploratory manner (Miles and Huberman, 1994) driven by our research question and themes of interest, which we already focused the interviews on (see section 4.2). Those were clustered into four main groups: 'institutional frame', 'expectancy structure', 'process and subject of profiling', and 'trends and strategic topics', which provide the basis for our framework (see also Figure 8.1). The 'institutional frame' cluster includes all information about who and what constitutes the frame (*relevant institutional players, their influence on the field*). The 'expectancy structure' cluster covers codes concerning the content of the expectation (*evaluation criteria, success factors, no-goes, must-haves*) as well as the clarity of them (*unambiguousness, predictability, understandability, subjectivity, objectivity*). The 'industry trends' cluster deals with themes, culinary styles and vogues that arise over time (*trends, culinary styles, evolution of the industry*). The 'process and subject of profiling' cluster is based around the evolutionary development process of the chefs over time, and its influence on their evaluation (*profile, concept, strategy, style, change, development, decisions*). Subsequent to the clustering, we tried to establish relationships between the different clusters over time. This resulted in our framework, which shows the interaction between the various levels of analysis over time and provides a description of the plasticity of the institutional frame and its influence on the space for creative action (see Figure 8.5). The results we were able to derive from the analysis will be discussed in the next section.

5. RESULTS

Our analysis of haute cuisine in Berlin reveals findings in two major areas:

1. Characteristics of the institutional frame, including the key players, the expectancy structure of the institutional context, as well as industry trends and strategic topics; and
2. The process of profiling.

All elements of Area 1 interact with one another and thus cause the plasticity of the institutional frame, which forces and enables the embedded actors to balance the paradoxical requirements by unfolding a process of profiling (Area 2) over time. The space for creativity effectively turns out to be highly subjective and dependent on the embedded actors' capability to creatively unfold and frame the paradoxical institutional frame. We will first present our findings in the respective areas, before discussing implications and possible future research.

5.1 Characteristics of the Institutional Environment

The institutional environment is comprised of key players (see section 5.1.1) that observe, judge and evaluate the actions of the embedded actors based on certain criteria. The evaluation of the actions always takes place in hindsight and is sent to the actors – i.e. the haute cuisine chefs – in the form of feedback. Through this interaction, the environmental players

Gault Millau	Michelin	Other critics
surprises + advancements	quality + continuity	novelty + descriptiveness

Figure 8.1 Key players in the institutional environment

have an influence on the constitution and evolution of the expectancy structure (section 5.1.2), and industry trends and strategic topics (section 5.1.3), which altogether shape the characteristics of the institutional environment.

5.1.1 Key players

As expected, we were able to confirm that the *Guide Michelin*, the *Gault Millau*, other national guidebooks such as the *Feinschmecker* and culinary journalists (the last two will be referred to as 'other critics') are the key players within the institutional environment. With regards to the relevance of their evaluation and their influence on the industry, the data revealed a clear hierarchical order: the *Michelin* is regarded as the most important external evaluator and is still described as 'the bible' (Interview 3: Quotation 13), followed by the likewise internationally renowned *Gault Millau*, then the nationally active *Feinschmecker* in conjunction with smaller gourmet guides and culinary journalists, who mostly act on a national or regional level. Within this overarching evaluation system, each of them plays a slightly different role, meaning they do not appear as a homogeneous entity (see Figure 8.1).

The *Michelin* is described as having a clear focus on quality, although it is not entirely clear whether this is purely with regards to the food – 'what is on the plate' (1:15) – or whether this includes the service and the entire ambience as well. In any case, reliability is the differentiating feature of the *Michelin*. This is due to multiple factors: First, all testers need to be learned cooks and thus experts in the field who truly understand the 'craft' and 'art' of haute cuisine cooking (1:15). Also, they almost entirely remain anonymous, granting them increased objectivity and legitimacy. Second, they are known to rate rather cautiously and conservatively. Being more interested in an incremental and continuous development of a restaurant/chef and the industry, they avoid unexpected up- or downgrading. Third, this goes along with a rigorous testing procedure: instead of going to a restaurant just once, a thorough investigation by the anonymous testers takes place before stars are awarded or taken away. This also includes various visits by different critics as well as intensive interviews with the restaurant manager about the fulfillment of 'other prerequisites' for an award (2:10). In short:

'They are just amazing when it comes to testing. They really spend money on it' (6:39). Besides the importance of continuity and quality, curiosity and merely speculations exist about their other criteria. This is also due to the fact that the *Michelin* does not justify or give explanations for why one, two or three stars are awarded. Undoubtedly though, certain norms need to be fulfilled for a respective number of stars. The *Gault Millau* – 'a different world' (3:13) – is much more interested in the aspects of change, advancement and differentiation than in continuity. It also often comes to unexpected and surprising evaluations, making it appear 'erratic' (6:42). The focus is not on solid and continuous ratings, but on sensations and surprises. One critic even calls the German head of *Gault Millau* 'a brilliant marketing strategist', who raises the interest in the new guide by praising or degrading someone where everybody goes – 'are you crazy?' (3:72). Thus, while quicker adjustments and changes in the ratings might yield more current evaluations, they also cause a touch of arbitrariness and reduce its influence compared to the *Michelin*. This is also due to divergent testing practices: first, testers do not need to know how to cook – it is sufficient to have shown interest and achievements in the gourmet industry (i.e. as journalists). They also do not remain as strictly anonymous as those of the *Michelin*. Second, the *Gault Millau* 'implicitly provides information about its criteria in each text' through extensive critiques, which are given for each restaurant/chef dyad included in the guide (7:43). Third, the *Gault Millau* tries to take more influence on the direction into which the industry evolves (see section 5.1.3) by actively sponsoring certain culinary styles or chefs. In this vein, decisions such as who becomes 'chef of the year' appear to be solely 'political' as in 'trend-setting', thus hampering its objectivity (2:35). Nevertheless, it is still regarded as the second most important external evaluator, yet far behind the *Michelin*. National guidebooks such as the *Feinschmecker* mostly play a role for a certain kind of clientele, but do not have such an overarching impact on the entire industry. One exception is annual awards such as 'chef of the year', which grants the winner media attention and much interest. Media attention is what gourmet journalists assure and on what they base their influence. We were surprised to find out that becoming a gourmet journalist and critic merely requires interest and passion in the subject and area but no specific training. Thus, various career paths exist which lead to a very diverse group of people. They again follow different principles and play a different role from the gourmet guides. For them, there is nothing greater than discovering someone or something new – in a positive or in a negative way. Since they do not have nearly as much budget for their testing activities as the professional gourmet guides, they are limited in their ability to continually and repeatedly go to different restaurants and form their opinion. Also, it is much harder, in particular on a regional level, to remain unknown and anonymous to the cooks. Big differences exist though as to how the journalists handle their relationships with the chefs. While some have a close and friendly relationship with a selection of chefs and openly talk about their 'favorites', others try to keep their distance in order to keep and protect their objectivity and legitimacy. We also found a diverse picture of their perceived influence on the industry. While some see themselves as promoters of certain chefs and restaurants, acting as guides and giving valuable advice to the chefs, others are more critical and regard their critique as a purely economic factor. Here, the old market principle holds true: 'Even bad critiques are good critiques' (3:46) – nothing is more important than continually getting public attention. In general, gourmet journalists follow the rules of mass media in their evaluations – looking for and promoting novelty that is still describable.

5.1.2 Expectancy structure

What results from this diverse and heterogeneous institutional environment with different roles and foci is a rather diffuse and paradoxical expectancy structure of the context (see Figure 8.2).

Instead of providing clear guidelines and criteria to which the chef/restaurant dyads have to adhere, we find that there are very few identifiable must-haves or no-gos. Instead, it appears as if the expectations rather represent norms than clearly defined standards. Due to the lack of specificity, the expectancy structure takes on a truly paradoxical form. We were able to identify four major and overarching paradoxical expectations that are sent to the embedded actors from the environment and with which they have to deal.

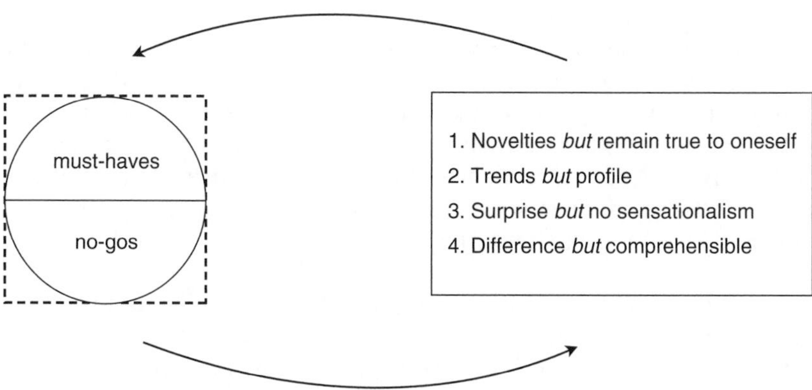

Figure 8.2 Expectancy structure

1. Novelties but remaining true to oneself Mutual agreement exists among all the major players that it is not enough to be excellent in order to gain and keep positive evaluations. Instead, a constant evolution and development is looked for. Some go so far as to claim that chefs are expected to reinvent themselves every six weeks. Others relativize this expectation but still put much weight on regular changes. Those changes mostly refer to small decisions like the menu. Bigger ones such as the interior or especially the location are much harder, if not almost impossible or only at high costs to influence and adapt. However, while needing to continually present novelties and progress, chefs are at the same time expected to develop and remain true to their culinary profile (4.2). Thus novelties are always evaluated against the background of former profile-forming behavior and should be perceived as compatible with it.

2. Responding to trends but in accordance with profile The same applies to the need to respond to industry trends and strategic topics, which shape the industry for a while before giving way to new orientations (section 5.1.3). It is not advisable to either follow everything that is en vogue 100 per cent, or to ignore contemporary dynamics completely (1:56). The best way seems to be to take in some influences and inspirations and adapt them to up to 10 per cent of the repertoire. This tactic is referred to as having a 'trend breeze' or 'trend hint' (3:65). Again, any decision to follow a trend in any way should be understandable for the environment and signal advancement without giving up any distinctiveness or individuality.

3. Surprising but no sensationalism In line with responding to trends is the need for chefs to surprise, for example by including novel dishes or elements, but not in an exaggerated manner. The key is to add an appropriate amount of humor to the dining experience without overwhelming or 'losing' the guest (4:53). Specifically young chefs try to impress their audience by adding as many surprising elements as possible to their menu. This might even trigger a 'wow-effect' (3:66) and raise interest in the first place. However: 'Then you have enjoyed this theatrical performance, it was funny . . . And everybody has said: It was great, a great opera, it was funny. But none of them goes a second time, since Wagner, seven hours, you do that once and that was it' (3:66).

4. Creating a difference but remaining comprehensible In general, haute cuisine chefs are expected to create a positive difference: to their competitors, to their past actions, to general expectations. However, this difference cannot be created at the expense of comprehensibility, i.e. it has to be made plausible. This need is also caused by the rising importance of authenticity. Additionally, not any difference can be created within an existing restaurant concept: Certain decisions are locked-in and hard to deviate from, or at least not without causing confusion.

5.1.3 Industry trends

In addition to the general expectations, the industry is shaped by trends. Those gain relevance for a period of time through adoption by a number of chefs and promotion by the key players. Trends are first and foremost culinary styles, which appear in waves. In the past, these were classical French cuisine, nouvelle cuisine, Italian cuisine, Asian cuisine, spice cuisine and more recently molecular cuisine. Currently popular is regional and vegetarian cuisine with a focus on local produce and authenticity. Other trends include the structure and wording of the menus. Formerly, lengthy descriptions were given about the cooking methods and the exact combinations in which ingredients are prepared, thus making it very explicit and obvious what to expect. Nowadays it is very much en vogue to just name the ingredients like 'Head. Foot. Asparagus'. With this wording it even remains unclear which aggregate state and size a dish has.

All past and recent cuisine styles and trends can be traced back to someone initiating it by performing a new style very successfully: nouvelle cuisine is directly linked to Paul Bocuse, molecular cuisine to Ferran Adriá, and regional and authentic cuisine to René Redzepi (chef of the *Noma* in Copenhagen). The success of those role models then entices more parties to assume and imitate tendencies of the trend and follow it as well. They do this by picking up new standards, including them in their repertoire and signaling to the guest and critics that they follow it, for example by being regionally conscious. Whether this holds true or not is at first hard to tell. According to one critic:

> This is only the basis, everybody has to do it, it gains more importance as a sign. Whether one can instill it with sense, whether one makes something out of it, or whether one even invents a new trend entirely depends on the personality. It is impossible to conduct this systematically (7:46).

Apparently, the pressure to follow certain movements is rather high, since it can be assumed that the institutional players reward the adaptation to a certain fashion. As with the general expectancy structure though, it is hard to tell in advance which directions gain popularity. For example, one Berlin haute cuisine chef started to specialize in regional and vegetarian cuisine about ten years ago and only now does he fit the overall trend. Also, tying back in with the paradoxical expectations, entirely concentrating on a new trend does not ensure long-term legitimacy and success: 'We have those waves and only those remain on top, who adopt these things, but transform them according to their personal style, which then becomes recognizable. Otherwise it is just playing and one already waits that the new wave also gets adopted' (7:46).

The boundaries and ends of such movements are characterized by the need for novelty that exists in the institutional environment. After a while, people get weary of repetitions and start to look for something new and fresh. Again, the critique takes a big stake in initiating decisions about future directions by demanding surprise and differentiation and strategically giving out awards. Also the 'Berlin Master Chef' is given to the candidate who portrays best the direction into which most of the critics want Berlin haute cuisine to evolve. The critics also regard it as their task to advance the movement and actively sponsor trends that they can identify with. Thus, trends interact strongly with the other elements of the institutional environment: They are promoted by critics and at the same time alter the content of their expectations. As with general expectations, they are signaled to the chefs who have to make sense of the signals and react in accordance with their profile, which also develops through an evolutionary process in interaction with the institutional environment (see Figure 8.3).

5.2 Process and Subjects of Profiling

As it became clear, being recognizable and differentiable is one criterion strongly demanded by the institutional environment. This refers to the need for chefs to develop a unique and robust profile over time which sets them apart from competitors and gives meaning to all their actions. The profile has to be distinguished from the 'concept', which is also referred to quite often by the critics. A concept incorporates the whole idea of the restaurant in which the chef works. It usually results from a conscious plan and includes the location, the equipment, the set-up, the service, the wine list, the menu and style of cuisine, etc. The concept is first developed

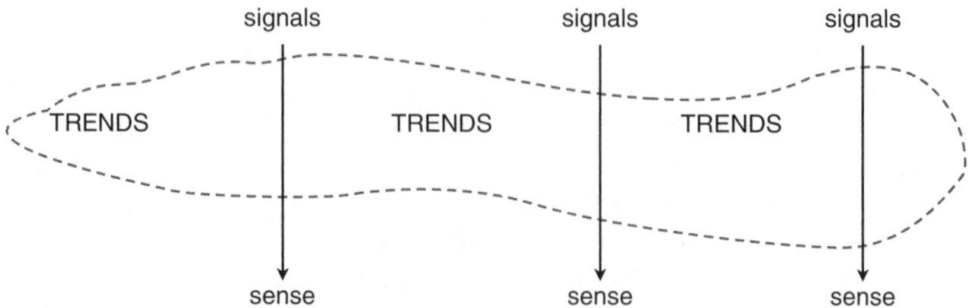

Figure 8.3 Industry trends

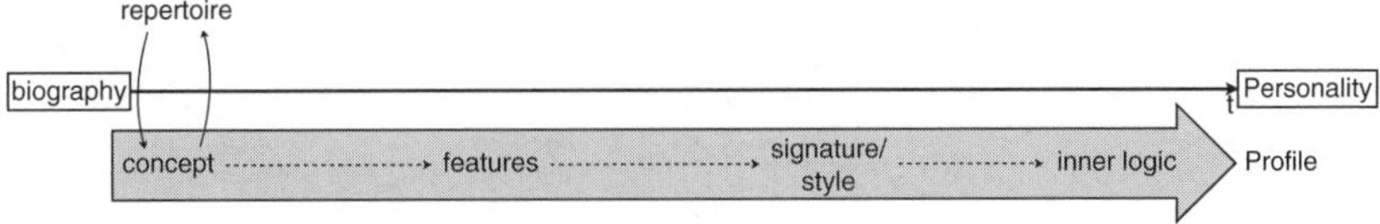

Figure 8.4 *Process of profiling*

prior to the opening of the restaurant and can be subject to changes, either due to a reorientation, a change in management or chef, a lack of success or an aim for higher goals that does not seem compatible with the current concept. The profile on the other hand is directly tied to the chef and develops over time in correspondence with the environment. The case of Pierre Gagnaire can exemplify the difference between the two terms: the French star chef has multiple restaurants around the world with different concepts, however he has developed an unmistakable profile as a chef.

We propose to view profiling as an iterative multi-step process which – if successful – leads to the development of a unique profile (see Figure 8.4).

At the start, chefs are influenced by their biography and their repertoire of what they have learned throughout their training. Usually they start out imitating what their teachers have done before developing their own style. Their background in the first place needs to be fitted into a concept. The first distinguishing step is adding certain features to the known repertoire, thus deviating a bit from the past. Those features should then over time develop into a characteristic signature or style. The highest level of profiling is to generate an inner logic, through which any action immediately makes sense and can be attributed and connected to the profile.

However, this process does not take place in isolation but is closely linked to the institutional environment through interpreting signals and responding to them, making it highly interactive (see Figure 8.5).

As indicated before, the chefs need to make sense of the expectancy structure and signals they receive from the institutional environment and act in a way that seems appropriate to them. By acting they send signals back to the institutional environment; those get evaluated according to the expectations and foci and feedback is returned. The institutional players thus play an active part in describing the profile, since they are constantly observing, judging and interpreting the actions in the quest for continuity, advancement or deviation from the past. The process is never truly complete, since the need for novelty and progression demands that the chefs remain open to the environment and make constant decisions about which signals to take in and into which direction to go. However, 'once a certain status is reached, it does not matter what you do – people will come anyway' (7:36).

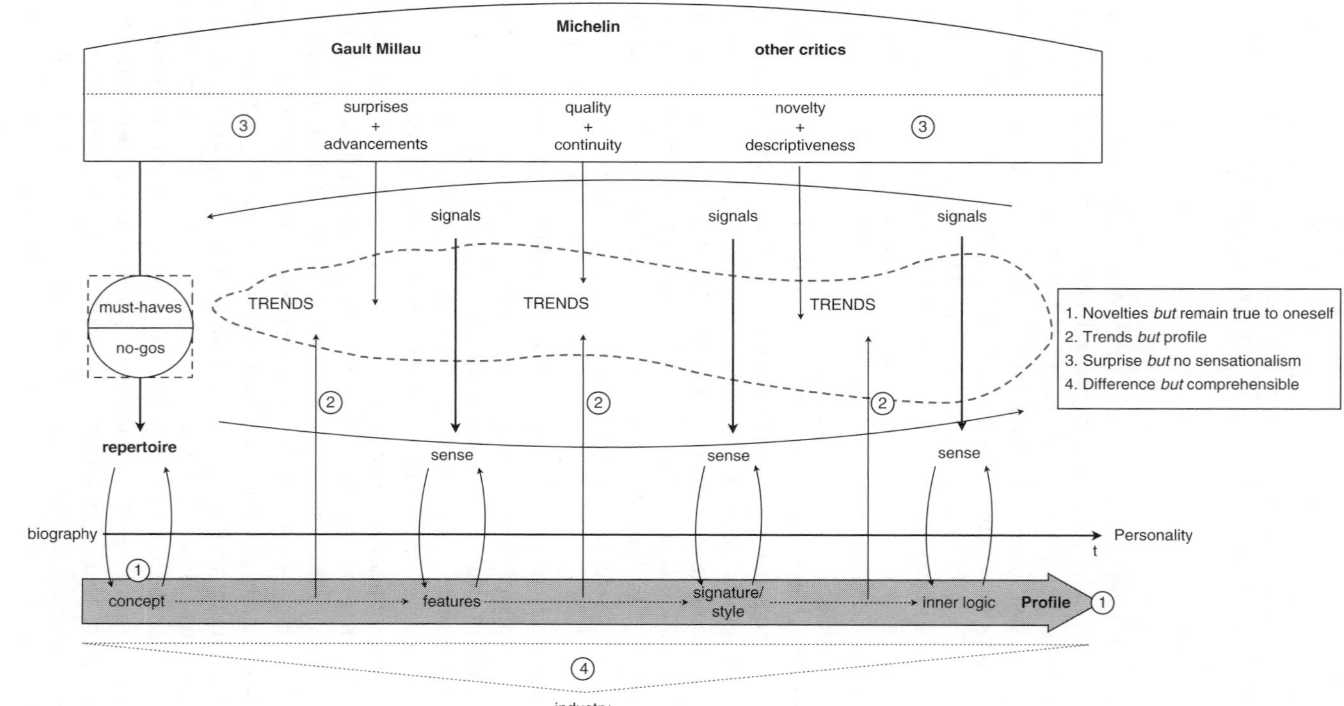

Figure 8.5 Profiling as an interactive process with the institutional environment

6. DISCUSSION AND IMPLICATIONS

Our study has revealed that haute cuisine, despite being a highly institutional field, does not provide actors with clear restrictions concerning their space for creative action and thus opens up room for organizational ingenuity in framing the frame. The institutional frame presents itself as very plastic, due to a number of key players with different roles as well as the lack of consistent and clear standards. Our results are thus in accordance with findings on cultural industries in general (Lampel et al., 2000), as well as with Johnson et al. (2005: 172), who attribute the *Guide Michelin* with 'vagueness'. This has the following implications: the expectations are so general and open that they can be interpreted in many different ways. Although one is supposed to have an appropriate location, a certain kind of product supply and wine list, a functioning service and kitchen team, a certain ambience and interior design, and a recognizable profile, which is still open to react to general industry trends and continually evolves, there are many possible ways as to how one could comply with these requirements. Since the understanding of what 'ambience', for example, means in the case of a gourmet restaurant varies among institutional players, there is no best way to describe what a one-star or a two-star restaurant ought to look like. The same holds true for other areas of expectation, such as the service. Upon passing a certain degree of requirement fulfillment (minimum standard), the norms appear to be pretty permeable and do not guide behavior. In general, it is hard for the restaurants to predict *ex ante* which steps to take in order to get positive feedback from the environment, since 'it all depends'. They can neither merely repeat their current actions due to the necessity for a creative evolution, nor can they ignore their profile development process, which sets boundaries to which steps are acceptable. Hence not any action is sensible, but action is necessary in order to get feedback from the environment. This renders the strategy making in restaurants a rather tentative process. This is true for the appropriate response to feedback as well: it might not be obvious what to change or what to keep stable in order to improve evaluations. This can result from either vague critiques in the case of the *Michelin*, which does not reveal much about its evaluation criteria, or from diverging critiques of the different players. All in all, this leads at first sight to trial-and-error based approaches of strategizing in interaction with the environment. At second sight, restaurants could and should differ a lot with regard to their processes of strategizing which is in turn the *genius loci* of strategic ingenuity. Nothing appears to be really set in stone as long as it all 'fits'. What remains are merely rough and flexible standards and it is up to the chefs to make decisions and take actions which they assume to be in line with the requirements and their personal interpretation of them. Not only though is the norm content subject to interpretation, their respective relevance varies over time and between critics. Depending on the focus the critics have, different features are of higher importance than others. All in all, the expectancy structure is highly subjective and dependent on interpretation.

While on the one hand this plasticity and 'vagueness' can be a source of uncertainty for aspiring chefs (Johnson et al., 2005), on the other hand it opens up opportunities for framing the frame and thus organizational ingenuity. As it turns out, through the lack of a clear and fixed institutional frame, actors can shape their space for creative freedom. This can be done through active engagement in what we term 'profiling', referring to the tentative and interactive process of acting and sensemaking. Only through interpreting given requirements in a *singular* way, taking into account the own repertoire and concept, trends and fashions and paradoxical expectations of the environment, and responding to them by creative action can actors over time build up a recognizable profile which becomes the benchmark for future evaluations.

Similar mechanisms can be identified in other studies. Svejenova et al. (2010) describe 'creative response' as crucial for achieving creative freedom (p. 420). According to them purposive action is key in the evolution of an individual business model (i.e. 'profile') (p. 424). Horng and Hu (2008) confirm a necessity for developing distinctive creative potential, as well as a mutual interaction between actors and business that opens up space for creative freedom: 'current fashion has an impact on the artist and he or she has an impact on current fashion, indeed (if creative enough) is capable of initiating a new fashion' (p. 223). Indications on how to build up 'situational identities' by the use of language can be found in the work of Fine (1996). He points to the use of 'occupational rhetorics' in order to relate work to a meaning system. By this process, which he calls similar to impression management, workers justify and explain their work and thus influence the factors by which they are judged (pp. 90ff.). What is required thus is the creation of a plausible story that aligns creative actions with the respective profile and creates a coherent identity.

Therefore, the key to success is not to comply with given (vague) standards, but to creatively unfold the paradoxical expectations of the environment. This means that actors continually need to balance contradictory forces and interpret them in ways granting them creative freedom. Consequently, the degree of freedom a chef has entirely depends on the steps taken along the path of profiling and how plausible they are made. According to our analysis and comparable results, there is no clear limit to creativity or renewal. Rather, creative freedom can be achieved through ingenious engagement with the institutional frame, resulting in a profile and thus personalized interpretation of the requirements. Organizational ingenuity thus means to act strategically in a way that does not appear to impede with the given profile and individuality (Paradox 1, 2 in Figure 8.5), remains comprehensible and describable (Paradox 4), creates a difference to the competitors and the past (Paradox 1, 4) and does not overwhelm. For this task, no 'best practices' exist. This is specifically due to the need to maintain a difference, forcing everyone to be the 'same but different'.

7. CONCLUSION AND FUTURE RESEARCH

Analyzing the institutional frame of haute cuisine reveals that the context is rather plastic due to different actors with divergent foci, industry trends and a complex paradoxical expectation structure. There are only a few issues clearly limiting the scope of creative action ('no-gos', 'must-haves'), but even these constraining factors are not considered absolute. Hence, even if a restaurant deviates from some of these 'no-goes' or 'must-haves', it would not immediately lose its legitimacy. Rather, this 'enabling limit' of the institutional environment has to be unfolded by the strategic agents creatively over time in an act of organizational ingenuity. This means that any deviation should be made plausible, i.e. a restaurant introducing something new has to have not only a convincing new 'product idea' but also a plausible story or narrative for this new and creative step. The degree of plausibility itself highly depends on the process of interaction between a strategic agent and the institutional context and is a result of dealing with the contradictory and flexible expectations ingeniously. In this vein, the plausible story or narrative has no single author but it is rather the product of an interactive 'sensemaking' process. This process of profiling entails both the expectation of having a clear profile (and hence an identity) and the expectation to deviate from that profile. Anchoring point for any creative development is the respective profile, which develops over time and upon which creative and innovative actions are contingent. Whereas young organizations are relatively unconstrained in their actions (Durand et al., 2007), the need to develop an identifiable profile can lead organizations to refrain from frame-breaking changes. From this it follows that history matters in the sense that interactions between the institutional frame and the strategic agents reciprocally frame the space for creative freedom over time. Due to the importance of continuous innovation and the contingent evaluation of it in light of past steps, chefs need to actively shape those through working on an authentic and flexible profile. As confirmed through our study, once a certain status is achieved, chefs are granted creative freedom ('*Narrenfreiheit*') and can move on to a 'quest for influence' and a 'quest for recognition' (Svejenova et al., 2010). In that sense we can say that chefs really are the architects of their creative freedom.

In conclusion, organizational ingenuity in our case means to balance the contradictory expectations of the institutional frame and interpret and respond to them in a way that increases the creative freedom for future development. Due to the plasticity of the given expectations, embedded organizations have an influence on the background against which they are evaluated. Hence, instead of trying to comply with seemingly strict standards, ingenious organizations should use the identified 'vagueness' and frame the frame by tentatively undertaking creative steps along their path of profiling. By signaling to the environmental players that what they do fits to their profile, considers trends and creatively deals with contradictory expectations, haute cuisine restaurants have the ability to maintain the creative freedom necessary for future scope of action and successful development. Thus, the space for creative action heavily relies on organizational ingenuity in form of an '*unfolding capability*' through the process of profiling and the plausible interpretation and reaction to environmental signals.

What can be inferred from our study is that ingenuity incorporates both sensing and shaping. Organizations need to gain a good understanding of the institutional frame and expectations they face, but more importantly use their ability to interpret and respond to them in a way that grants them creative freedom. This is most easily done by interactively building up a recognizable and distinct profile, which is then used as a benchmark for evaluating future actions. Depending on what profile is built up over time, the degree of creativity possible varies. This, however, is not due to fixed environmental constraints, but rather a result of the interaction process and influenced by the actors themselves. Thus, when facing contradictory and plastic environmental expectations, organizational ingenuity means to sense and shape them in a creativity enhancing way, leading to an increased scope of action and creative freedom.

In this chapter we have focused on the context in the first place with regard to the general mechanisms enabling and constraining the agents in the field, contributing to an understanding of the complex interaction process and the paradoxical expectations. Future research should focus on the embedded systems, i.e. the restaurants, and analyze to what extent they are aware of the paradoxical expectations, which of those they regard as relevant, and what acts of organizational ingenuity they use to create space for creative unfolding and framing the frame. We agree with

Stierand and Lynch (2008) that this should include exploring the belief systems and perceptions of the chefs as well as the consequences they face for their innovative actions. Another point of interest is whether the process of profiling leads to path dependence and into a lock-in situation (Sydow et al., 2009).

NOTES

1. http://www.berlin-partner.de/?id=1331&L=1 (last accessed 27 September 2013).
2. We refrained from including data from the subsequently held interviews with the Berlin star chefs in this chapter and analysis, since we first aim to understand the institutional frame and the view from the outside before extending the perspective to the systems embedded in the context. This will be the second step of our project.

REFERENCES

Barley, S.R. and P.S. Tolbert (1997), 'Institutionalization and structuration: studying the links between action and institution'. *Organization Studies*, **18**(1), 93–117.

Battilana, J., B. Leca and E. Boxenbaum (2009), 'How actors change institutions: towards a theory of institutional entrepreneurship'. *The Academy of Management Annals*, **3**(1), 65–107.

Bouty, I. and M.-L. Gomez (2010), 'Dishing up individual and collective dimensions in organizational knowing'. *Management Learning*, **41**(5), 545–59.

Bouty, I. and M.-L. Gomez (2013), 'Creativity in haute cuisine: strategic knowledge and practice in gourmet kitchens'. *Journal of Culinary Science & Technology*, **11**(1), 80–95.

Bouty, I., M.-L. Gomez and C. Drucker-Godard (2013), 'Maintaining an institution: the institutional work of Michelin in haute cuisine around the world'. *Research Center ESSEC Working Paper 1302*, available at: http://halshs.archives-ouvertes.fr/hal-00782455/ (accessed 1 May 2013).

Christensen, B.T. and J. Strandgaard Pedersen (2011), 'Evaluative practices in the culinary field'. *Creative Encounters Working Paper # 69*, available at: http://openarchive.cbs.dk/handle/10398/8393 (accessed 1 May 2013).

Dacin, M.T., J. Goodstein and W.R. Scott (2002), 'Institutional theory and institutional change: introduction to the Special Research Forum'. *The Academy of Management Journal*, **45**(1), 43–56.

Drori, I. and D. Landau (2011), *Vision and Change in Institutional Entrepreneurship: The Transformation from Science to Commercialization*. Oxford, New York: Berghahn Books.

Durand, R., H. Rao and P. Monin (2007), 'Code and conduct in French cuisine: impact of code changes on external evaluations'. *Strategic Management Journal*, **28**(5), 455–72.

Fauchart, E. and E. von Hippel (2008), 'Norms-based intellectual property systems: the case of French chefs'. *Organization Science*, **19**(2), 187–201.

Ferguson, P.P. (1998), 'A cultural field in the making: gastronomy in 19th-century France'. *American Journal of Sociology*, **104**(3), 597–641.

Fine, G.A. (1996), 'Justifying work: occupational rhetorics as resources in restaurant kitchens'. *Administrative Science Quarterly*, **41**(1), 90–115.

Harrington, R.J. (2005), 'Part I: the culinary innovation process – a barrier to imitation'. *Journal of Foodservice Business Research*, **7**(3), 35–57.

Horng, J-S. and M-L. Hu (2008), 'The mystery in the kitchen: culinary creativity'. *Creativity Research Journal*, **20**(2), 221–30.

Johnson, C., B. Surlemont, P. Nicod and F. Revaz (2005), 'Behind the stars: a concise typology of Michelin restaurants in Europe'. *Cornell Hotel and Restaurant Administration Quarterly*, **46**(2), 170–87.
Karpik, L. (2000), 'Le Guide rouge Michelin: Michelin's red guidebook'. *Sociologie du travail*, **42**(3), 369–89.
Lampel, J., H. Benson and I. Drori (2011), 'Discovering creativity in necessity: organizational ingenuity under institutional constraints'. *Organization Studies*, **32**, 1305–8.
Lampel, J., T. Lant and J. Shamsie (2000), 'Balancing act: learning from organizing practices in cultural industries'. *Organization Science*, **11**(3), 263–9.
Meyer, J.W. and B. Rowan (1977), 'Institutionalized organizations: formal structure as a myth and ceremony'. *American Journal of Sociology*, **83**, 340–63.
Miles, M.B. and A.M. Huberman (1994), *Qualitative Data Analysis: An Expanded Sourcebook*. Thousand Oaks, CA: Sage.
Ottenbacher, M. and R.J. Harrington (2007a), 'The culinary innovation process'. *Journal of Culinary Science & Technology*, **5**(4), 9–35.
Ottenbacher, M. and R.J. Harrington (2007b), 'The innovation development process of Michelin-starred chefs'. *International Journal of Contemporary Hospitality Management*, **19**(6), 444–60.
Powell, W.W. and P. DiMaggio (1991), *The New Institutionalism in Organization Analysis*. Chicago: University of Chicago Press.
Rao, H., P. Monin and R. Durand (2003), 'Institutional change in Toque Ville: nouvelle cuisine as an identity movement in French gastronomy'. *American Journal of Sociology*, **108**(4), 795–843.
Scott, W.R. (2001), *Organizations and Institutions* (2nd edn). Thousand Oaks, CA: Sage.
Sewell, W.H. (2005), *Logics of History: Social Theory and Social Transformation*. Chicago: University of Chicago Press.
Stierand, M. and P. Lynch (2008), 'The art of creating culinary innovations'. *Tourism and Hospitality Research*, **8**(4), 337–50.
Stierand, M. and J. Sandt (2007), 'Organising haute-cuisine service processes: a case study'. *Journal of Hospitality and Tourism Management*, **14**(1), 24–36.
Stierand, M., V. Dörfler and J.C. MacBryde (2009), 'Innovation of extraordinary chefs: development process or systemic phenomenon?' Paper presented at BAM 2009, Brighton, UK, 17–19 September 2009. Available at: http://strathprints.strath.ac.uk/id/eprint/14430 (accessed 1 May 2013).
Svejenova, S., C. Mazza and M. Planellas (2007), 'Cooking up change in haute cuisine: Ferran Adrià as an institutional entrepreneur'. *Journal of Organizational Behavior*, **28**(5), 539–61.
Svejenova, S., M. Planellas and L. Vives (2010), 'An individual business model in the making: a chef's quest for creative freedom'. *Long Range Planning*, **43**(2–3), 408–30.
Sydow, J., G. Schreyögg and J. Koch (2009), 'Organizational path dependence: opening the black box'. *Academy of Management Review*, **34**(4), 689–709.
Yin, R.K. (2009) *Case Study Research: Design and Methods*. Thousand Oaks, CA: Sage Publications.
Zucker, L.G. (1977), 'The role of institutionalization in cultural persistence'. *American Sociological Review*, **42**(5), 726–43.

9 Creating innovative solutions in microfinance and the role of organizational ingenuity

Ana Cristina O. Siqueira, Sandra R.H. Mariano, Joysi Moraes and Gregory Gorse

INTRODUCTION

We examine organizational ingenuity in Brazil by addressing the context of community banks in areas with severe economic constraints. The concept of 'organizational ingenuity' refers to the 'ability to create innovative solutions within structural constraints using limited resources and imaginative problem solving' (Lampel et al., 2011). The ability to create ingenious solutions with limited resources is often shaped by the social and institutional environment where individuals live and work.

We investigate conditions under which organizational ingenuity has emerged in specific low-income communities or shantytowns in Brazil. Such bottom-of-the-pyramid communities typically face institutional constraints including poverty, deficient sanitation and basic infrastructure, violence, and low levels of education. However, specific entrepreneurs in some areas have emerged as change agents.

We also examine how specific individuals have built organizations that have transformed the social and institutional environment where they operate. The concept of organizational ingenuity refers also to the notion of creative reformulation of institutional constraints (Honig et al., 2012). Most individuals have to operate within existing norms and boundaries to create solutions to existing problems. Yet the work of some individuals and the organizations they build can alter the institutions in which they operate.

In this study we explore the following research questions:

Research Question 1: How have leaders of community banks operating in areas with severe economic constraints in Brazil demonstrated organizational ingenuity?

Research Question 2: How is the social and institutional environment of community banks in low-income communities coordinated and transformed by leaders' organizational ingenuity?

ORGANIZATIONAL INGENUITY IN MICROFINANCE

Sustainability

The sustainability of microfinance leaders and their firms could be another integral part in creating a unique need for organizational ingenuity in the field of microfinance. One reason is the size of the microfinance market. Almost 4 billion people worldwide earn less than US$2 000 each per year. This equates to about 65 per cent of the world's total population (Prahalad and Hammond, 2002). Using this amount as a possible representation of the microfinance market shows how large it may potentially be. A market of this size could implicate that leaders in microfinance would need to find ways to make the best use of their limited resources so that these populations may survive. Organizational ingenuity might be such a way to accomplish this. It is possible that organizational ingenuity can serve as a mechanism through which microfinance leaders address the overwhelming size of the microfinance market, so that they are able to achieve sustainability for both their community members and their firms.

One well-known example of a leader using creative and resourceful skills to become sustainable while affecting the microfinance market is the case of Muhammad Yunus and his Grameen Bank. Yunus' Grameen Bank can be considered the most prominent in terms of combining several models of microlending to form its business model for reaching the microfinance market. Both for-profit and not-for-profit organizations alike have begun to implement elements of Grameen Bank's structure and processes. Today, Grameen Bank is considered the world's largest microlending organization (Bruton et al., 2011). Yunus' organizational ingenuity may have been a key contributing factor for the ability of Grameen Bank to be sustainable.

Culture

Another contributing factor towards organizational ingenuity playing a role in microfinance is the environment of microfinance itself. That is, microfinance environments are often different from the markets in which non-microfinance businesses operate (Nielsen and Samia, 2008). For example, in some areas of Africa, small businesses that have been started as a result of microlending are often run by women. However, in some instances the businesses are either started by men then left to the women to run, or they are started by women but still culturally considered to be owned by their husbands (Khavul et al., 2009). This could create an issue for microlenders in determining who to target through their business operations. As a result of situations such as these, microfinance leaders operating in these types of environments might need to use organizational ingenuity in order to develop more creative and resourceful solutions.

Additionally, it is also important to consider how microloan customers generally behave (Wood et al., 2008). Microfinance leaders often need to develop creative ways of gathering information about these customers. For instance, individuals who seek microfinance in bottom-of-the-pyramid areas may not have documents that could serve as proof of residency or formal registration of their business. The need for this creative information gathering could be derived from the size and culture variation of microfinance environments.

Outreach

Microfinance leaders who exhibit organizational ingenuity may stretch their ability to be creative and resourceful to greater lengths in order to accomplish their goals. These lengths may be unattainable unless certain conditions existed that helped in enhancing these leaders' organizational ingenuity. One such enhancing condition to enhance organizational ingenuity may be the leaders' involvement and outreach in local communities. Outreach at the micro-level may assist leaders in fully realizing the needs of the people that they are trying to aid. For example, Martin Burt's work with Paraguayan communities, through his organization Fundacion Paraguaya, may have enhanced his organizational ingenuity to the point that his organization was able to positively impact the lives of many Paraguayans by enabling and teaching them to be self-sustainable (Maak and Stoetter, 2012). If Burt had not been involved closely with this local community, he might not have been able to achieve the level of organizational ingenuity enhancement necessary to accomplish his goals. This points towards the possibility of involvement in local communities as being an integral enhancer in leaders' development of their organizational ingenuity.

Another condition that may enhance leaders' organizational ingenuity is their perception of social disparities. Often, social leaders tend to share a common goal of wanting to create a better world for those who live in it. This unprecedented challenge tends to lead to responsible leadership approaches (Ferdig, 2007). As a result these leaders may use imaginative problem solving, while having limited resources, to carry out their mission. Leaders' perception of social disparities could also enhance their organizational ingenuity by allowing them to explicitly recognize opportunities intended to generate different kinds of value to specific sectors of society (Kuratko, 2005). Examples of this particular organizational ingenuity enhancer at work may include bottom-of-the-pyramid and microfinance business strategies. When a leader's social disparities perception comes into play, they might be driven to discover innovative solutions for reaching their intended outcomes. Outcomes of which may include reducing poverty, promoting education, or feeding the hungry (Murphy and Coombes, 2009). Outreach towards serving the poor in itself has potential to be an enhancer of leaders' organizational ingenuity. It could give social entrepreneurs a unique outlook on their surroundings. This in turn can allow them to learn and create new collaborative efforts that are able to create social value for those who need it most. At the same time it may lead to leaders' efforts being financially beneficial for them (Seelos and Mair, 2005). With limited resources, satisfying the needs of so many individuals sometimes is only possible through imaginative problem solving and innovative solutions. In this way, a well-established perception of social disparities seems an essential contributing factor for the development of organizational ingenuity.

Power

In the context of microfinance, a concern with multiple stakeholders and responsible leadership also has the possibility of being an enhancer for leaders' organizational ingenuity. Leaders may have the ability to show proper use of responsible leadership techniques by operating in the best interests of their multiple stakeholders. It is important that microfinance leaders show this best interest because of the power they hold due to the impacts their decisions will have on these multiple stakeholders. The concept of responsible leadership could be classified through the use of three different components: wholeness of values and virtues; wholeness in the sense of being part of something larger than the person; and wholeness as a person in the sense of aligning thinking, feeling, and acting. In addition to these components, it is possible that responsible leadership may be shown through a leader's responsibility for themselves, for others, and for issues in the economic, social, natural, and political environment (Pless, 2007). All of these aspects of responsible leadership may contribute to the responsible use of power that could be necessary in order for leaders' actions to have the greatest impact. In this way, microfinance leaders' organizational ingenuity may encourage a focus on the best interest of multiple stakeholders.

Institutional Constraints in Brazil

There are important institutional constraints involving access to credit in Brazil. One key obstacle for individuals who live in bottom-of-the-pyramid communities to open a traditional bank account is providing proof of residency. For low-income populations who live mainly in the periphery of large cities, the lack of formal documentation of their property is an obstacle not only because individuals lack proof of residency, but also because they cannot use their property to secure credit (Ushizima, 2008). In this case, residents have a permanent address, but they have no proof of residency. Additionally, the fees charged by traditional banks may correspond in some cases to the amount low-income individuals would be able to spend in their monthly repayment of the loan, and thus are often too high for low-income populations.

Another institutional constraint for individuals to access credit includes the need to document income from a formal activity or demonstrate repayment capability. Nonetheless, 'how can low-income individuals have access to credit if they do not have documented repayment capability?' (Ushizima, 2008, p. 1). Credit may not be provided to those who need it most, such as 'individuals who lack social protection, unemployed, women who need to complement family income or support single-parent families' (Toscano, 2004, p. 6).

Major constraints for businesses to open a business account include the need to present a document showing legal registration of the business, according to Brazil's Central Bank (BACEN, 2012). Nonetheless, many low-income individuals operate informal businesses. 'Informal businesses' are those not registered with a governmental authority (International Labour Organization, 1993). To help residents of bottom-of-the-pyramid communities overcome this constraint, certain community banks like Paju Bank in the state of Ceara have offered microfinance to informal businesses.

METHODS

Setting

A sector that has developed strategies to serve low-income populations is the industry of microfinance organizations such as community banks.

Community banks 'are an associational initiative, involving residents in a specific community who seek a solution to concrete public problems related to their daily life condition, by promoting the creation of socio-economic activities' (Franca Filho, 2008, p. 118). Currently, Brazil has 51 community banks, of which 37 are located in the Northeast, the country's poorest region. Within the Northeast region, 28 community banks are located in the state of Ceara, the home of Brazil's first community bank, Palmas Bank.

DATA COLLECTION

To examine our research questions, we use a case study approach (Eisenhardt, 1989; Eisenhardt and Graebner, 2007). For our case studies, we focus on two Brazilian community banks that serve low-income communities: Palmas Bank (*Banco Palmas*), which was launched in 1998, and Paju Bank (*Banco Paju*), which was launched in 2006, both in the state of Ceara. We selected these banks due to their influence on the development of microfinance in Brazil and the relevance of their operations for their respective communities. Palmas Bank has developed a methodology to create community banks in Brazil, which has influenced the creation of Paju Bank. These community banks have their own currency, which circulates only within their communities. Palmas Bank is located in the city of Fortaleza and serves the *Conjunto Palmeiras* neighborhood, while Paju Bank is located in the city of Maracanau and serves the *Pajucara* neighborhood. Palmas Bank offers microfinance to *individuals* in its low-income community, while Paju Bank offers microfinance to formal or informal *businesses* in its low-income community. In this way, Palmas Bank and Paju Bank represent two information-rich environments in different stages of development, contributing to rich insights in this study.

The case study method is a preferred strategy to investigate the types of research questions that start with 'how', which focus on current phenomena involving real world problems (Yin, 2005, p. 19), such as the questions we investigate in this study. In a case study approach, cases need to be contextualized by specific, unique, and well-delimited characteristics (Stake, 1995). Accordingly, we seek to examine each bank within its specific social context.

Secondary data. For our case studies, we use both primary and secondary data. Our secondary data come from a detailed review of published sources about these community banks and their social impact.

Primary data. Our primary data come from direct observation and interviews. Direct observation was an additional method we used to capture specific details of the context under study (Trivinos, 1987).

This method enabled us to have a direct experience with the research subjects and a broader view of the phenomenon by visiting the community banks. A benefit of direct observation is knowledge about the context (Patton, 2002), which allowed us to become familiar with the actual daily operations and people involved in the activities of the bank.

We conducted two in-depth interviews, one with the founder of Palmas Bank and the other with the founder of Paju Bank. The goal of these two interviews was to understand the interviewee's perspective and generate insights beyond those based on secondary data (Patton, 2002). The interviews were conducted in the summer and fall of 2012. This study was approved by US Institutional Review Board research procedures. We conducted interviews in Portuguese, the interviewees' native language, lasting about 2 hours. The interviews were audio-recorded and transcribed. Our semi-structured interviews were designed to include essential questions while offering ample opportunity for further inquiries during the interview (Merriam, 2002; Trivinos, 1987). The interviews addressed topics involving the history and the foundation of the banks, the context of microfinance in these specific bottom-of-the-pyramid areas, constraints and challenges faced by the banks, and impact of microfinance on the community.

Data Analysis

We conducted our analysis informed by grounded theory methodology (Glaser and Strauss, 1967; Corbin and Strauss, 2008). We identified themes that emerged from the data connected to the research questions (Corbin and Strauss, 2008). Our first research question asked: How have leaders of community banks operating in areas with severe economic constraints in Brazil demonstrated organizational ingenuity? For this research question, we categorized two themes in our data: involvement in local communities and perception of social disparities.

Additionally, we categorized three themes that emerged from the data in connection with our second research question. Our second research question asked: How is the social and institutional environment of community banks in low-income communities coordinated and transformed by leaders' organizational ingenuity? Consistent with case study and qualitative research guidelines (Eisenhardt and Graebner, 2007; Stinchfield et al., 2013), we did not share a priori sensitizing concepts with our informants. We categorized the following themes: creative solutions with diverse stakeholders, transcending the individual level to the community level, and pursuing an impact at the national level. Finally, we formulated a set of theoretical statements. The goal is that the abstract statements can be transferable and serve as a guide for investigating other samples (Auerbach and Silverstein, 2003).

FINDINGS

Examining Research Question 1: How Leaders have Demonstrated Organizational Ingenuity

Our interviewee demonstrates organizational ingenuity by viewing himself as a person who 'provides ideas' and is 'kind of a dreamer':

> *Founder and President, Paju Bank*: I provide ideas [about projects for the community bank]. I am kind of a dreamer, sometimes I come up with odd ideas and they end up working well.

Additionally, our data suggests that leaders demonstrate organizational ingenuity by pursuing a greater level of involvement in local communities, exhibiting a greater perception of social disparities, and formulating creative solutions with diverse stakeholders. We address these themes in the following sections.

Involvement in local communities

Our interviewee highlights that he had been involved with the community many years prior to founding the microfinance organization:

> *Founder and President, Paju Bank*: Since very young, I have always been worked with social and community issues. We founded the Association of Residents. I have always had, in the community, a strong relationship based on dialogue, on participation in several activities . . . to combat hunger and poverty . . . so people see me as the community bank, as part of the history of the community.

Leaders' involvement in the community, and particularly one with a long-term orientation, may stimulate organizational ingenuity by organizing the community for action and entrepreneurship. Our interviewee makes a connection between community organization and entrepreneurship:

> *Founder and President, Palmas Bank*: At that time people would not talk about entrepreneurship, but in practice it was entrepreneurship . . . An organized community and getting things done.

First, involvement in the community may promote organizational ingenuity by enabling leaders to uncover not only what the community needs, but what the community hopes for. For instance, one leader emphasizes personal contact and dialogue in a process that is 'very human':

> *Founder and President, Paju Bank*: There is this whole process of dialogue with the community . . . A community bank is very human, physical contact, . . . personal contact.

Second, involvement in the community may facilitate organizational ingenuity by enabling leaders to gain the trust of community members, which is essential to gain support for the implementation of innovative ideas. Our interviewee highlights the connection between involvement in the community and the development of trust:

> *Founder and President, Paju Bank*: Management is very . . . humanized, face to face. The Bank has gained the trust of residents in the community.

One example of organizational ingenuity resulting in an innovative project was the 'mapping of production and consumption' developed by the founder of Palmas Bank:

> *Founder and President, Palmas Bank*: 'Several initiatives here we carried out based on our proactivity. We created them. Practically without an external reference, we conduct the mapping of consumption in the community . . .

This example of organizational ingenuity was a project to better understand the reasons of poverty in the community. This project was developed in 1997 in the *Conjunto Palmeiras* neighborhood. The founder of Palmas Bank had a hypothesis that, in order to alleviate poverty, it would be essential not only to provide microfinance as a means to stimulate entrepreneurship, but also to encourage consumption within the neighborhood as a means to maintain wealth inside the community. The solution was a face-to-face survey that mapped whether residents were spending their money outside the community by visiting residents and collecting data in a somewhat informal way. This solution was created based on the leader's proactivity and 'practically without an external reference', which suggests it was a creative solution demonstrating organizational ingenuity.

Theoretical statement 1: *Leaders' involvement in their local communities will increase the likelihood of organizational ingenuity*. Involvement in local communities may play a role in facilitating the development of leaders' organizational ingenuity. Greater involvement in communities may enable leaders to develop creative solutions based on their greater knowledge of the community needs and the trust they have gained from community members.

Perception of social disparities

One interviewee describes his perception of a 'contradiction between rich and poor':

> *Founder and President, Paju Bank:* I started as a construction worker. Then, I became a door man in one of the buildings I helped build . . . Then, I gave up my professional life, because I gave up what I had to come here, taking the risk. But, in fact, financially I was living better there, but I was not a happy person, because did not like that contradiction between rich and poor. Although I was trusted, I was not an insider, I was always the one who came from the slum. And this bothered me a lot, see this contradiction.

A greater perception of social disparities may encourage the development of organizational ingenuity in leaders because it can serve as a motivation or driver to change. Indeed, our interviewee felt that social disparities 'bothered [him] a lot', to the point that he left a stable job to found a microfinance organization.

Additionally, the perception of social disparities may promote organizational ingenuity by encouraging leaders to develop solutions to alleviate poverty. For example, the question 'Why are we poor?' was a central motivation for our interviewee to start a microfinance organization.

> *Founder and President, Palmas Bank:* When we had the idea of founding the community bank, there were many questions . . . 'Why we are poor?' People would say 'we are poor because we do not have money' . . . In fact, I had the belief that the community was poor because residents would buy outside [so their wealth would not circulate or stay in the community].

The founder of Palmas Bank played a pivotal role in the creation of a system in which community banks offer their own 'social currency' in Brazil. A *social currency* is a currency issued by a community bank, which circulates only within a specific low-income community. Brazil's Central Bank requires that a community bank's currency is totally backed by Brazilian *reais*, such that for each issued social currency banknote, there is a corresponding one in Brazilian *real*. Moreover, Brazilian community banks use banknotes that are produced with security measures, such as banknote paper, watermark, bar code, and serial number to avoid falsification. The purpose of the social currency is to increase the proportion of wealth that remains inside the community.

The perception of social disparities was a central motivation for our interviewee to develop initiatives such as the creation of a 'social currency'. Consequently, a greater perception of social disparities may be connected to initiatives that demonstrate organizational ingenuity.

Theoretical statement 2: Leaders' perception of social disparities will increase the likelihood of organizational ingenuity. Organizational ingenuity may be influenced by leaders' perception of social disparities. Individuals with a greater perception of social disparities may be more likely to develop their own organizational ingenuity because they may see a greater need for a change.

Examining Research Question 2: How the Social and Institutional Environment is Transformed by Leaders' Organizational Ingenuity

Organizational ingenuity addresses the creative reformulation of one's social and institutional environment (Honig et al., 2012). While most individuals operate within existing boundaries to create solutions to current problems, in turn they may also change the institutions in which they operate. We examine how specific individuals have built organizations that have transformed the social and institutional environment where they operate.

Creative solutions with diverse stakeholders

Our interviewee describes how he developed a leading-edge public-private partnership:

> *Founder and President, Palmas Bank:* How do you get powerful institutions like Mastercard, Redecard [for-profit company that processes electronic payments], Vivo [for-profit mobile telecommunications company], and *Caixa Economica* [governmental bank], and build a relationship with the community bank, and the community bank says: 'I do not feel I am dominated by them.' I am the protagonist. I did a single, small, humble negotiation, and it was vital.

The founder of Palmas Bank demonstrates organizational ingenuity by creating a pioneering initiative that has established a 'virtual social currency', which aligns the interests of diverse stakeholders. The 'negotiation' to which our interviewee refers was his insight to ask Vivo Telecommunications to enable purchases only inside his specific community. The goal was to stimulate production and consumption within that bottom-of-the-pyramid community as a means to alleviate poverty. This initiative enables purchases via cell phone inside the community. The buyer easily transfers the payment from his account to the account of a local entrepreneur via the phone. The purpose is that the wealth of residents remains inside the community. The founder of Palmas Bank coordinates the partnership between Caixa Bank, Vivo Telecommunications, Mastercard, and Redecard with Palmas Bank to enable the virtual social currency.

The virtual social currency initiative aligns the interests of different companies (e.g., Caixa Bank, Vivo Telecommunications, Mastercard, and Redecard), who have certain benefits such as access to new customers or increased revenues. At the same time, this initiative emphasizes benefits for small entrepreneurs and low-income customers, who pay reduced fees or use a service for free. Additionally, the community bank has the benefit of reinforcing its central role by coordinating the enrollment of small businesses and low-income customers in its premises. The virtual social currency initiative represents an example of the founder's organizational ingenuity:

> *Founder and President, Palmas Bank:* This is the secret . . . I allow any company to meet with me. I negotiate on my own terms . . . We are the protagonists . . . This is a result of our capacity of organization.

Theoretical statement 3: Leaders' organizational ingenuity will increase the likelihood of solutions that creatively reconcile the interests of diverse stakeholders. A key mechanism through which leaders' organizational ingenuity influences the social and institutional environment may be the ability to envision win–win relationships with diverse stakeholders. Microfinance in a bottom-of-the-pyramid community provides a unique scenario to observe this phenomenon because it is even more complex to align the interests of for-profit organizations with the social mission of the microfinance organization.

Transcending the individual level to the community level

In his interview, the founder of Paju Bank not only highlights his personal conduct, but also his initiatives to develop a leadership team of other individuals who prioritize the community level.

> *Founder and President, Paju Bank:* If you want to be a leader, if you want to take care of your community and do your best, you have to dedicate yourself to the community. It is like priesthood. Your family is the bank, your family is the community. Your personal life is secondary. The priority is the collective good . . . This is a concern of mine because I was the founder, I am still here today, but I am not eternal. So . . . I want to form a team so that they can, if I leave tomorrow, continue the community bank's operation without my presence. Perhaps in the beginning, if this happened today, they would have more difficulty in letting go their personal life to emphasize the collective good.

Complementarily, the founder of Palmas Bank also maintains an emphasis at the community level:

> *Founder and President, Palmas Bank:* People will not overcome poverty individually! People need community action.

Leaders' organizational ingenuity may alter their social environment by understanding the knowhow of the community and mobilizing a large number of individuals for action. Our interviewee highlights that Palmas Bank has been able to develop several actions, such as vocational training, which other organizations have not been able to implement because they do not have ties with the community as strong as the microfinance organization.

> *Founder and President, Palmas Bank:* It is much easier for a woman who is a resident here to do sewing training in the community, than going to SESI [Social Service of Industry] or SEBRAE [Brazilian Service of Support for Micro and Small Enterprises] . . . because she has little empowerment, she has little direction for her life, she comes here because she leaves her child in the school, she comes, she leaves, picks up her child on her way home, because another woman from the community brings her . . .

Notably, leaders' organizational ingenuity may enable a greater impact at the community level than the government in certain situations when the government does not have close ties with community members:

> *Founder and President, Palmas Bank:* There are many things that only a community does . . . If the government wants to do it, it is too expensive, it will not happen, people will not attend it . . . Truly stable is . . . a nation that is established from the bottom up, from reorganization, from the knowhow of the community.

Consequently, Palmas Bank has been able to transform its social environment by encouraging not only individual entrepreneurship, but also collective actions such as community fairs and cooperatives:

> *Founder and President, Palmas Bank:* . . . ordinary people here, selling in the market fair, creating relationships, trying to sell together, form a cooperative together, produce together. For us this is the only way out.

Another way in which community banks have supported local businesses is by encouraging community members to buy locally and fostering the circulation of their social currency within the community (Palmas Bank, 2012). The circulation of the social currency takes place in the local market, and shoppers who use the social currency generally receive a discount from local merchants and producers to stimulate the use of the social currency in the community. Any producer or merchant registered with a community bank can exchange the social currency for Brazilian *reais* (R$) if she needs to make a purchase or payment outside the community. These measures aim at ensuring that the wealth generated by residents stays in the community (Palmas Bank, 2012) and thereby generate a positive impact on the community.

Theoretical statement 4: Leaders' organizational ingenuity will increase the likelihood of creative solutions that can have an impact at the community level. Organizational ingenuity may increase the likelihood of high impact entrepreneurship because ingenious individuals may pursue the creation of innovative solutions that in turn alter their social environment. One way to change their social and institutional environment is by pursuing emphasis at the community level.

Pursuing an impact at the national level

Our interviewee describes the impact of his microfinance organization on people's lives, which started to be featured in the media nationwide:

> *Founder and President, Paju Bank:* Here the community bank brought three main benefits. First is the economic one, because people who have access to credit improve their lives . . . [Second,] citizenship, because people have access to the service without . . . owning somebody's a favor. Imagine a woman from the slum, who never had access to a bank and today is able to have a bank account, have access to credit. For her, this is high self-esteem. Third, . . . 'change the subject' . . . sometimes entrepreneurs, people who do beautiful things . . . nobody shows them . . . Rede Globo [a major Brazilian TV channel] came here to tape a positive video showing our projects. We are also featured in *Epoca Magazine*, which circulates nationwide. We were featured in the magazine of the Ministry of Labor and Employment, which highlights the community's entrepreneurship.

Leaders' organizational ingenuity may be associated with solutions that are more innovative and impactful and therefore have a greater likelihood of being featured in initiatives with a national scope. For example, our interviewee explains that the quality and accountability of the social results accomplished by his community bank were the primary reason to gain visibility in a central event with a national scope:

> *Founder and President, Paju Bank:* We went to the closing ceremony of the United Nations Development Program, which was held here in Fortaleza, and *Caixa* [governmental bank with nationwide presence] presented their . . . social results and our project was the first. Of the projects they approved, ours was the first in terms of points, results from the last year, in terms of accountability.

A key mechanism through which leaders' organizational ingenuity may increase the likelihood of an impact at the national level is by promoting mobilization and association at the country level. For example, the founder of Palmas Bank explains the system he has been building in Brazil, which can enable several community banks in Brazil to increase their joint impact:

> *Founder and President, Palmas Bank:* I am building a democratic system of social management, in which each community bank has managerial autonomy and at the same time shares . . . the responsibility. There are 28 community banks [in Brazil] . . . [A community bank who wishes to be part of this system] has to use a software, has to track their operations . . . if a bank does not take care of its delinquency or nonpayment rate and let it increase, it will be detrimental to the 28 banks collectively . . . Today my bank is the anchor because we are the most experienced.

Palmas Bank has also coordinated the Network of Community Banks in the country. This network emphasizes the role of community banks as 'targeting the generation of employment and income from the perspective of reorganizing local economies, based on the principles of solidarity economy' (Brazilian Network of Community Banks, 2007, p. 12). Additionally, with the objective of managing knowledge and disseminating practices related to solidarity economy, Palmas Bank has developed a methodology for the creation of community banks, which has influenced the creation of many other community banks in the country.

Theoretical statement 5: Leaders' organizational ingenuity will increase the likelihood of creative solutions that can have an impact at the national level. Leaders' organizational ingenuity may play a central role in promoting an impact at the national level. A mechanism to pursue an impact at the national level may be gaining visibility nationwide. Another mechanism may be promoting mobilization and association at the country level.

DISCUSSION

We contribute to the literature on organizational ingenuity by addressing how the perspective of organizational ingenuity can be valuable to better understand the creation of innovative solutions in the context of microfinance in low-income communities. There is a growing interest in the role of leaders for the success of microfinance organizations (Galema et al., 2012). Social entrepreneurs have highlighted that the success of social enterprises (Javetski, 2009), including various types such as leveraged non-profits, hybrid not-for-profits, and hybrid for-profits (Hartigan, 2006), depends on leadership. In this study, we focus on how an analysis of founders and leaders in the context of microfinance in low-income communities can be enhanced by using the perspective of organizational ingenuity.

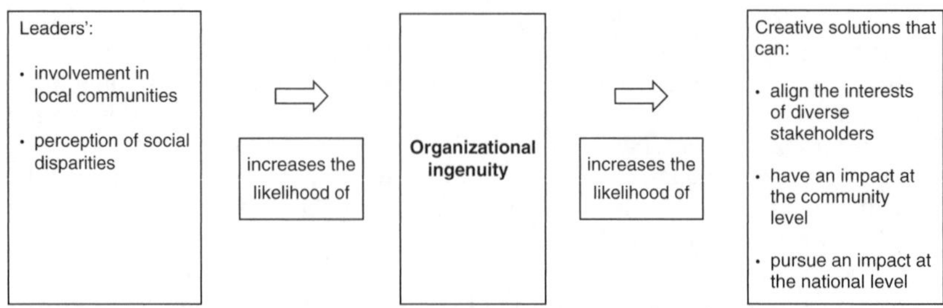

Figure 9.1 Conceptual framework: organizational ingenuity in microfinance

Theoretical Implications of this Study for Organizational Ingenuity in Microfinance

Based on our findings, we develop a conceptual framework for organizational ingenuity in microfinance, shown in Figure 9.1.

According to Figure 9.1, the first implication is that leaders' involvement in local communities and their perception of social disparities are both elements that can increase the likelihood of organizational ingenuity in microfinance. This observation is shown in Theoretical Statements 1 and 2. These insights add to the literature on leaders in microfinance (e.g., Galema et al., 2012). These statements provide us with a better understanding of what elements are helpful for microfinance leaders to be able to develop organizational ingenuity.

The second implication is that leaders' organizational ingenuity can increase the likelihood of creative solutions that reconcile the interests of diverse stakeholders, have an impact at the community level, and have an impact at the national level. This insight stems from Theoretical Statements 3, 4, and 5. These observations extend the literature on institutional change in the context of microfinance (e.g., Khavul et al., 2013). These statements suggest how organizational ingenuity can enhance the outreach and survival of microfinance lending firms. Organizational ingenuity allows these entities to make the most efficient use of their resources through creative means to achieve the greatest amount of outreach possible. This outreach is necessary in order to meet the interests of diverse stakeholder groups at the community level or nationally. Using organizational ingenuity as a means for stakeholder outreach also helps to drive the survival of microfinance leaders and their firms by allowing them to optimally address the large size of the microfinance market. Overall, these implications show us first what increases the likelihood of organizational ingenuity, and then how organizational ingenuity increases the likelihood of creative solutions for stakeholder interests at both community and national levels.

Practical Implications of Organizational Ingenuity in Microfinance

Our findings generate practical implications for the literature on organizational ingenuity. First, our study suggests that organizational ingenuity has a greater likelihood to emerge when leaders pursue a greater level of involvement in local communities and exhibit a greater perception of social disparities. The leaders in our case studies have pursued a long-term involvement in their communities in order to understand the needs of two bottom-of-the-pyramid communities in Brazil. These leaders were able to develop creative solutions such as the 'mapping of production and consumption', the issuing of a 'social currency' that circulates only inside the community, the provision of microfinance to individuals and formal or informal businesses, the promotion of local entrepreneurship, and the coordination of diverse stakeholders including for-profit organizations to achieve goals aligned with the community bank's social mission. Thus, microfinance leaders may benefit from their involvement in local communities and perception of social disparities to enhance their organizational ingenuity.

Second, this study highlights how microfinance leaders have coordinated and transformed their social and institutional environment by using organizational ingenuity. The area of microfinance has faced institutional changes including a dominant development logic, a shift to market logic, and a conflict over regulatory logic, with entrepreneurs often serving as 'change agents' (Khavul et al., 2012). For instance, Palmas Bank has developed innovations such as a methodology to create new community banks in Brazil, which has influenced the launching of many other community banks in Brazil. Consequently, Palmas Bank has not only developed its local community, but also inspired microfinance initiatives in several other communities in the country.

Our case studies indicate that Palmas Bank has become a reference and main coordinator of the Brazilian community bank model, adopted as a public policy initiative by the Brazilian government. Palmas Bank was the first community bank of Brazil and its first goal was to organize and encourage its 30 000 inhabitants to produce and consume in their own neighborhood. With the expansion of initiatives to create community banks in Brazil, Palmas Bank started to provide a community bank certification. Palmas Bank is responsible for 'certifying the creation of a new community bank and communicating to Brazil's Central Bank the creation of a new social currency' (Melo Neto Segundo and Magalhaes,

2009, p. 25). Although other participating organizations in the Brazilian Network of Community Banks also provide training and assistance to community banks in different states, Palmas Bank is an exemplar case of creative reformulation of social and institutional environments through organizational ingenuity. Therefore, microfinance leaders may increase the likelihood of creative solutions for stakeholder interests at both community and national levels by developing their organizational ingenuity.

Future Directions

Our findings extend the literature on organizational ingenuity by addressing the creative reformulation of institutional constraints through the impact of the community banks on their social and institutional environment. Organizational ingenuity transcends other perspectives that deal with entrepreneurs' creativity in resource-poor environments such as the perspective of bricolage (Baker and Nelson, 2005) by addressing to a greater extent how some entrepreneurs and their organizations can alter the institutions in which they operate. When a social entrepreneur's resources can be so scarce and limited, bricolage may help entrepreneurs determine the most optimal solution for the utilization of these resources. Yet organizational ingenuity goes beyond the scope of the bricolage perspective by addressing how entrepreneurs reformulate their social and institutional environments. Our findings indicate that organizational ingenuity increases the likelihood that the actions of microfinance entrepreneurs will reconcile the interests of diverse stakeholders and have a greater impact at the community and national levels.

Limitations of this study represent opportunities for future research. Our study focuses on two microfinance organizations in Brazil. New studies should examine other microfinance organizations in different countries to assess the influence of diverse environments. Future research should also investigate other leaders of microfinance organizations to explore how the effectiveness of microfinance can be facilitated by organizational ingenuity. Our study suggests that leaders' involvement in local communities and perception of social disparities can increase the likelihood of organizational ingenuity, and that leaders' organizational ingenuity can increase the likelihood of creative solutions with an impact at the community and national levels. This study suggests that our understanding of the context of microfinance can be enhanced by using the perspective of organizational ingenuity.

REFERENCES

Auerbach, C.F. and L.B. Silverstein (2003), *Qualitative Data: An Introduction to Coding and Analysis*, New York: New York University Press.

BACEN (Central Bank of Brazil) (2012), 'FAQ – Contas (abertura, encerramento, e bloqueio)' ['FAQ – Checking accounts (opening, closing, and blocking)']. Available at: http://www.bcb.gov.br/pre/bc_atende/port/servicos1.asp#2 (last accessed 26 September 2013).

Baker, T. and R.E. Nelson (2005), 'Creating something from nothing: resource construction through entrepreneurial bricolage', *Administrative Science Quarterly*, **50**(3), 329–366.

Brazilian Network of Community Banks [Rede Brasileira de Bancos Comunitarios] (2007), *Memoria – II Encontro da Rede Brasileira de Bancos Comunitarios* [*Memory – 2nd meeting of the Brazilian Network of Community Banks*]. Caucaia, CE, Brazil: Sage Publications, available at: http://www.inovacaoparainclusao.com/uploads/4/2/2/8/4228830/relatrio_do_ii_encontro_rede_de_bancos-1.pdf (last accessed 26 September 2013).

Bruton, G.D., S. Khavul and H. Chavez (2011), 'Microlending in emerging economies: building a new line of inquiry from the ground up', *Journal of International Business Studies*, **42**, 718–39.

Corbin, J. and A. Strauss (2008), *Basics of Qualitative Research: Techniques and Procedures for Developing Grounded Theory*, Thousand Oaks, CA: Sage Publications.

Eisenhardt, K.M. (1989), 'Building theories from case study research', *Academy of Management Review*, **14**(4), 532–50.

Eisenhardt, K.M. and M.E. Graebner (2007), 'Theory building from cases: opportunities and challenges', *Academy of Management Journal*, **50**(1), 25–32.

Ferdig, M.A. (2007), 'Sustainability leadership: co-creating a sustainable future', *Journal of Change Management*, **7**(1), 25–35.

Franca Filho, G.C. (2008), 'Consideracoes sobre um marco teorico-analitico para a experiencia dos bancos comunitarios' ['Considerations about a theoretical and analytical milestone for the experience of community banks']. In J.T. Silva Jr, R.T. Masih, A.C. Cancado, and P.C. Schommer (eds), *Gestao social: Praticas em debate, teorias em construcao* (pp. 113–24). Juazeiro do Norte, Brazil: Universidade Federal do Ceara. Available at: http://gestaosocial.paginas.ufsc.br/files/2011/07/Livro-1-Cole%C3%A7%C3%A3oEnapegsV1_Gest%C3%A3oSocialPraticasDebatesTeoriasConstru%C3%A7%C3%A3o.pdf (last accessed 26 September 2013).

Galema, R., R. Lensink and R. Mersland (2012), 'Do powerful CEOs determine microfinance performance?', *Journal of Management Studies*, **49**(4), 718–42.

Glaser, B.G. and A.L. Strauss (1967), *The Discovery of Grounded Theory: Strategies for Qualitative Research*, New York: Aldine de Gruyter.

Hartigan, P. (2006). 'It's about people, not profits', *Business Strategy Review*, **17**(4), 42–5.

Honig, B., J. Lampel and I. Drori (2012), 'Each in its own way: national culture and organizational ingenuity', Professional development workshop at the meeting of the Academy of Management, Boston, MA.

International Labour Organization (1993), *Fifteenth International Conference of the Labour Statisticians: A Report of the Conference*. Geneva: International Labour Organization.

Javetski, B. (2009), 'Developing entrepreneurs among the world's poorest: an interview with Acumen Fund's founder', *McKinsey Quarterly*, (3), 136–42.

Khavul, S., G.D. Bruton and E. Wood (2009), 'Informal family business in Africa', *Entrepreneurship Theory and Practice*, **33**(6), 1217–36.

Khavul, S., H. Chavez and G.D. Bruton (2013), 'When institutional change outruns the change agent: the contested terrain of entrepreneurial microfinance for those in poverty', *Journal of Business Venturing*, **18**(1), 30–50.

Kuratko, D.F. (2005), 'The emergence of entrepreneurship education: development, trends, and challenges', *Entrepreneurship Theory and Practice*, **29**(5), 577–98.

Lampel, J., B. Honig and I. Drori (2011), 'Discovering creativity in necessity: organizational ingenuity under institutional constraints', Call for papers for the Special Issue on 'Discovering Creativity in Necessity: Organizational Ingenuity under Institutional Constraints', *Organization Studies.*, **32**(5), 715–17.

Maak, T. and N. Stoetter (2012), 'Social entrepreneurs as responsible leaders: Fundacion Paraguaya and the case of Martin Burt', *Journal of Business Ethics*, **111**(3), 413–30.

Melo Neto Segundo, J.J. and S. Magalhaes (2009), 'Bancos comunitarios' ['Community banks']. *Mercado de Trabalho*, **41**, 21–6, available at: http://www.ipea.gov.br/sites/000/2/boletim_mercado_de_trabalho/mt41/10_Eco_Bancos.pdf (last accessed 26 September 2013).

Merriam, S.B. (2002), *Qualitative Research in Practice: Examples for Discussion and Analysis*, San Francisco: Jossey-Bass.

Murphy, P.J. and S. Coombes (2009), 'A model of social entrepreneurial discovery', *Journal of Business Ethics*, **87**(3), 325–36.

Nielsen, C. and P.M. Samia (2008), 'Understanding key factors in social enterprise development of the BOP: a systems approach applied to case studies in the Philippines', *Journal of Consumer Marketing*, **25**(7), 446–54.

Palmas Bank [*Banco Palmas*] (2012), 'Rede de bancos comunitarios' ['Network of community banks'], available at: http://www.bancopalmas.org.br/oktiva.net/1235/secao/9963 (last accessed 26 September 2013).

Patton, M.Q. (2002), *Qualitative Research and Evaluation Methods*, Thousand Oaks, CA: Sage Publications.

Pless, N.M. (2007), 'Understanding responsible leadership: role identity and motivational drivers', *Journal of Business Ethics*, **74**(4), 437–56.

Prahalad, C.K. and A. Hammond (2002), 'Serving the world's poor, profitably', *Harvard Business Review*, **80**(9), 48–58.

Seelos, C. and J. Mair (2005), 'Social entrepreneurship: creating new business models to serve the poor', *Business Horizons*, **48**, 241–6.

Stake, R.E. (1995), *The Art of Case Study Research*, London: Sage.

Stinchfield, B.T., R.E. Nelson and M.S. Wood (2013), 'Learning from Levi-Strauss' legacy: art, craft, engineering, bricolage, and brokerage in entrepreneurship', *Entrepreneurship Theory and Practice*, **37**(4), 889–921.

Toscano, I. (2004), 'Bancos populares de desenvolvimento solidario' ['Popular banks for solidarity development']. Instituto Polis [Polis Institute], available at: http://www.saopaulo.org.br/download/27.pdf (last accessed 26 September 2013).

Trivinos, A.N.S. (1987), *Introducao a pesquisa em ciencias sociais: pesquisa qualitativa em educacao* [*Introduction to Research on Social Sciences: Qualitative Research in Education*], Sao Paulo, SP: Atlas.

Ushizima, L.Y. (2008), 'Acesso ao sistema financeiro: alternativas a exclusao – o caso dos bancos populares' ['Access to the financial system: alternatives to the exclusion – the case of popular banks']. Forum Brasileiro de Economia Solidaria, available at: http://www.fbes.org.br/index2.php?option=com_docman&task=doc_view&gid=775&Itemid=1 (last accessed 26 September 2013).

Wood, V.R., D.A. Pitta and F.J. Franzak (2008), 'Successful marketing by multinational firms to the bottom of the pyramid: connecting share of heart, global "umbrella brands", and responsible marketing', *Journal of Consumer Marketing*, **25**(7), 419–29.

Yin, R.K. (2005), *Estudo de Caso* [*Case Study*], Sao Paulo, SP: Bookman.

10 Acting ingeniously: opportunity development through institutional work

A.M.C. Eveline Stam, Ingrid A.M. Wakkee and Peter Groenewegen

INTRODUCTION

This chapter proposes an integrated perspective on institutional work as ingenious actions of entrepreneurs for the purpose of opportunity development. Ingenuity refers to actions undertaken by entrepreneurs who work within 'the institutional setting and uses this embeddedness by co-opting, coordinating, and whenever possible, drawing on organizational slack to effect change' (Lampel et al., 2011: 715). Ingenuity involves bringing together thinking and acting, individually and collectively, to take advantage of opportunities or to overcome problems. As such, ingenuity denotes the space between a challenge and a solution (Homer-Dixon, 2000). We present day-to-day actions of an ingenious entrepreneur who combines opportunity development and institutional work. The concept of opportunity development is central in the study of entrepreneurship (Shane and Venkataraman, 2000). Recognizing opportunities, entrepreneurs develop initial ideas of new combinations of products, processes, markets and organizing to be offered to the market. Opportunity development requires the dynamic and interactive process to prepare a resource base and create an organizational structure for opportunity exploitation, and interactions with a diverse set of stakeholders in the institutional environment and understand their conflicting needs and motivations and to perceive possibilities resulting from these (Slotte-Kock and Coviello, 2010).

As much as agency, interest and rent-seeking behavior and defying from structure come naturally to entrepreneurship studies (e.g. Alvarez and Barney, 2004, 2005), in institutional theory the argumentum for institutions to come about and be maintained is due to actors' costs associated with moving away from taken-for-granted patterns of practice, technologies and rules (DiMaggio and Powell, 1983; Scott, 2001). To explain institutional change and the role of actors herein, the concept of 'institutional entrepreneurship' came about focusing on 'activities of actors who have interest in particular institutional arrangements and who leverage resources to create new institutions or to transform existing ones' (Maguire et al., 2004: 657; Battilana et al., 2009). Institutional entrepreneurs create or change institutions by engaging in 'projective agency' (Dorado, 2005). They focus on offering future visions or remedies to resolve current organizational or social problems on the societal level. Institutional entrepreneurship studies typically focused on the influence of single actors on the change of institutions and consequently were criticized for neglecting the ongoing and day-to-day practices of embedded actors. The concept of institutional work tried to fill this void and extend the range of individuals in organizational fields to 'the purposive action of individuals and organizations, aimed at creating, maintaining and disrupting institutions' (Lawrence and Suddaby, 2006: 215; Lawrence et al., 2009).

Engaging with the environment is recognized as being critical for the entrepreneur in both research areas but the questions of how institutions serve as a source of opportunity and how entrepreneurs engage institutional forces in shaping opportunity structures remain somewhat neglected. Efforts of entrepreneurs to recognize and open up 'locked-in' opportunities in the institutional setting remain relatively unaccounted for. Rather, research on opportunity development invokes mostly the tweaking and adapting of the front end of opportunity development and idea generation to fit market conditions. Thus entrepreneurs seemingly take institutional settings for granted or as a settled condition, even if innovative entrepreneurship frequently evokes the image of the creative destruction of existing markets. However, how entrepreneurs seek to destroy the status quo to allow for the development of opportunities is scarcely addressed. This study therefore seeks answers to the following research questions: How does opportunity development relate to institutional work? Why, and in what way, can institutional work be both a strategy and a set of tactics of ingenious entrepreneurs?

By examining action patterns of an ingenious entrepreneur in the changing Dutch homecare field, this chapter contributes to the integration of entrepreneurship and institutional work scholarships in three ways. First, we describe which entrepreneurial actions can be interpreted as institutional work and seek to understand how these actions are aimed at influencing institutional processes. Specifically, we seek to understand which social skills and set of tactics form the base of entrepreneurial institutional work. Thus we show that institutional constraints are not static and that understanding the underlying dynamics helps entrepreneurs to influence institutional processes. Second, we extend and modify the view on opportunity development by showing how ingenious entrepreneurial engagement in institutional processes simultaneously creates latitude in opportunity structures for new ventures. Third, we contribute to the insight in institutional work regarding both delegitimating or destructional work and legitimating or creational work of entrepreneurs in densely interconnected organizational fields. Institutionalized fields produce administered markets with ancillary organizational arrangements, interdependent resource relations, and increased dependencies between organizations and controlling agencies (Scott et al., 2000). These controlling agencies, accrediting bodies, regulatory organizations and governance associations are 'legitimating organizations' established to maintain institutional arrangements by conferring legitimacy on other social actors and establish mechanisms of compliance and membership (Lawrence, 2004) and form additional barriers to field structuring mechanisms (Trank and Washington, 2009). Consequently, entrepreneurs not only compete with cognitive norms but also with those organizations purposely designed to actively preserve these.

Viewing the institutional setting from an opportunity perspective fits with increasing attention to dynamic aspects of organizational fields as witnessed by an increase of studies on how institutions change over time (Dacin et al., 2002). For this, we elaborate on the concept of 'entrepreneurial ingenuity': the combined mental orientations and skills that entrepreneurs utilize to achieve innovative solutions in organizational fields, and at the same time engage in entrepreneurial activities to exploit existing opportunity structures. We demonstrate that, whereas entrepreneurship has been described as requiring primarily an individual capability or skill, institutional work is considered the combination of repetitive actions and small changes; the combination pictures day-to-day processes. With daily effort entrepreneurs influence the destruction of institutional status quo and the creation of windows of opportunity towards relevant resource base alignment allowing for start-up and the implementation of innovations in complex and densely regulated environments. It requires a degree of ingenious entrepreneurial mental orientation and subsequent actions to understand the interconnected institutional processes in an organizational field characterized by different time horizons and spaces, and opportunity recognition and development. Researchers increasingly focus on organizational actors' 'practical evaluative agency' in relation to institutional work (e.g. Battilana and D'Aunno, 2009). Even institutional work and the quest over legitimacy on the intra-organizational level is documented (e.g. Drori and Honig, 2013). More insight is nevertheless necessary on how and why, under conditions of resource constraint, independent entrepreneurs, as peripheral and marginalized field actors, ingeniously engage simultaneously into both destructional and creational work.

Our analysis is grounded in the case of a nurse who creates and organizes activism towards the field for the introduction of a care cooperative of freelance homecare nurses. His initiatives have a recognizable effectuating logic in that he does not begin with a predetermined exogenous opportunity of a predicted future goal but with a given set of means and fluid future ends (Sarasvathy, 2001; Sarasvathy and Dew, 2005). Being frustrated about the institutional barriers for operating as a freelance homecare nurse, and lacking the resources to break these barriers he starts blogging and is surprised by how he manages to create a support network of other aspiring freelance homecare nurses. The network serves both as means and an end to exploit the opportunity recognized from it: selling his knowledge on how to start as an independent homecare nurse. His blog forms the base of his institutional work to gain support from field and political and policy actors. As he becomes a knowledgeable expert, he is sought out as a strategic partner by other field actors to build other health care ventures and activism coalitions. These new means render the opportunity of the care cooperative for freelance nurses. Through effectuating additional coalitions he exercises normative influence towards regulatory changes that are necessary for cooperatives to be eligible within reimbursement structures and to allow this previously inconceivably new organizational form to enter Dutch health care.

The remainder of this chapter discusses the concepts employed, the methods of our embedded case design and the results of our findings. We finish with the conclusions drawn from the analysis, and the discussion of these findings as well as the limitations of our research.

OPPORTUNITY DEVELOPMENT AND INSTITUTIONAL WORK

Institutional work suggests that even maintaining institutions requires effort (Lawrence and Suddaby, 2006). Efforts vary in both degree and kind however, and therefore a range of actions can function as institutional work. Three main categories of institutional work represent these efforts: the development of institutional political activities; the creation of new identities with supporting norms and relations; and efforts propagating new institutional conditions (Lawrence and Suddaby, 2006). Institutional work provides a natural fit with entrepreneurial action as advocated by Greve (2001) and McMullen and Shepherd (2006). Their pragmatic approach encourages the study of entrepreneurial action by distinguishing between an entrepreneur's intentions (the creative ideas) as they form, and whether subsequent entrepreneurial behavior (the implementation of creative ideas) occurred. In this way, analogies between the perspectives on entrepreneurial action and opportunity development on the one hand, and institutional work on the other, suggest individuals acting with intentionality and displaying actions accordingly.

Organizational Fields and Institutional Work

We situate entrepreneurial actions and institutional work at the organizational field level. An organizational field is 'a community of organizations that partakes of a common meaning system and, whose participants interact more frequently and fatefully with one another than with actors outside the field' (Scott, 2001: 56). The field includes actors such as government agencies, critical exchange partners, professional and trade associations, special interest groups and external audiences (Fligstein and McAdam, 2012: 3). DiMaggio and Powell (1991) assert that an organizational field consists of four critical elements: (1) network structures from interaction among field organizations; (2) principles ordering field actors according to sharp interorganizational patterns of domination and coalition; (3) a culture and practice shared by field members; (4) a shared identity of organization members. Research on the dynamics of organizational fields suggests that actors whose interests are not served by existing institutional arrangements will seek to disrupt the extant set of institutions (DiMaggio, 1991). Also Bourdieu (1993) argues that the differential allocation of capital based on institutional structures induces conflict in organizational fields. Actors compete to gain privileged positions or disrupt institutions restricting their access to capital. In Abbott's (1988) analysis of professional fields inter-jurisdictional conflicts in which professional groups compete for the right to engage in particular forms of activity are highlighted. Thus organizational fields form arenas for action and pacification as well as unstable backdrops against which actions and effort take place.

Because of their relation to entrepreneurial processes actions, we focus on two types of institutional work: disruptive institutional work and creational work. Disruptive institutional work involves attacking or undermining the mechanisms that lead members to comply with institutions (Lawrence and Suddaby, 2006). Actors disrupt institutions primarily by redefining, recategorizing, reconfiguring, abstracting, problematizing and, generally, manipulating the social and symbolic boundaries that constitute institutions. Lawrence et al. (2009) further divided disruptive institutional work into three categories. The first is often judiciary in character and consists of the state-apparatus disconnecting sanctions and/or rewards. The second category is destructive work by means of disassociating the practice, rule or technology from its moral foundation as appropriate within a specific cultural context. The third category of destructive work covers effort undermining core assumptions and beliefs. Moving away from institutions and differing from taken-for-granted patterns through introducing innovations, technologies and/or new practices comes with risk and costs (Scott, 2001). Moving away from existing templates becomes cost-effective when underlying beliefs and assumptions of old templates are dismantled. For Lawrence and Suddaby (2006), two types of work emerged which undermine core assumptions and beliefs. These are through innovations that break existing, institutional assumptions and the more gradual undermining of current core assumptions and beliefs through contrary practice.

Gradual undermining of institutions by organizations is often accompanied by the alternating between coupling and decoupling (Meyer and Rowan, 1977). In essence, through close coupling organizations create and maintain direct congruence between their formal structure and their actual task accomplishment and productivity, which ensure external recognisability, and actively display intentions of good faith (Bromley and Powell, 2012). On the contrary, strategic *de*coupling of organizations, even labeled institutional 'dirty work' when politically controversial means and goals are involved (Hirsch and Bermiss, 2009), consists of the intentional creation of electorally preferred frames often developed from existing financial and cultural resources enabling legitimized actions of powerful political actors. Strategic decoupling thus creates institutional buffers ensuring – relative – stability permitting the maintenance of influence structures. Decoupling can also be developed as a strategic response to institutional processes and pressure, when organizations engage in concealment; the disguising of nonconformity of their activities; the reduction of external inspection and evaluation of activities (Oliver, 1991). Although coupling and decoupling are concepts often bridging across the conceived categories of institutional work of creating, maintaining and disrupting institutions (Lawrence et al., 2009: 23) it ultimately may contribute to gradual institutional changes.

On the other side, actions that actors initiate in order to create new institutional arrangements can be categorized in three broader types (Lawrence and Suddaby, 2006). The first, creational institutional work involves the reconstructing of rules, property rights and boundaries that define access to material resources. It involves the practices and tactics of 'vesting', 'defining' and 'advocacy' in which actors actively engage. The second type of creational work takes on the shape of the reconfiguration of belief systems. It consists of practices emphasizing 'constructing identities', 'changing norms/normative association' and 'constructing normative networks'. The final type of creational institutional work focuses on the alteration of abstract categorizations in which meaning systems are rooted and consists of activities entailing ‘mimicry’, ‘theorizing’, and ‘educating’ (Lawrence and Suddaby, 2006).

Independent actors, such as entrepreneurs, however typically have no back up of a formal organizational structure, are short of means, and operate outside formal organizational structures for endowing legitimacy. Thus they cannot fall back on organizational power or buffering to decouple strategically. *How* independent entrepreneurs, under full-blown resource constraints and with marginalized field positions, ingeniously engage with and perform institutional work, is thus an open question. Thus institutional maintenance and change lead and work aimed at disrupting institutions is potentially more effective when it reduces or even removes those costs by facilitating replacement templates, or decreases the perceived risks of innovation and differentiation (Lawrence and Suddaby, 2006).

Understanding this requires us to empirically explore and understand how and why entrepreneurs concomitantly put effort into institutional work and how this maps onto the three categories of institutional work just discerned. For this we need to understand how ingenious practices influence the construction of windows of opportunity in the institutional setting permitting resource structure alignment supporting future entrepreneurial opportunity. In our view, we need to understand what it is and how entrepreneurial ingenuity enables us to demonstrate how institutional work takes shape and leads to the undermining of current institutional arrangements and the reshaping of the organizational field to favor resource structure alignment in relation to opportunity development. These actions are imbedded in day-to-day practices and actions of entrepreneurs and aligned to the process of venturing and requires the use of social skills as described by Fligstein (1997). These include the ability to induce cooperation in others, to exert authority, frame arguments, act creatively in bringing together unusual components; and, engage in bargaining and brokering based on mental orientations and accompanying social tactics.

Similarly, Ansell (1997) calls upon an entrepreneur’s cultural skills which involve the ability to frame the envisioned institutional setting in a way that is appealing to a wider audience via the building of specific normative attitudes and the creation of common identities. Additionally the literature suggests that ingenious entrepreneurs require the ability to develop abstract models and representations of the institutional setting, i.e. they must possess strong analytical skills (Strang and Meyer, 1993). Such analytical skills are closely connected to the mental orientations entrepreneurs possess. Our focus is to understand the relevance of ingenuity as a set of skills and orientations on how starting entrepreneurs act on future projections and beliefs, while facing resource constraints, coercive institutional barriers and negotiating with institutional preserving actors. Thus, how ingenuity contributes to the generation of social acceptance through legitimacy attribution of actors towards both the introduction of creative ideas and implementation compliance of other institutionalized actors is accomplished is our foremost objective.

RESEARCH DESIGN

Our empirical research consists of an embedded case study based on the gathering and use of multiple sources of data and analysis techniques (Yin, 1994). We longitudinally map the behavior of a health care entrepreneur over the period January 2008 (start of his weblog) until June 2011. Legislative changes marked the end when the Covenant for Care Cooperatives, aiming to facilitate such cooperatives under the reimbursement of the 'Exceptional Medical Expenses Act' (AWBZ), was signed in June 2011. In our investigation we explore both field level and venture level actions undertaken by this entrepreneur in relation to opportunity development and connect these to critical events occurring at the field level.

Site and Case Study Selection

The case is set in the Dutch health care context, which is characterized by a number of issues such as an aging population, limited financial resources and rapid legislative changes. In the Netherlands, the AWBZ provides funding for homecare for the elderly, physically challenged or those otherwise incapacitated. Under this law patients may receive a personal care budget (PGB). The PGB permits care receivers to purchase care services from different care providers. Patients in need of homecare may benefit from using the services of unincorporated self-employed nurses or flexible workers as they provide more flexible options compared to the services from an established homecare organization, thus potentially improving care quality, affordability and accessibility. The institutional positions differ as homecare organizations pay freelance nurses similar wages as tenured workers, thus incurring fewer costs because freelance nurses need to pay insurance, pension and social security fees from their income. This results in a lower net salary for a freelance nurse compared to tenured workers. This disparity is a key motivator to change legislation on the tax and social security regulation for freelance nurses. Additionally the tax code and quality of care regulation prevents freelance nurses from organizing themselves collectively.

The case was purposely selected because of the longitudinal data in the weblog, for the relevancy of both the entrepreneur's actions and the issue of freelance nurses in Dutch health care. The entrepreneur was aged 34 when he started his blog in 2008. He was educated as a hospital nurse between 1996 and 2000. After having worked in hospitals and care institutions from 2000 to 2005, he went on to work as a nurse in the offshore industry (2005–10). His work schedule and the seclusion in the offshore industry allowed him the time necessary to reflect on a more entrepreneurial career. He felt he did not devote enough time and attention to his patients in the hospital context and his job no longer met his personal interests. At the same time, he felt challenged by the opportunities arising from the structural changes in the health care market. He envisioned the prospect of working as a freelance entrepreneurial nurse in the Dutch homecare system and recognized institutional constraints to this personal goal.

Data Sources and Data Collection

The principal source of information consists of the entrepreneur's weblog over the period March 2008–July 2011. Weblogs, or 'blogs', refer to websites that contain a series of frequently updated posts usually written by a single author (Bar-Ilan, 2005; Hookway, 2008). Blogs offer substantial benefits for social scientific research by providing a publicly available, low cost and instantaneous technique for collecting substantial amounts of data (Hookway, 2008). The archived nature of blogs and the textual form of data makes blogs open to examine social processes over time. Altogether, using blogs as a data source helps overcome issues of accessing unsolicited diaries of entrepreneurial actions, while they are not 'contaminated' by the particular interest and possibly biased search for data by a researcher (Hookway, 2008). This particular weblog was launched early in 2008 and, with a short interlude in 2009, continued to be updated until and beyond June 2011. It includes 215 entries, many of which include references and hyperlinks to policy documents, newspaper articles and other materials which are also incorporated in the study. A semi-structured interview with the entrepreneur focused on his life story and networking behavior. The interview, which was taped and later transcribed, lasted 2 hours and 15 minutes. Two interviews with the entrepreneur published in health care expert magazines were also used. The blogs and interviews were complemented with a substantial amount of secondary materials obtained from position papers, policy archives, white papers written by the entrepreneur, discussion forums, newspapers, and professional health care magazines.

Data Preparation, Analysis and Interpretation

The preparation and analysis of the data was conducted in several phases and steps. The first phase concerned the creation of a story line. First and pertaining to the 215 blog posts, the first author manually and inductively reduced the data by identifying 99 issues regarding the venture and the field. The issues were supported by relevant quotes from the entrepreneur's blog and include activities undertaken by the entrepreneur as well as his observations about the field and field level events. The entrepreneur was the only author of the blog, thus the issues are a reflection of his views and constitute a selection of actual events suiting his personal needs and goals. Therefore, to ensure the trustworthiness and accuracy of these issues, the first author cross-referenced these issues with data obtained from other sources. Next, the second author re-examined the resulting data portfolio for consistency. After that, and following mutual discussion between all three authors these 99 issues were aggregated and grouped into eight inductively emerging field level actions and nine inductively emerging venture level actions.

To counter potential bias and to distance ourselves from a 'myopic view' on the success of the intentionality of actors in changing institutions (DiMaggio, 1988; Aldrich, 2010) and to abstain from attributing our entrepreneur with 'supernatural social skills', we framed the entrepreneur's efforts in relation to two other processes. In the first place we describe critical events in the Dutch homecare field occurring in conjunction with the entrepreneur's actions and practices and relate these where possible. Included are policy and political changes and those initiated by other key institutional stakeholders, such as health insurance companies and professional associations. These critical field events are included into the framework of Figure 10.1 at the top and serve as a background against which the story line unrolls.

Second, interviews were conducted with the entrepreneur and two field expert actors. The first was an influential Member of the House of Representatives (MP 2007–2010 and 2012–present) who initiated a freelance workers union in the Netherlands in 2005. The second field expert was a senior consultant in the Dutch health care sector who, though ultimately unsuccessful, tried to introduce a care cooperative, approximately eight years earlier in 1994 when being employed as a homecare manager. By comparing his experiences to that of the entrepreneur we explore how current field conditions may impede successful implementation of a new organizational form at one moment in time while allowing for it several years later. These interviews were taped and lasted 1 hour and 1 hour 45 minutes respectively and were aimed at identifying and cross-checking relevant field level critical incidents and understanding the difference in the field forces for the acceptance of the care cooperative.

To develop insight into the extent, nature and relation of the entrepreneurial actions and the institutional work performed by the entrepreneur, we analyzed our data repeatedly, discussing the emerging picture jointly. We use Lawrence and Suddaby's (2006: 221) categorization as discussed in the conceptual background as a frame for the different types and roles of institutional work the entrepreneur conducted. To understand the entrepreneur's role in terms of entrepreneurial action and opportunity development, we focus on his practices targeted at *disrupting* the current inefficient institutional arrangement in tax and social security legislation for freelance nurses and his practices involving creation of new institutions in relation to: (1) identity building of 'entrepreneurial nurses'; (2) a more efficient tax and social security arrangement fitting their needs; and (3) the freelance nurse cooperative in a way that fits his intended opportunity. We search for patterns based on the actor's motivation and logic of action developing over time within each of the field and venture level activities separately and across the entire timeline, thus connecting the activities to the critical field events as well.

RESULTS

Case Descriptive: Field and Venture Level Actions

Field level actions refer those efforts initiated by the entrepreneur to influence or alter the institutional setting. Included are responses to the actions of other players in the field and self-initiated actions undertaken with the intent to benefit future entrepreneurial activities. Venture level actions refer to those efforts the entrepreneur initiates in direct relation to his current business activities and to generate more immediate revenue streams. Each of these efforts, which differed in duration, was then mapped on the case study timeline, which is presented in Figure 10.1.

Whereas the distinction between field level and venture level actions was derived from theory, the entrepreneur recognizes and also distinguishes between these. Regarding his current business he says:

> With our company we try to achieve innovative solutions for safety and quality as well as providing and placing medical service professionals in technical industrial settings. We work with specialized nurses and our company is growing and offering extra services such as safety courses and medical care delivery courses for industrial employees (B/11/10/2009).

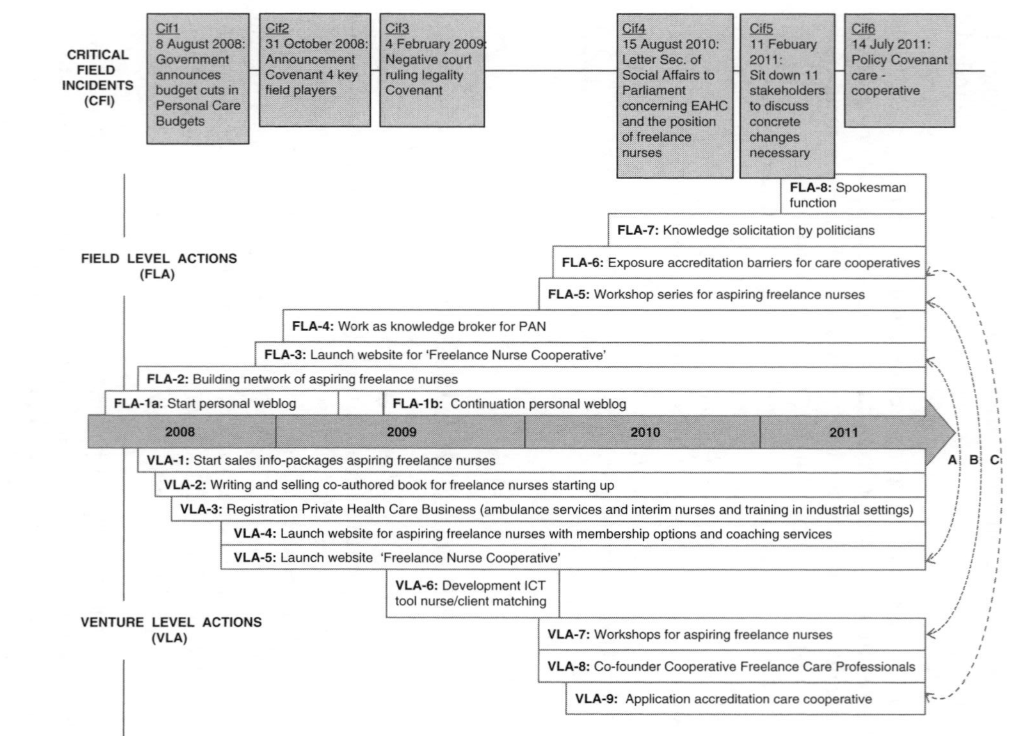

Figure 10.1 Entrepreneurial actions at the Field Level (FLA) and at the Venture Level (VLA), and Critical Field Incidents (CFI), 2008–11

Yet that is not the only thing he does, he states in a summer 2010 blog:

> . . . I have focused myself on understanding and organizing the overarching interests of freelance health care nurses. This has taken on the form of a broad network that I use for a higher goal: to anchor the position of freelance nurses in Dutch health care. (. . .) I think it is a highly relevant issue for our current society where health care staff in general and domiciliary staff in particular needs to be addressed (. . .) (B 25/3/2008).

It is important to notice that the entrepreneur undertakes his field level work voluntarily and compares himself and some colleagues interested in cooperatives – to 'white raven', 'just some people that, without any direct economic benefit, nor the guarantee that we ever will, spend much time and effort to organize ourselves'.

From here we discuss the key events from Figure 10.1 starting with the *Field Level Actions* (hereafter FLA) 1–9 followed by the *Venture Level Actions* (hereafter VLA) 1–9 and analyze these in terms of the entrepreneurs' action, their institutional work characteristics and relation to opportunity development. Table 10.1 illustrates these key events with the most representative quotes we distilled from our data. Both the field level and venture level activities can directly or indirectly be linked to the key events presented in Figure 10.1: the acceptance of the political covenant of July 2011 is the endpoint.

Attending to Field Level Opportunity Development

As of March 2008, the entrepreneur started to invest time in researching entrepreneurship and opportunity identification for freelance nurses in health care in a broad sense. He pushed himself to understand what being a freelance nurse entailed in the light of the market development, the shortage of health care workers, and the need for flexibility in labor relations in the Dutch health care sector. First he noted that freelance and flexible workers are currently locked out of working in the AWBZ sector, which accounts for approximately 50 per cent of all health care costs in the Netherlands. Second, he understood that most freelance nurses hardly knew anything about the complex issues and found it important to educate them in this respect (I/3/2010:7). Thus he decided to start blogging (FLA-1a/1b and FL-2: Figure 10.1) to keep multiple parties informed of news and developments on the issue. A third issue he noted was that freelance nurses were dependent on intermediary organizations to match them with care customers.

The entrepreneur's weblog content and actions demonstrate a distinct learning curve in dealing with the field position of freelance nurses in

Table 10.1 Field level and venture level actions, 2008–11: relevant quotes

Field level actions 2008–11	
FLA-1a	*Start personal weblog, and launch information/coaching service website.* 'I write for aspiring freelance homecare nurses, fellow health care entrepreneurs, nurse/client matching organizations health care managers and chairmen, those who work as policy agents in the tax and social security area and politicians in charge of health care issues.' (BHP/portfolio)
FLA-1b	*Continuation of personal weblog.* Around the summer of 2009 there is a four-month long silence; comparing the activities of the entrepreneur at the venture level, we notice that around that time he initiated four new venture activities. Afterwards, the entrepreneur excuses the interruption and continues frequent blogging.
FLA-2	*Building network of aspiring freelance nurses.* 'To convince the freelance nurses of the potential of working in the specialized care sectors was another issue, to do this we have to organize ourselves if we want to achieve anything on the regulatory front is an important mission.' (I/3/2010:7) 'I have put my mind to understanding their situation which is equally my own, with all the problems, limitations as well as its potential.' (BHP/portfolio)
FLA-3	*Launch websites on information/coaching service for aspiring nurses and on 'Freelance Nurse Cooperative'.* 'Deciding on your own working schedule, choosing for whom you want to work, higher salaries, it all sounds too good to be true. In reality it is, organizing to become and function as a flexible nurse is difficult, therefore the concept of a cooperative of flexible nurses is a very good solution. All together this means we need a collective force to join in and on the one side warrant continuity to patients and agencies we as flexible workers work for and on the other side to keep the profits of being a flexible worker. This is why I research the possibility of founding a cooperative of freelance nurses.' (24/11/2009)
FLA-4	*Work as knowledge broker for professional association of nurses.* 'There is a lot to be done I noticed, this is why I started to network and I have developed good contacts with professional associations like the professional organization for nurses and caregivers, the Chambers of commerce, and innovation networks. I use this knowledge to generate attention for our issue.' (1/10/2008)
FLA-5	*Launch workshop series for aspiring freelance nurses.* 'In this way I hit three targets at once; (informs about the potentials of independent nursing, attracts more nurses for his independent nursing network and the nurses get credits for the seminar).' (B/12/02/2009)

Table 10.1 (continued)

Field level actions 2008–11	
FLA-6	*Exposure accreditation barriers for care cooperatives.* 'I try to construct a sustainable cooperation within our group and between our group and other actors in working towards clear quality norms for freelance nurses.' (B/13/10/2008)
FLA-7	*Knowledge solicitation by politicians.* 'This is the beginning; we are being noticed by the government! This is a start and yes the very resilient governmental agencies start viewing our point.' (B/17/03/2010)
FLA-8	*Spokesman function.* 'Last week I was contacted by the directors of the Ministerial Department of long term care and extraordinary care. They asked me to come to The Hague and talk about the market position of independent nurses. I look back at a very successful meeting. In discussing our issue I found out that the Ministry of Health sees a strong potential market position for flexible workers in the delivery of homecare.' (B/29/03/2010)
Venture level actions 2008–11	
VLA-1	*Start sales of info-packages for aspiring freelance nurses.* 'I have organized all necessary information on how to work independently, it is great, please contact me and for the total amount of 30 Euros I will send it to you. However, at the same time I am working to unite freelance nurses in health care, please visit my website for freelance nurses.' (B/18/04/2008)
VLA-2	*Writing and selling a co-authored book for freelance nurses starting up.* 'I have organized all necessary information on how to work independently, it is great, please contact me and for the total amount of 30 Euros I will send it to you. However, at the same time I am working to unite freelance nurses in health care, please visit my website for freelance nurses.' (B/18/04/2008)
VLA-3	*Registration of private health care business (ambulance services and interim nurses and training in industrial settings).* 'Amongst the group of freelance nurses that registered, I met my current business partner, we had a similar affinity with nursing within industrial environments. Together we took on an assignment on a large building site where medical assistance was needed and we organized the job. Because of our productive cooperation we founded a business.' (BHP/portfolio). 'Since about three months I have met a business partner with whom I work on a totally different initiative; (the starting of the ambulance care company red.) it has nothing to do with homecare; but it is an entrepreneurial opportunity in health care that is highly feasible.' (B/30/03/2009)

Table 10.1 (continued)

Venture level actions 2008–11	
VLA-4	*Start coaching website for aspiring freelance nurses.* 'My last initiative of this year is the launch of another website. For the lasts months I have advised future independent nurses on how to start their own nursing company and be self supporting. This website is a place where I give professional support and personal coaching to independent nurses that need information on how to start their independent working business.' (B/06/12/2008)
VLA-5	*Launch website 'Care Nurse Coöperative' with information on, and membership options for, cooperative.* 'One way of doing this (organizing a market position for freelance nurses red.) is by starting a cooperative for freelance nurses for the structural cooperation amongst them and to force new labor options for them. On the other side, we ourselves have to deal with the quality issue and find a solution for this. The network aspect is important in here and it is crucial that we really cooperate for this and unite our position and opinion.' (B/07/04/2009). 'On the other hand, I see it as an investment; if in the long term, I want to be able to operate this cooperative legally and practically, I need to invest in [it] right now.' (I/2010/p12)
VLA-6	*Development of ICT tool for freelance nurse/care client matching.* 'Without sharing the concept (ICT application for freelance nurse matching labor demand for patients with available freelance nurses, we have been working on developing this application. I now see the amount of time this application needs and deserves and unfortunately this cannot be combined with running my other private company. It is a shame and I do not hope I will refrain from further development but now I just do not have sufficient time.' (B/04/02/2010)
VLA-7	*Workshop for aspiring freelance nurses.* 'In this way I hit three targets at once; (informs about the potentials of independent nursing, attracts more nurses for his independent nursing network and the nurses get credits for the seminar).' (B/12/02/2009)
VLA-8	*Co-founder of Cooperative Freelance Care Professionals/Application accreditation care cooperative.* 'Alternatives such as the [xxx] concept for a collectively organized administrative solution for freelance nurses are available' or 'I spend the day yesterday with [Mr B] in discussing the concept of [xxx]. The concept in arranging a working structure for freelance nurses is interesting, however the functional part of the organization structure is not what truly independent nurses want (. . .)' (B/6/06/2008) 'However, many of you will see that I share less and less details on what the initiative of our cooperative is all about because it is not

Table 10.1 (continued)

Venture level actions 2008–11	
	smart to share all of this information in a market where there is so much hostility about organizing our interests. However we really want to make this work and not give away our recipe (smiley.)' (B/31/03/2010)

Note: Coding for data sources is as follows: BHP – Blog Home Page; I/month/year/page – Interview; (B/31/03/2010) – Blog/day/month/year; PP/month/year – Policy Paper; PP/actor/month/year – Position Paper/Actor.

homecare work. He works on discrediting the institutional arrangements surrounding their marginalized situation. His writing and actions are targeted at undermining current institutional core assumptions and beliefs which he perceived as obstructing collective goals and his own potential opportunities. Disassociation of moral foundations occurs by exposing management practices of large organizations in homecare, such as bureaucratic billing and excessive executive salaries. Moreover, he discredited illogical and irrational policies and policy implementation of heath care governmental agencies and large care organizations with concrete examples of how efficiency programs in large homecare organizations actually lead to increased workload and pressure. At the same time, the entrepreneur described in detail substandard classification practices causing patients to be incorrectly indicated for the care they need and thus receiving less money to spend on care. Equally he reports on fraud and scam practices of large homecare organizations. Furthermore, he discussed the 'dysfunctional HRM practices' of large care organizations in hiring (or not hiring) freelance nurses indicating the institutional rigidity of incumbents and exposing their strategic decoupling practices. Finally, through his posts he sought to expose barriers for obtaining an official accreditation for care cooperatives.

Analysis of the discourse in the weblog shows the reasoning of an actor with explicit field level awareness and consideration for stakeholder pluralism. Content and discourse are targeted at the different stakeholders and exhibit a bottom-up approach. This is demonstrated through the opportunities he develops from his own aspirations for becoming a freelance nurse. Recognizing how client-nurse matching and administrative burden hinder freelance nurses, he studies and reports on the potential efficiencies of a care cooperative. His intentions are exposed already in his early blogs and displayed great frustration with the current solutions for homecare in the Netherlands. In seeking ways to operate as a freelance homecare nurse and getting reimbursed for it he encountered the limitation of the institutional system. This was his single motivation in sharing his story with others. Personal relegations of the current institutional setting of Dutch home-care delivery as reported in a cyber diary therefore formed the base of his uncalculated and organically grown entrepreneurial activism.

Building Alliances

Realizing how his online exposure and his aptitude for knowledge translation could facilitate the organization of his peers, he continued to develop and apply his social skills to address different stakeholder groups. Through his blog he persuaded and enabled aspiring freelance nurses to contact him. The numbers of blog visitors grew from about 100 individual visitors per week in April 2008, to 750 visitors per month after 2.5 months and 2 000 visitors per month after one year. By May 2010 the blog was visited 23 000 times and just over 600 aspiring freelance nurses had become members. To educate, inform and build critical mass among them for his care cooperative, the entrepreneur also launched a website with information offering a coaching service for aspiring freelance nurses: 'Freelance Nurse Coöperative' (FLA-3: Figure 10.1). Due to his increasing experience and public exposure, the Dutch Professional Organization of Nurses and Caregivers approached him and asked him to work as a knowledge broker and to represent them in stakeholder events on the issue (FLA-4: Figure 10.1). Subsequently, this professional organization and a representative from the Chamber of Commerce, whom he already knew from a previous project designing a series of workshops for aspiring freelance nurses (FLA-5: Figure 10.1), contacted the entrepreneur to understand the ins and outs on starting up.

Wanting to move forward with the care cooperative, the entrepreneur realized that freelance nurses need a way to acquire a professional license permitting individual care delivery. In addition he recognized that the care cooperative would require official accreditation as a health care organization in order to be eligible for reimbursement, something which was impossible at the time. He realized that this could undermine the potential success of care cooperatives as it would prevent these from gaining access to necessary resources including funding. To that end, the entrepreneur scanned and scrutinized the environment for existing initiatives and new ideas to solve these issues. In his weblog, he inserted hyperlinks to relevant sources and comments on the pitfalls of alternative solutions offered by other entrepreneurs in order to build a case for his own solution. For instance he positioned some of the competing ideas as 'flawed franchising system for independent nurses' (B/17/06/2008).

Legitimating at the Field Level

As his own opportunity crystallized, the entrepreneur and his business partners decided to simply start the full accreditation process and, hence, began to openly confront and problematize the current institutional barriers (FLA-6: Figure 10.1). The entrepreneur forecasted in his posts that this deliberate act, the information obtained in the process and the media attention it would generate may well be usable as 'ammunition' in showcasing how institutional barriers prevent innovation in homecare.

One of the workshop announcements, as described under FLA-5, drew the attention of a Member of the House of Representatives who decided to attend the entrepreneur's presentation on the issue of freelance nurses with a colleague. The information obtained in the presentation and a short additional talk to the entrepreneur led them to ask official questions to the responsible Minister on how freelance nurses are inhibited to solve labor demand and quality issues in homecare due to current regulations (FLA-7: Figure 10.1). The topic was thus brought to the political agenda and this forced the government to clarify, and take a stance on, the issue of freelance entrepreneurial workers in health care. Ultimately, in 2011, the entrepreneur was invited multiple times to speak and contribute to political debates on this issue in the Chamber of Deputies (FLA-8: Figure 10.1). Also, the Social and Economic Advisory Board (SER) invited him to speak on the topic as their representative. These events display his journey from developing an initial idea in his off-duty hours while working as a nurse in the offshore industry to becoming a knowledge worker informing politicians about the current shortcomings in freelance nurse legislation.

In relation to how the entrepreneur worked on the 'undermining of core assumptions and beliefs' we observed that the blog contained posts redefining and reframing the labor shortage in Dutch homecare as a result of institutional barriers rather than as a labor supply issue. He argued that improvements in tax and social security regulations would make the freelance nurse occupation a more attractive career option. The undermining of core assumptions and beliefs is furthermore displayed in the posting of multiple essays on the general goals and inefficiency of current homecare policy versus 'how it could or should be'. Furthermore, the entrepreneur sought to expose the arbitrary role of the Professional Association of Health Care Entrepreneurs' (Actiz) from whom he hoped to obtain support since they represent all health care entrepreneurs but who gave negative advice on allowing freelance nurses to work in homecare due to pressure of incumbents.

Attending to Venture Level Opportunity Development

After the first few months of blogging, the amount of reactions he obtained from aspiring freelance homecare nurses showed the value of the resource structure the entrepreneur had initiated. He envisioned its economic potential in relation to venture level opportunities. From here he started selling info-packages on his blog on how to start up and what the current limitations are (VLA-1: Figure 10.1), providing further indication of an effectuation logic. Parallel, and owing to the demonstration of his knowledge on the issue and his writing capacity, early in 2008 the entrepreneur was invited through his blog by an informed author to jointly write a handbook on freelance nursing practices and start-ups. Once written, the book could be purchased via the weblog (VLA-2: Figure 10.1). While continuing to operate part time as a nurse offshore, he began to explore entrepreneurship in health care in a broad sense later on in 2008. After meeting his future business partner earlier in the year (again through the blog) he noticed that the two of them could be a forceful team. Within three months they started a private health care business (VLA-3: Figure 10.1) offering industrial medical support services such as the providing and operating of ambulances in industrial areas, and the educating and the provision of specialized nurses in industrial settings. Only after being contacted by aspiring freelance nurses in search of information following on from the blog did the entrepreneur start to think about another initiative, namely, the selling of private coaching packages for aspiring freelance nurses (VL-4: Figure 10.1). This can be considered his second entrepreneurial venture.

Legitimation on Both Levels

Soon after, and while studying the possibilities to efficiently relieve administrative burdens and increase care client/nurse interaction, he realized that the organizational form of care cooperative permits a collective organizational (umbrella) structure above individually freelance operating nurses (VLA-5: Figure 10.1). Promoting care cooperatives constituted both a venture level and a field level activity, as the entrepreneur recognized that he had to attend to these cooperatives at both levels (point B: Figure 10.1). Around the same time as starting the coaching website, the entrepreneur worked on the launch of an ICT platform to match freelance nurses to patients in need of domiciliary care (VLA-6: Figure 10.1). However, due to a work overload he put this initiative on hold. Shortly after, he was approached by the Dutch Professional Organization of Nurses and Caregivers and the Chamber of Commerce on the designing and giving of a series of workshop for aspiring freelance nurses (VLA-7: Figure 10.1).

Several actions in particular can be seen as examples of legitimating efforts: first the registration of the Cooperative for Freelance Care Professionals at the Chamber of Commerce together with his business partners (VLA-8: Figure 10.1); and second their decision to 'just start going through the full accreditation process' thus permitting them to openly confront and problematize the faced barriers (VLA-9: Figure 10.1). Both actions were used as means to generate the required legitimacy for this new organizational form underlying both his own opportunity and to allow others in the field to follow suit with similar rent-seeking behaviors. Effort in combining obvious personal interest with field level interest vested in the rearrangement of these institutional conditions is indicated by the dotted line 'C' in Figure 10.1.

Resource Building and Organizing: Creational Work

Concurrently the entrepreneur realized that he needed the acceptance and collaboration from other aspiring freelance nurses to eventually exploit his own venture of a care cooperative where he would occupy a director function. For this he engaged intensively with the group of 600+ members of his online network. He substantiated it with options for the nurses to obtain factual knowledge and coaching material and educating aspiring freelance nurses on 'how to start' and 'Entrepreneurship for freelance nurses in practice series'. The educating of the network members occurred through translating complex information that allowed freelance nurses to provide AWBZ care. In Figure 10.1, the dotted line 'A' represents the collision of the venture level and field level work of the entrepreneur into a symbiotic movement of actions targeted at both opportunity development and institutional work. The blog started out as an information source with timelines and multiple stakeholder group issue descriptions. Soon the digital platform became a matching point for supply and demand because aspiring nurses emailed him for information and support, and homecare clients wanted to know where they could find freelance nurses that met their personal care budgets. However, this was never its main intention; information giving was more important (BHP/portfolio).

While providing aspiring freelance nurses with ideas and information on how to proceed with a start-up, he equally started reporting on his personal transition from being a nurse in large health care institutes to being an independent health care entrepreneur. This was dialoging because he invited others to share his stories with him, although he did not expose them on his blogs. Addressing the aspiring nurses personally, they formed the base from which the insurrection of an entrepreneurial nurse identity was created and templated. Thus, he actively tried to reconstruct the nurses' identity from a large care organization of 'working bees' with all their organizational, financial and personal constraints, to the 'independent entrepreneurial freelance nurse' in charge of their own destiny. First he tried to boost the self-confidence of nurses into 'entrepreneurial nurses' by demonstrating the advantages and benefits of working as a freelancer. Second, by targeting their knowledge and a belief system and introducing the label 'entrepreneurial nurse' he permitted them to consider an independent future with more work satisfaction and higher income. Equally, the normative network provided a social group from which collective identity construction evolved.

Similarly, the normative network and the weblog visitors were educated through the entrepreneur's content, abstracting and translating results of scientific research agencies and governmental bodies on domiciliary care research. The same went for the translating and commenting on policy discussion, news coverage on happenings and news in the area of domiciliary care towards care patients. The entrepreneur also targeted blog content towards the effort of educating and convincing homecare patients holding a PGB of the unique selling points of hiring freelance care nurses. Alongside, he presented the benefits of the new organizational form of the care cooperative to aspiring freelance nurses. This indicates transparency in the entrepreneur's communication and addressing multiple stakeholder groups simultaneously in explaining the need and benefit of the care cooperative. He created the image of the entrepreneurial nurse who is flexible, helpful, and creative with time and resources and is personally attentive and thus delivers care at a higher quality level.

Addressing and problematizing current administrative issues, complaint issues and fee issues for PGB holders formed an equally important tactic in this process. While diffusing and fueling the new identity of the strong entrepreneurial nurse among patients, he informed homecare patients concomitantly on 'how to find, hire, judge and pay flexible freelance nurses with your PGB'. Both tactics were aimed at changing PGB holders' norms and beliefs regarding how homecare should be organized. As such he presented a new organizational model of homecare delivery that not only brought better service but did so at lower costs to both individual patients and society at large. Doing so at a time when efficiency, resource conservation and cost reduction were at the center of the public debate shows an ingenious bricolage of stakeholder and self-interest that significantly helped in winning support for this model.

Field Level Events and Logics

We now pay attention to critical incidents and field logics in the field of Dutch homecare. For this, we describe field level change arguments contributing towards the signing of the covenant. This data permits insight into a field level social change explanation to which our entrepreneur focused his efforts. We present specific event-time relations in understanding why these field level change arguments contribute to the coming about of the policy covenant allowing care cooperatives of freelance nurses in 2011 and why it failed seven years earlier.

In checking our story line with the secondary data of the homecare sector and the personal political experiences of the MP and the senior consultant, we were able to frame time-dependent field logics influencing the decision to change legislation. The covenant in July 2011 (CFI-6; Figure 10.1) allowed homecare to be carried out by freelance nurses in the new organizational form of care cooperatives and to be eligible for reimbursement. Around the time of the development of the covenant, two major Dutch homecare organizations went bankrupt. The Member of Parliament noted: 'it opened up the eyes of many to realize that something is really wrong in the way we organize homecare'. This caused the loss of thousands of nursing jobs and was subject to massive media exposure to underwriting the fraud and malpractices such as those elaborated upon by our entrepreneur about two years earlier in his weblogs. The MP witnessed the importance of these bankruptcies in going back to the 'small is beautiful' way of organizing homecare: 'new forms of organizing, small-scaled care initiatives within close proximities of self-organizing teams of freelance nurses are hot now, it's cheaper and the quality is evidence-based better'. At the same time, the proportion of freelance workers in the Dutch labor market was close to 1 million unincorporated workers. Thus:

> You cannot negate their positions anymore (. . .) two years after the covenant signing everyone here in the political capital understands that the quality demands cannot be met by large bureaucratic operating health care organizations (. . .). This brings openings for entrepreneurship and for people to jump in. It is not as tight as it used to be six or seven years ago, you can feel it.

Thus, specific time-related characteristics such as the bankruptcies of some homecare organizations due to fraud and malpractice, the economic crisis and the increasing health care costs and changing health care quality norms have played important roles bringing the resource constraint of the current institutional setting on the table, thus creating awareness and consideration among the institutions' persevering actors building alignment in the signing of the covenant.

We also wanted to understand what field conditions and logics were in place at the time of the failed initiative of the introduction of the organizational form of a care cooperative in 1994. Our field expert responsible for the design and implementation of the 1994 effort noted: 'with our care cooperative we were slightly too late or too soon depending on how you analyze it'. He explained that the introduction of the AWBZ back in 1968 brought a coercive, state-initiated game-changer and was then considered to introduce locally organized innovative homecare. In those days, economies of scale and market efficiencies were the only impetus for action.

> We needed larger organizations to cut overhead and bring the least amount of higher educated nurses to the neighborhood. Initiatives of small-scale care organizations were not considered as added value because within the new AWBZ all sorts of homecare were anticipated for and reimbursement only went along those lines.

While elaborating on the different field logics emphasizing and prioritizing economies of scale in the 1994 initiative, this demonstrates the coercive social boundary the AWBZ represented: homecare services would not be reimbursed if the organizational form was not recognized and justified within the AWBZ.

CONCLUSIONS AND DISCUSSION

This chapter set out to understand how, in organizational fields, institutional work of individual entrepreneurs can be related to opportunity development and why and how, institutional work can be both an entrepreneurial strategy as well as a set of ingenious tactics. To understand how ingenuity is translated into action and results on both the venture level and the institutional level we performed an embedded case study examining the synchronized manifestations of destructional and creational efforts of an ingenious entrepreneur in the changing Dutch homecare. We reported on an entrepreneur's effort in the disclaiming of the current institutional arrangements, the creation of new identities of entrepreneurial freelance homecare nurses and activities on the propagating of new tax and social security conditions. In particular we paid attention to the dynamics of this process.

We exposed how destructional and creational work, as an entrepreneurial strategy are different sides of the same coin called opportunity development and that influencing, proposing and contributing to efforts leading to destroying and creating institutions is, inherently, institutional work. However, it also is dependent on the skilful ability to learn from experiences and adapt to the circumstances at hand, i.e. processes of effectuation. Thus we showed in particular the requirement for actors to adapt mental representations, social skills and sets of tactics as put forward by Fligstein (1997) and Fligstein and McAdam (2011). In the early days of his blog, our entrepreneur was a peripheral actor with start-up ambitions to incorporate as a freelance homecare professional and lacked informal influence or legitimate authority. During the search for a foundational principle he adapted his quest to the field structure, his efforts were oriented towards the framing of the institutional limitations. At the same time, his efforts were targeted at creating a field position by both organizing critical mass within a normative network of freelance nurses, which he empowered with a new entrepreneurial identity and the prospects of new resources. This building of a normative network of aspiring freelance nurses generated a community of feeling which is referred to as 'cool mobilization' for 'hot causes'; or the challenge for market rebels in forging a collective identity and mobilizing support that arouses emotion and creates a community of members signaling identity and sustaining commitment (Rao, 2009: 12). Just as in social movements, the entrepreneur's cool blogging mobilization strategy engaged network members in collective experiences wherein participants actively installed and started living according to new meanings and the values associated with them (ibid.: 13).

From this we can recognize the three ingeniously paralleled efforts of blogging and occupying a spot in cyber space and tactical efforts in our entrepreneur's institutional work; venture activity and field development in which he shows particular acumen in bricolage with field dynamics. The weblogs show rhetoric reflections outright denouncing institutional constraints, but also the identifying of the gap between the current institutional setting and the ingenious entrepreneurial solution. Key to this is how our actor, at the center of his institutional work, displays the ability to reflect on the disparities between own means-end relations in venturing and overarching prerequisite organizational field changes for the creation of new ones. This permitted him to recognize, categorize and label the barriers and dysfunctional field institutions and mechanisms against the organizing of ventures for entrepreneurial nurses. As such, this case shows an entrepreneur engaging in institutional work, who activated mental orientations based on reflective organizational field level awareness. The destructional institutional work yielded options to construct ingenious new solutions and was characterized by translating of multifaceted institutional barriers into actionable options for value creation. This critical skill of 'translation' was the ability to not only move ideas and resources from a particular system, but also to adapt them into appropriate forms for the receiving system as well as opportunistic behavior vis-à-vis new possibilities (Lawrence and Suddaby, 2006).

Besides the display of the cognitive aptitude in the learning, mapping and externalizing of his personal field level insights, the entrepreneur displayed concentrated focus and drive, resulting in the documentation of his insights in his weblog and the persuading of others on the necessity of policy changes. While blogging, the entrepreneur reported on who he met and how he built both the normative network of aspiring freelance nurses and private health care businesses. Thus, he displayed himself as a social chameleon employing social skills in engaging and networking with field groups and actors. Without the purposeful intention to introduce a new organizational form or organizing activism and policy change the entrepreneur took advantage of emerging opportunities (Sarasvathy, 2001). His 'ingenuity on the fly' is distinct from deliberate and premeditated implementation of a great creative idea for potential value creation.

Further, we found that, noticeably, the entrepreneur dexterously employed rhetoric to work within both destructional and a creational discourse, even simultaneously. Using rhetoric has been associated both with entrepreneurial resource acquisition strategies (Aldrich, 1999; Suchman, 1995; Zimmerman and Zeitz, 2002; Cornelissen, 2005) and actors contributing to institutional change processes (Suddaby and Greenwood, 2005; Jones et al., 2010). While the entrepreneur profiled himself as an altruistic actor engaging in and claiming missionary efforts to defend both his personal and the collective stakes of aspiring freelance nurses, he also showed an opportunistic way of adapting to external dynamics such as the shifting political priorities. We showed that, in this specific case, institutional setting and dynamics thereof helped the entrepreneur to gain leverage in influencing institutional change processes. As such, in our case, institutional work became an entrepreneurial tool in the orchestrating of an opportunity resource base under conditions of uncertainty, permitting rent generation and appropriation (Alvarez and Barney, 2004, 2005).

In the literature, institutional work directed at the destruction of institutions is mostly seen as coercive actions exercised by the state or professional groups (Lawrence and Suddaby, 2006) involving the defining and redefining sets of concepts (Suchman, 1995) in ways that reconstitute actors and reconfigure relationships between actors. Employing the concept of institutional work in relation to opportunity development offers new ways of studying entrepreneurial effort under institutional constraint. We showed that normative and effectuated actions by peripheral entrepreneurs contribute to agenda setting in two ways: first by destructing current institutional arrangements through the unveiling and exposing of its inefficiencies and unproductive practices of institutionalized actors both leading to resource scarcity.

Second, through translation, negotiation and persuasion the entrepreneur effectuated new resource bases in the form of a network of aspiring nurses wanting to exploit their own skilled resources within a more effective new organizational form of a care cooperative. Thus, individual institutional work and opportunity development can equally be seen as mechanisms partaking in or working towards large-scale, revolutionary change (Greenwood and Hinings, 1996; Lawrence and Phillips, 2004). We showed how an entrepreneur not only caused waves but also demonstrated the ability to ride them and direct them to new ingenious solutions.

Our independent starting entrepreneur could not rely on 'organizational slack' providing a formal organizational structure with means and resources. Confronted with even higher levels of resource constraints, he exploited the one resource he had most of: time. He started a compelling weblog in the hours between his shifts in the offshore industry. Without grand intentions to change the environment, and much to his surprise, his engagement into the low cost influencing technique of blogging gave way to the mobilizing of a networked community of mostly aspiring freelance nurses. While the network's members consisted mostly of members adhering to his ideas and intentions as witnessed in 'normative networks' (Lawrence and Suddaby, 2006), we viewed that this network also served as 'opportunity accelerator' by providing a new resource base through contact with new care clients giving way to environmental feedback from other, often more central, field actors.

Through the mobilizing of a networked community of aspiring freelance nurses our nurse was maybe no 'market rebel' (Rao, 2009), but characteristics of a 'field rebel' can be attributed to him. The improvisational and insurgent character of the low cost influence technique of blogging used was an important part of the resource constrained form of 'ingenuity on the fly'. Through the mobilizing of a community network, new identities of the aspiring freelance nurses were developed and signaled, and commitment of its members sustained. From this, our entrepreneurial nurse created new opportunities, formed new venturing teams and structured new coalitions permitting influential interaction with central field actors. With incessant motivation for effort, intuitive knowledge leaps and progressive insight through ordinary but striking occurrences, the entrepreneur developed, accelerated and exploited opportunities. These were situated across different levels of the Dutch health domiciliary care field and together they built up towards the field introduction of a new organizational form. Originally situated in a peripheral field position, our independent entrepreneur thus blogged and effectuated network connectivity as well as the transfer and translation of knowledge transfer. As a result of this 'entrepreneurial ingenuity on the fly', exploitable externalities at both the venture level and field level came about wherein efforts of institutional work and opportunity development were mutually constitutive and influencing.

Finally, our research delineates the prerequisite and context-dependent nature of concurrences between institutional work and entrepreneurial opportunity development. As noticed, administered markets coincide with densely interconnected institutionalized fields. Opportunity development and institutional work are hence likely to co-occur when the institutions targeted are apparent and obvious, or even as in our research, of coercive nature. At the same time this prevailed reiteration of two potential shortcomings of institutional work studies concerning: (a) the taken-for-grantedness and under problematizing of the actors' identities and interests; and (b) the under theorizing of institutions (Hwang and Colyvas, 2011). We stuck to the shortest relation possible between work and a delimited, coercive institution and referred back and forth between the actor's identity and interests at multiple levels in specific relation to the institutions' effects.

Limitations

Rich in material and findings, our embedded case study offered unique opportunities to engage in inductive theoretical work. However, our research setting and approach carry inherent limitations relating to methodological and contextual constraints and, as such, limit the generalizability of our findings. The entrepreneur and his practices represent a particular instance of an agent's field work in an administered market. His drive and motivation to extensively and longitudinally map his actions in a transparent and accessible way, combined with the additionally collected data, provided a fairly complete idea of the internal validity and genuineness of our findings. Although we sought to avoid bias by reporting on the ongoing field changes in Dutch homecare and the field position of freelance nurses, and sought the views of two additional expert field actors, we still followed only one actor. Thus, our findings and conclusions may be predisposed in different ways and susceptible to over-interpretation. Therefore, additional studies researching the relation between opportunity development and institutional work in field settings with dissimilar levels of institutionalization and bureaucratic culture are necessary. At the same time, studies where 'ingenuity on the fly' ended up creating difficulties for the entrepreneurs involved are equally of great importance for its understanding.

ACKNOWLEDGMENTS

The authors thank Rabobank NL and the Amsterdam Centre of Entrepreneurship@VU University for financial support to this research and Syntens, Department of Human Health for their generous field access and cooperation. We equally thank the editors for their constructive comments.

REFERENCES

Abbott, A. (1988). *The System of Professions*. Chicago: University of Chicago Press.

Aldrich, H.E. (1999). *Organizations Evolving*. London: Sage.

Aldrich, H.E. (2010). Beam me up Scott(ie): institutional theorists' struggles with the emergent nature of entrepreneurship. In W.D. Sine and R.J. David (eds), *Institutions and Entrepreneurship* (pp. 329–64). Bingley, UK: Emerald Group Publishing.

Alvarez, S.A. and Barney, J.B. (2004). Organizing rent creation and appropriation: toward a theory of the entrepreneurial firm. *Journal of Business Venturing*, **19**(5): 621–35.

Alvarez, S.A. and Barney, J.B. (2005). How do entrepreneurs organize firms under conditions of uncertainty? *Journal of Management*, **31**: 776–93.

Ansell, C.K. (1997). Symbolic networks: the realignment of the French working class, 1887–1894. *American Journal of Sociology*, **103**(2): 359–90.

Bar-Ilan, J. (2005). Information hub blogs. *Journal of Information Science*, **31**(4): 297–307.

Battilana, J. and D'Aunno, T. (2009). Institutional work and the paradox of embedded agency. In T.B. Lawrence, R. Suddaby and B. Leca (eds), *Institutional Work. Actors and Agency in Institutional Studies of Organizations* (pp. 31–58). Cambridge: Cambridge University Press.

Battilana, J., Leca, B. and Boxenbaum, E. (2009). How actors change institutions: toward a theory of institutional entrepreneurship. *Academy of Management Annals*, **3**: 65–107.

Bourdieu, P. (1993). *Sociology in Question*. London: Sage.

Bromley, P. and Powell, W.W. (2012). From smoke and mirrors to walking the talk: decoupling in the contemporary world. *Academy of Management Annals*, **6**(1): 483–530.

Cornelissen, J.P (2005). Beyond compare: metaphor in organization theory. *Academy of Management Review*, **30**: 751–64.

Dacin, M.T., Goodstein, J. and Scott, W.R. (2002). Institutional theory and institutional change: introduction to the special research forum. *Academy of Management Journal*, **45**: 45–56.

DiMaggio, P.J. (1988). Interest and agency in institutional theory. In L.G. Zucker (ed.), *Institutional Patterns and Organizations: Culture and Environment* (pp. 3–22). Cambridge, MA: Ballinger.

DiMaggio, P.J. (1991). Constructing an organizational field as a professional project: US art museums, 1920–1940. In W.W. Powell and P.J. DiMaggio (eds), *The New Institutionalism in Organizational Analysis* (pp. 267–92). Chicago: University of Chicago Press.

DiMaggio, P.J. and Powell, W.W. (1983). Iron cage revisited: institutional isomorphism and collective rationality in organizational fields. *American Sociological Review*, **48**, 147–60.

DiMaggio, P.J. and Powell, W.W. (1991). Introduction. In W.W. Powell and P.J. DiMaggio (eds), *The New Institutionalism in Organizational Analysis* (pp. 1–38). Chicago: University of Chicago Press.

Dorado, S. (2005). Institutional entrepreneurship, partaking, and convening. *Organization Studies*, **26**(3): 383–413.

Drori, I. and Honig, B. (2013). A process model of internal and external legitimacy. *Organization Studies*, **34**(3), 345–76.

Fligstein, N. (1997). Social skill and institutional theory. *American Behavioral Scientist*, **40**: 397–405.
Fligstein, N. and McAdam, D. (2011). Toward a general theory of strategic action fields. *Sociological Theory*, **29**(1): 1–26.
Fligstein, N. and McAdam, D. (2012). *A Theory of Fields*. Oxford: Oxford University Press.
Greenwood, R. and Hinings, C.R. (1996). Understanding radical organizational change: bringing together the old and the new institutionalism. *Academy of Management Review*, **21**: 1022–54.
Greve, W. (2001). Traps and gaps in action explanation: theoretical problems of a psychology of human action. *Psychological Review*, **108**: 435–51.
Hirsch, P.M. and Bermiss, Y.S. (2009). Institutional 'dirty' work: preserving institutions through strategic decoupling. In T.B. Lawrence, R. Suddaby and B. Leca (eds), *Institutional Work. Actors and Agency in Institutional Studies of Organizations* (pp. 262–83). Cambridge: Cambridge University Press.
Homer-Dixon, T. (2000). *The Ingenuity Gap*. New York: Knopf.
Hookway, N. (2008). 'Entering the blogosphere': some strategies for using blogs in social research. *Qualitative Research*, **8**(1): 91–113.
Hwang, H. and Colyvas, J.A. (2011). Problematizing actors and institutions in institutional work. *Journal of Management Inquiry*, **20**, 62–6.
Jones, C., Livne-Tarandach, R. and Balachandra, L. (2010). Rhetoric that wins clients: entrepreneurial firms' use of institutional logics when competing for resources. Institutions and Entrepreneurship. *Research in the Sociology of Work*, **21**, 183–218.
Lampel, J., Honig, B. and Drori, I. (2011). Call for papers for a special issue on discovering creativity in necessity: organizational ingenuity under institutional constraints. *Organization Studies*, **32**: 715.
Lawrence, T. (2004). Rituals and resistance: membership dynamics in professional fields. *Human Relations*, **57**(2): 115–43.
Lawrence, T.B. and Phillips, N. (2004). From Moby Dick to Free Willy: macro-cultural discourse and institutional entrepreneurship in emerging institutional fields. *Organization*, **11**(5): 689–711.
Lawrence, T.B. and Suddaby, R. (2006). Institutions and institutional work. In S.R. Clegg, C. Hardy, T.B. Lawrence and W.R. Nord (eds), *The Sage Handbook of Organization Studies* (pp. 215–54). Thousand Oaks, CA: Sage.
Lawrence, T.B., Suddaby, R. and Leca, B. (2009). Introduction: theorizing and studying institutional work. In T.B. Lawrence, R. Suddaby and B. Leca (eds), *Institutional Work: Actors and Agency in Institutional Studies of Organizations* (pp. 11–59). Cambridge: Cambridge University Press.
Maguire, S., Hardy, C. and Lawrence, T. (2004). Institutional entrepreneurship in emerging fields: HIV/AIDS treatment advocacy in Canada. *Academy of Management Journal*, **47**(5): 657–79.
McMullen, J. and Shepherd, D. (2006). Entrepreneurial action and the role of uncertainty in the theory of the entrepreneur. *Academy of Management Review*, **31**: 132–52.
Meyer, J.W. and Rowan, B. (1977). Institutionalized organizations: formal structure as myth and ceremony. *American Journal of Sociology*, **83**(2): 340–63.
Oliver, C. (1991). Strategic responses to institutional processes. *Academy of Management Review*, **16**(1): 145–79.
Rao, H. (2009). *Market Rebels: How Activists Make or Break Radical Innovations*. Princeton, NJ: Princeton University Press.
Sarasvathy, S. (2001) Causation and effectuation: toward a theoretical shift from economic inevitability to entrepreneurial contingency. *Acadamy of Management Review*, **28**(2): 243–63.
Sarasvathy, S. and Dew, N. (2005) New market creation through transformation. *Journal of Evolutionary Economics*, **15**: 533–65.
Scott, W.R. (2001). *Institutions and Organizations*. Thousand Oaks, CA: Sage.
Scott, W.R., Ruef, M., Mendel, P.J. and Caronna, C.A. (2000). *Institutional Change and Healthcare Organizations*. Chicago: Chicago University Press.

Shane, S. and Venkataraman, S. (2000). The promise of entrepreneurship as a field of research. *Academy of Management Review*, **25**(1): 171–184.
Slotte-Kock, S. and Coviello, N. (2010). Entrepreneurship research on network processes: a review and ways forward. *Entrepreneurship Theory and Practice*, **34**(1): 31–57.
Strang, D.B. and Meyer, J.W.B. (1993). Institutional conditions for diffusion. *Theory and Society*, **22**(4): 487–511.
Suchman, M.C. (1995). Managing legitimacy: strategic and institutional approaches. *Academy of Management Review*, **20**: 571–611.
Suddaby, R. and Greenwood R. (2005). Rhetorical strategies of legitimacy. *Administrative Science Quarterly*, **50**: 35–67.
Trank, C.Q. and Washington, M. (2009). Maintaining an institution in a contested organizational field: the work of the AACSB and its constituents. In T.B. Lawrence, R. Suddaby and B. Leca (eds), *Institutional Work: Actors and Agency in Institutional Studies of Organizations* (pp. 236–61). Cambridge: Cambridge University Press.
Yin, R.K. (1994). *Case Study Research: Design and Methods*. London: Sage.
Zimmerman, M.A. and Zeitz, G.J. (2002). Beyond survival: achieving new venture growth by building legitimacy. *Academy of Management Review*. **27**: 414–31.

11 Constraints and ingenuity: the case of Outokumpu and the development of flash smelting in the copper industry

Janne M. Korhonen and Liisa Välikangas

INTRODUCTION

Prior work on ingenuity under constraints has focused on whether resource constraints and operating limitations help or hinder creativity, innovation, and organizational performance. Although the evidence is somewhat inconclusive, as discussed below, an emerging consensus seems to be that constraints may trigger cognitive and organizational mechanisms that promote creativity, innovativeness and other desirable characteristics. However, what is missing in most prior studies is a detailed 'autopsy' of constraint-induced fruits of ingenuity, detailing the roots and antecedents of the innovation and thereby placing the constraint-induced technological departure in a broader context. Thus, we remain somewhat in the dark about the processes of innovation that constraints tend to promote. This study uses a detailed, longitudinal case study of copper smelting to describe how constraints promoted innovativeness in this well-documented but rarely fully analyzed case. We show that the adoption of existing technological ideas, here diffused worldwide, is an important part of constrained innovation.

A major reason why the relationship between constraints and innovations attracts interest is because constraints can theoretically act as 'focusing devices' (Bradshaw, 1992; Hughes, 1983; Rosenberg, 1969) that attract inventive attention to a specific problem. Some extant research suggests that such focus may produce solutions that would have not been invented otherwise (Goldenberg et al., 2001; Moreau and Dahl, 2005; Ward, 2004), that are functionally superior to non-focused efforts (Gibbert and Scranton, 2009; Popp, 2006) or, at least, produce similar results but faster and with fewer resources (Gibbert and Scranton, 2009; Gibbert et al., 2007; Hoegl et al., 2008; Lampel et al., 2012; Weiss et al., 2011). The premises underlying these arguments are that inventions, or at least important ones, require specialized inventive efforts, such as R&D investments, and involve a conceptual leap of ingenuity that separates the innovation from the existing state of the art. The latter premise is implicitly accepted in all research that defines innovations as something novel, because novelty is generally defined as being something unusual to the context under consideration. Typically, innovations are also seen to involve more or less radical breaks from the traditions of the field. For example, frequently quoted works use words and phrases like 'discontinuities' (Foster, 1985), 'changes in core technological architecture' (Henderson and Clark, 1990), 'launching new technology versus making progress along established paths' (Christensen and Rosenbloom, 1995), or that 'radical exploration builds upon distant technology that resides outside of the firm' (Rosenkopf and Nerkar, 2001).

Flash smelting, the technology we are studying in this chapter, is often mentioned (e.g. Habashi, 1993, 1998; Särkikoski, 1999; King, 2007) as an example of a breakthrough or radical innovation in the classical sense of the word (Henderson and Clark, 1990). We wish to add the historical perspective to the emergence of this technological innovation. Hence, our case study spans more than 100 years and details the development of copper smelting around the world including a company not well known outside the mining industry that became a major innovator in copper smelting: Outokumpu in Finland.

CALIBRATING THE ROLE OF RESOURCE CONSTRAINTS IN INNOVATION

As stated previously, constraints can be thought of as focusing devices that direct attention to a particular problem. The often-unstated assumption of research on constrained innovation is that constraints allow or force conceptual leaps of ingenuity to happen. Presumably, these leaps would not happen without the constraints, so the resulting innovation cannot, *ipso facto*, be obvious to a member of the community of knowledge. However, if we accept that innovations tend to be incremental improvements of existing state of the art, the view that constraints induce innovation and, specifically, that constraints induce *radical* innovation or 'out of the box thinking' may need a slight rethink. We do not need to discard the concept of focusing devices, and we are not arguing that constraints *do not* focus attention to particular problems or promote ingenuity on occasion. Instead, we argue for two things: first, for a more nuanced and historically informed study of examples used in innovation literature. Second, we extend the argument made by, for example, Hoegl et al. (2008) and Weiss et al. (2011) regarding financial constraints to the domain of non-financial constraints and argue that constraints by themselves are largely a neutral factor in the development of innovations: they have the capability both to help and to hinder technological development, but what actually happens depends on contingent factors.

This dichotomous nature of constraints on innovation is clearly visible from wildly diverging research results, some claiming that constraints generally hinder innovation (e.g. Amabile, 1996; Damanpour 1991; Nohria and Gulati, 1996) and others assigning them a more ambivalent or even positive role (e.g. Arthur, 2009; Gibbert and Scranton, 2009; Gibbert et al., 2007; Gibbert and Välikangas, 2004; Hoegl et al., 2008, 2010; Katila and Shane, 2005; Lampel et al., 2012; Mone et al., 1998; Nonaka and Kenney, 1991; Välikangas and Gibbert, 2005; Weiss et al., 2011). We believe that an important reason why such contrasting viewpoints can be empirically supported is because these studies have only rarely accounted for the 'prehistories' of technology examples they use to make their case. What appears to have the decisive influence is the nature of technological development *prior* to the imposition of a constraint, and in particular, whether such early development has 'primed' the innovation environment with parallel, multiple, relatively small incremental advances – a dynamic bearing some similarity to 'punctuated equilibrium' models of organizational change (e.g. Romanelli and Tushman, 1994). Our claim is therefore that constraints generally spur innovation if and only if the components required for the 'constraint-induced innovation' already exist and are widely known within the community of knowledge. Thus, constraints may not be very good at producing completely new-to-the-world technological innovations, and hence they are not very good drivers of technological advance *per se*: partly because constraints by definition cause a problem that needs to be solved relatively quickly, they tend to be poor at inducing time-consuming and fundamentally risky, highly exploratory R&D on novel solutions.

However, constraints can be powerful incentives for overcoming switching costs, first-mover disadvantages and other barriers to adoption of already developed but less than completely adopted technologies (e.g. Montalvo, 2008; Nemet, 2009; Popp et al., 2010). In this way, constraints can act as very efficient distributors of technologies that have already been commercialized, or are near commercialization, and it is primarily in this sense that constraints drive the evolution of technology. In the rest of this chapter, we present a historical case study that seeks to understand one particular constraint-induced radical innovation with the help of its prehistory.

To examine our theoretical proposition, we now turn to an extended case study of copper metallurgy. The twentieth century saw extensive developments in the technologies used to wrest usable copper in greater and greater amounts from increasingly poor ores (Figure 11.1). By 1990, the technologies that had dominated copper smelting for over 200 years

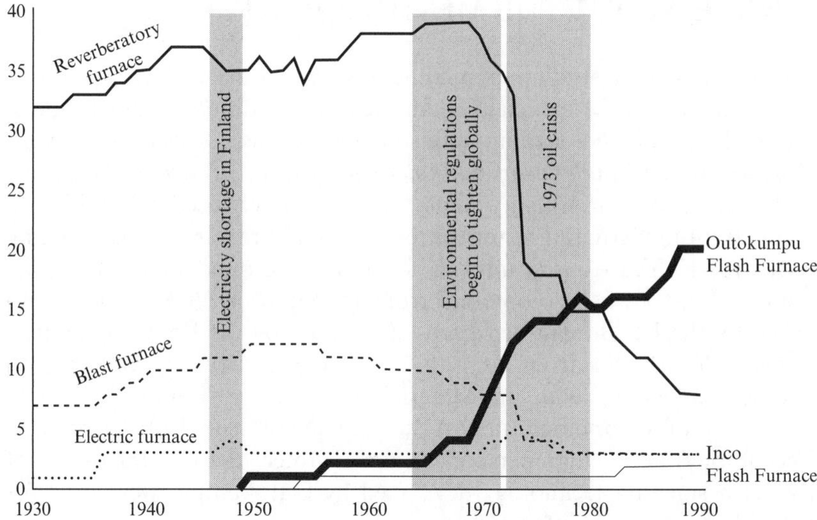

Note: Figures refer to number of smelters, not the number of smelting furnaces, as individual smelters often used multiple furnaces, sometimes of different types.

Source: Data from US Bureau of Mines *Mineral Yearbooks* 1931–1990 and from metallurgy textbooks (Biswas and Davenport, 1976; Davenport et al., 2002; Newton and Wilson, 1942).

Figure 11.1 Furnace types used by the world's copper smelters

were all but dead and replaced with technologies largely brought to use during the second half of the century. Several of these innovations were and still are hailed as landmarks in metallurgy. In this case study, we focus on what has been sometimes lauded as one of the most important metallurgical breakthroughs of the twentieth century (Särkikoski, 1999): the technology for flash smelting of sulfide ores. Flash smelting was first realized on a commercial scale in 1949 by Outokumpu, a small copper mining company in Finland, and a few years later in a slightly different form by nickel giant Inco in Canada.

The *effects* of Outokumpu's and Inco's advances were certainly significant, even radical. Replacing the previously dominant technology while simultaneously greatly improving energy efficiency, increasing production *and* reducing harmful emissions is no mean feat. But were these technologies really radical breaks from existing practice, requiring significant ingenuity, as argued by some (Särkikoski, 1999; Habashi, 1993)? Or were they just logical culminations of steady, incremental accumulation of knowledge?

A NOTE ABOUT THE METHODOLOGY

The historical case study presented here is based on a variety of sources, of which only some are cited in the text. In addition to cited works, the archives of the leading industry periodicals *Journal of Metals*, *Transactions and Bulletins of the American Institute of Mining Engineers*, and *Engineering and Mining Journal* were extensively used, in addition to a variety of modern and historical textbooks and research articles dating back to 1847. Year-to-year information and background about the copper mining industry was also obtained from annual *Minerals Yearbooks* published by the US Bureau of Mines (after 1995, by the US Geographical Survey). Yearbooks from 1931 to 1990 were particularly valuable. In addition, interviews with the Materials Engineering faculty of Helsinki University of Technology (now Aalto University School of Science and Technology) and senior personnel from Outotec, the current owner of the flash smelting technology developed by Outokumpu, were used for valuable background information.

The case was originally researched because flash smelting provided a rich history of an innovation developed as a direct response to resource constraints; it was not specifically chosen to support the argument outlined in this chapter. Although restrictions on length prevent us from providing more examples, it should be noted that several other technologies studied by the authors show similar dynamics (e.g. Korhonen, 2013).

THE DEVELOPMENT OF AUTOGENOUS SMELTING: THE ACCUMULATION OF INNOVATION INCREMENTS IN COPPER SMELTING OVER TIME

The prehistory of flash smelting of copper reaches back to the mid-1800s, when the spreading electrification greatly increased the world's appetite for this ductile, corrosion resistant and conductive metal. By 1880, the process of industrial scale copper production was relatively standardized across the planet: the ore was first crushed to a suitable size, then usually roasted in a special roasting furnace to remove most of the sulfur, then smelted in a smelting furnace to remove gangue and iron, and produce low-grade copper (so-called copper matte), and finally converted (purified) in a third furnace (Figure 11.2).

By the turn of the century, two types of smelting furnace provided the majority of the world's copper (Peters, 1898a). The older vertical shaft or 'blast' furnace, dating from the middle ages, was well suited for relatively rich ores that yielded solid lumps of copper-bearing sulfites. Its

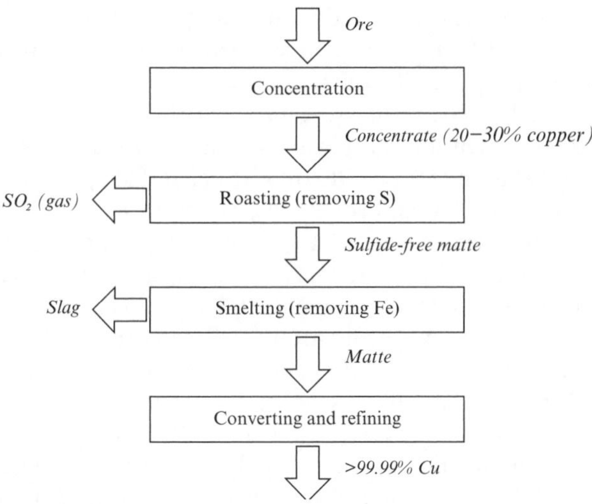

Figure 11.2 Basic copper smelting process

somewhat younger competitor, the horizontal or 'reverberatory' furnace, could process finely ground ore concentrates that were increasingly what the poorer copper mines could produce, and did not require purified and expensive coal-derived coke that blast furnaces depended upon.

To metallurgists, however, the choices were far from satisfactory. Both furnaces used considerable amounts of fuel, averaging at about 60 per cent of the total cost of copper smelting (Peters, 1898b). What made this particularly galling was the fact that the copper ore itself seemed to contain a significant source of energy: theoretically, the sulfur in the ore could be burned in the furnace, instead of wasting its heat content in the pre-furnace step of roasting. As one textbook on copper metallurgy, originally from 1877, lamented:

> In treating ordinary sulphide ores . . . we are at great pains to burn and destroy, more or less perfectly, the very sulphur and iron that form nature's fuel to melt the ore itself. . . . It is as though we employed the contents of our coal-bins to burn up a large portion of our coke-pile, so that we could get at the residue of it more conveniently. (Peters, 1898b, p. 376)

In about 1866, a Russian engineer named Semennikov made the first recorded recommendation for using sulfur as fuel (Sticht, 1898, p. 400). In this, he was inspired by the example of the steel Bessemer process, dating from 1855 and utilizing similar principles in the production of steel. As an example of either rapid spread of ideas or independent discovery, a mention of the idea can be found in an American textbook published in 1870 (Särkikoski, 1999). In due course, experiments were made in utilizing the sulfur's potential. The first concrete proof of possibility came in 1878 (Sticht, 1898). The result, so-called 'pyritic smelting' was a logical, straightforward extension of blast smelting process: a blast furnace 'scarcely original to the process,' to quote Sticht (1898, p.417), was loaded with ore rich in sulfides, and some coke added to help ignite the sulfur. The method promised to save up to 40 per cent of the fuel costs, which was a considerable advance. Furthermore, crushing and roasting processes and their associated equipment and personnel, costing anything between 17 to 50 per cent of the total smelting cost, could be dispensed with as well (Peters, 1898b). First industrial scale plants were erected already in 1881, and the process understandably attracted considerable interest (Sticht, 1898). However, it was difficult to control, yielded poor-quality copper matte and, most damningly, was very particular in regard of the ore used. Ore of suitable chemical composition and rich enough to be mined in lumps was yielded by some mines, but these ore bodies were largely exhausted by 1930. As a result, the pyritic smelting and its derivatives gradually fell out of use (Mäkinen, 1933). Therefore, from *c.* 1890 to 1950, the reverberatory furnace was increasingly the go-to method for smelting copper, although a few electric smelters were built in areas where abundant electricity was cheaply available (Newton and Wilson, 1942, p.146).

The idea of using sulfur as fuel did not vanish, however. In 1897, the first patent explicitly describing a method for 'autogenous' smelting of finely ground ores was granted in the United States (Bridgman, 1897; Inco, 1955). The patent, granted to a Chicagoan inventor Henry L. Bridgman, describes a largely conventional horizontal reverberatory furnace, where finely ground ore concentrate is ignited in air in place of conventional fuel.

It is probably no coincidence that Bridgman's patent appeared in the 1890s. Because ore bodies, by and large, were getting poorer and poorer, novel techniques for pre-processing copper ore had to be developed at the same period. Especially relevant to Bridgman's invention, one of these concentrating techniques under development at the time, froth flotation, could produce extremely fine, almost dust-like and relatively rich concentrate. Igniting dust-like concentrate would be significantly easier than igniting more or less gravel-like ore from earlier mines; in fact, it seems that Bridgman's patent was granted mere years after the first adoption of copper production processes that could support it!

Nevertheless, one major difficulty remained. Compared to pyritic smelting, too much heat was lost in flue gases. The process could thus only reduce fuel use, not eliminate it altogether. What is more, the fuel savings may have been as little as 5 per cent (Newton and Wilson, 1942). This and Bridgman's death in 1900 seem to be key reasons why there is no mention of Bridgman's autogenous furnace ever being used on an industrial scale.

Only a few years later, a group of metallurgists working for Anaconda Copper Company in Montana were granted another patent on autogenous smelting (Klepinger et al., 1915). The key improvement over Bridgman's patent was an addition of a heat exchanger: this re-used heat from the furnace flue gases to preheat incoming blast air and thereby greatly improved the heat balance of the process. Again, the invention was scarcely radical. The developer of pyritic smelting, John M. Hollway, had suggested in 1879 that heat of flue gases could be used to preheat the blast (Sticht, 1898), and a thorough discussion of pyritic smelting in a textbook published one year after Bridgman's patent (1898) noted that 'all metallurgists are well aware of the great effect produced by even a slight increase of temperature [achievable by preheating the blast] on the rapidity and energy of chemical reactions' (Peters, 1898b, p. 383). In a similar vein, Sticht writes in 1893 that:

> Locally, much has been made by the uninitiated, in the regions where the [pyritic smelting] process is in use, of this application of a preheated blast as a startling novelty in the smelting of the more precious metals [i.e. copper]. In truth, however, *hot blast, in the course of its nearly sixty years of uninterrupted employment for metallurgical purposes in Europe, has long ago found its way into the smelting of the more valuable metals and, in fact, is theoretically urged for introduction wherever the nature of the metals treated permits of its use.* (Sticht, 1898, p. 412, emphasis added)

Although the text does does not explicitly mention Bridgman's horizontal furnace design, the principle of 'recycling' heat from the flue gas to preheat the air blast would have been self-evident to any competent metallurgist. Preheating blast air had been invented in 1828 by James Beaumont Neilson in Scotland (Gale, 1967, pp. 55–8). E.A. Cowper and Sir Carl Wilhelm Siemens took the logical step of recovering heat from flue gases in the 1850s, with patents for 'regenerators' granted in 1856 and 1861 respectively (Gale, 1967, pp. 74–7, 98–100). In other words, what Klepinger et al. created was not so much a novel invention, but rather the logical improvement for the earlier process. It may be surmised that Bridgman might have himself made the same improvement, had he not died just three years after patenting his original furnace design.

As patent citations were a later innovation, it is difficult to obtain concrete proof that Klepinger et al. were aware of Bridgman's patent. However, Bridgman was not an isolated inventor, but an active member of the American Institute of Mining Engineers (AIME), the premier professional body for mining and metallurgy (see, for example, mentions of his work in *Engineering and Mining Journal*, October 10, 1891). Furthermore, the Bulletin and Transactions of the AIME and other periodicals like *Engineering and Mining Journal* covered extensively any new developments and patents relevant to the industry. It seems very highly unlikely that knowledge of prior art and archives of these periodicals were not available to Klepinger's group, working as it did for one of the largest copper mines in the United States. In general, the tightly knit, relatively small network of professional metallurgists, bound together by mutual interests and often known by name to each other, and the speed with which ideas in the mining and metals industry spread across the world seems to have precluded truly independent developments. Instead, the inventions were more or less collective (Allen, 1983): even competitors exchanged information about their processes and experiments relatively freely (see also Särkikoski, 1999, for examples of this network in the copper industry).

WORLDWIDE EXPERIMENTATION

There is no evidence that even Klepinger et al.'s furnace or the later, slightly different design by Horace Freeman (Freeman, 1932) were ever realized on a commercial scale (Bryk et al., 1958). The idea that fine concentrate dust produced by froth flotation could in principle be ignited and used to smelt the ore was well established, however, and several important developments occurred during the 1930s. First, in the spring of 1931, Frederick Laist and J.P. Cooper, again from Anaconda Copper Company, performed an experiment with a down-draft roasting shaft mounted over a small reverberatory furnace (Cooper and Laist, 1933). The concentrate was fed to the vertical roasting shaft and ignited in air. The fuel consumption was decreased by 60 per cent, and the experimenters reported that fitting shaft roasters to standard reverberatory furnaces would be reasonably expected to increase the smelting capacity by 50–100 per cent without increases in fuel consumption.

Another experiment along the same lines was conducted in the Soviet Union. Bryk et al. (1958) refer to a series of tests performed in 1935, in which concentrate-air suspension is injected into a shaft furnace built over a reverberatory-type vertical settler bath. However, wear of furnace bricks was intense and molten flue dust had tended to clog up the flue. In around 1937, a French-owned Societe Mines du Bor also tried autogenous smelting with vertical shaft furnace in Bor, Yugoslavia (Bryk et al., 1958; Särkikoski, 1999, p. 113). However, the results were negative and the outbreak of the Second World War ended the experiments. Although the former experiment was apparently reported in Russian periodicals, it seems that these experiences were largely ignored in North America.

Perhaps the most important development of the decade, however, was the publication of an article by T.E. Norman in 1936 (Norman, 1936). Working for nickel giant Inco in Canada, he discussed the theoretical aspects of smelting copper concentrates without the use of extraneous fuel. This research confirmed that the heat balance of the process would not suffice, unless remedial measures were taken: too much heat was absorbed by inert nitrogen introduced by the air blast. If, however, the concentrate were burned in an atmosphere of 40–95 per cent oxygen, then there would be less material to absorb the heat evolved. A higher temperature could be attained, and sulfide ores would melt. The theoretical discussion was accompanied with calculations based on smelter feeds at three large smelters in North America, showing that fully autogenous smelting was, at least in theory, entirely possible.

By that time, the result seems to have been almost self-evident to most metallurgists. The benefits of oxygen were well known even in the 1800s (Davis, 1923), the only problem having been affordable tonnage production of the gas. However, around the turn of the century major steps in solving the problem were taken by almost simultaneous developments in the air liquefaction process by Linde in Germany and Claude in France (Greenwood, 1919). The nitrogen produced from liquefied air had important uses in munitions and fertilizer industries, and tonnage production of oxygen was an established industry by the 1930s. As a metallurgy textbook published in 1942 states in an almost offhand discussion of Norman's research, 'Provided that a suitable furnace could be designed which would be as satisfactory as present equipment, the factor that would be of primary importance would be the relative costs of fuel and oxygen.' (Newton and Wilson, 1942, p. 161).

No mention is made of any insuperable difficulties: autogenous smelting was now seen as little more than an exercise in practical engineering, to be put into use once oxygen is cheaper than alternative fuels. In the latter half of the 1930s, several inventors had indeed experimented with different technologies and patented air- and oxygen-based furnaces (Haglund, 1940; Zeisberg, 1937).

But copper smelting technology was not the only force on the march. On September 1, 1939, the Second World War broke out in Europe. As industries round the world geared for war, productivity and reliability were what mattered. Uncertain, untested technologies had to wait, as long as 'there was a war to win' (Queneau and Marcuson, 1996, p. 14).

POST-WAR DEVELOPMENTS: THE BIRTH OF AUTOGENOUS SMELTING IN FINLAND

After the war, a small Finnish copper mining company Outokumpu found itself in dire straits. Founded in 1914 to exploit an eponymous copper deposit discovered in 1910 in eastern Finland and 100 per cent state-owned since 1924 (Kuisma, 1985), it had completed the world's largest electric copper smelting furnace in 1936, utilizing electricity from a new, close-by Enso hydropower plant (Newton and Wilson 1942, p. 147; Särkikoski, 1999, p. 36). The smelter produced valuable copper until summer 1944, when a Soviet attack forced it to relocate to the western coast of the country. The facility had almost been brought back to full operation when the armistice between Finland and Russia went into effect. One of the stipulations of the armistice required Finland to cede major parts of eastern Finland, including the important Enso and Rouhiala hydropower plants, to the USSR. In one fell swoop, one third of the electricity generating capacity of the country was lost (Kuisma, 1985, p. 124). Combined with the demobilization of the army, the gradual return to civilian life and, in particular, the exacting war reparations (amounting to 61 per cent of total exports in 1945, for example) that put heavy strain on the domestic industry to increase its production, the inevitable end result was a steep rise in electricity prices. Outokumpu's electric smelter, which alone accounted for some 3 per cent of the country's electricity use (Kuisma, 1985, p. 162), naturally suffered as its electricity costs saw an almost five-fold increase between 1946 and 1948 (Kuisma, 1985, p. 124).

A state-owned company producing not just an extremely important export product and perhaps even more important war reparatory deliverable (copper), but also some useful raw material for domestic industry (iron) and a necessary ingredient for paper and fertilizer manufacturing (sulfur) might have been able to commandeer the scarce electricity by arguing that its wares were more important than heat and lighting. However, the company's directors did not press the issue. In fact, even before the State Electricity Commission formally enquired whether Outokumpu could limit its electricity use in late 1945, a crash program to develop another smelting method had been started (Särkikoski, 1999, pp. 99–100). In just three months of work by a core group of three to four engineers, the work produced an alternative. Back-of-the-envelope calculations suggested that flotation concentrate from the eponymous Outokumpu copper mine could be ignited in air, and if the heat from the combustion could be recycled back as preheated air, the energy balance of the process would stay positive – or at least good enough so that the difference could be economically made up with scarce fuel oil. The hunt for scarce construction materials began in 1946, and in February 1947, less than two years after the start of the project, the pilot plant roared to life (Särkikoski, 1999; Kuisma, 1985). After a few days of adjustments, true autogenous smelting was achieved: as long as the finicky heat exchanger worked, the plant did not need extraneous fuel. The Outokumpu flash smelter, whose direct descendants would eventually be responsible for smelting more than 60 per cent of the world's primary copper, was thus born. Planning for a full-scale furnace began in September 1947, and building commenced three months later (Kuisma, 1985, p. 126). By 1949, the method was in commercial use, although an old electric furnace accounted for a third of the production as late as in 1953 (Kuisma, 1985, p. 127).

The pre-war developments, the speed with which Outokumpu's project proceeded despite post-war difficulties in obtaining even basic construction materials, and the fact that there is no evidence of the company's directors even seriously considering other options such as demanding priority for electricity or import credits for coal, strongly suggest that the discovery of the reverberatory furnace was 'in the air' at the time. Outokumpu had had no history to speak of in the field of technological innovation, but all the evidence suggests that its engineers were very well-informed of developments in copper smelting elsewhere. The company had been importing advanced mining and metallurgical technologies and skilled personnel since its founding in 1914, and its culture, partly a product of conscious attempt to develop its own skill set and partly a product of wartime necessities that forced it to do so, emphasized learning and improvising to keep the wheels turning. Särkikoski (1999, p. 71) summarizes the philosophy of Outokumpu's long-time managing director, Eero Mäkinen (state-appointed controller since 1919 and managing director from 1921 to 1953), as 'one should strive to do the best oneself, but if this does not succeed or if there is not enough time, it is not shameful to acquire the expertise from someone else'.

The managing director of Outokumpu and the éminence grise of the flash smelter, Eero Mäkinen, had a PhD in geology, an MSc diploma in mining engineering, and had worked for years to forge connections between Finland's fledgling mining engineering community and centers of learning abroad (Kuisma, 1985, 2008). He saw the development of Outokumpu from a simple commodity ore mine to value-added finished goods producer where 'every gram of ore promotes Finnish production' as a part of a larger production program for the industrialization of Finland, which was a poor agrarian country at the time (Särkikoski, 1999, p. 71). In his view, developing the company was synonymous with patriotism, and lack of ardor in pursuing increased domestic capability was tantamount to treason. His attitude may be best illustrated by a private letter to a key member of parliament in 1924, when the government's financial affairs committee contemplated renting the Outokumpu copper mine to foreign interests that would probably have had little interest in building refineries in Finland: in a vituperative letter, he threatened to 'gather all the dynamite he could find in the mine to the [financial committee's office] basement and blow the committee sky high' if the mine had been rented out (Kuisma, 1985, p. 62).

However, he also emphasized that there were no shortcuts to knowledge (Särkikoski, 1999, p. 71) and acted accordingly, taking an active part in the higher education of mining engineers and metallurgists. While serving as a managing director, Mäkinen went as far as to author one of the first comprehensive treatments of mining and metallurgical technology in Finnish (Mäkinen, 1933). (The text includes a brief but well-informed account of pyritic smelting of copper.) Before the war, Outokumpu gave financial aid to hire foreign specialists to teach in the Helsinki University of Technology, and engineers and geologists from Outokumpu were regularly sent abroad to learn from best available practices (Kuisma, 2008; Särkikoski, 1999). The experience of Petri Bryk, the first inventor in the key patents granted to flash smelting, is illustrative. When he joined the company as a fresh MSc graduate in 1938, the 25-year old's very first assignment, by Mäkinen's orders, was a two-year learning tour of copper refineries in North America (Särkikoski, 2008). This was a remarkable journey at the time, and it is an indication of Outokumpu's (and Mäkinen's) connections within the industry.

These connections and the 'can-do' attitude proved their worth when developing flash smelting. The fruit of Outokumpu's labors resembled the Laist-Cooper furnace tested in 1931 in the United States (Cooper and Laist, 1933).[1] A vertical reaction shaft was attached on top of a horizontal settler, greatly resembling the traditional reverberatory furnace. Ore concentrate and preheated air would be injected from the top and ignited. A flue collector at the other end of the settler would collect the gases and pass them to the heat exchanger. This, however, would prove to be a very troublesome piece of equipment: in the mid-1950s, the unreliable heat exchanger was in practice eliminated in favor of oil-fired preheaters (Särkikoski, 1999).

As mentioned previously, none of these developments were really novel: besides ideas already introduced within the copper industry, many of the components were in widespread use in other industries, and were consciously modeled after these practical examples (Särkikoski, 1999). Particulate reactions in air suspension were understood thanks to work in coal dust-fired power plants, sulfide burning in suspension was already practiced in the manufacture of sulfur and sulfur dioxide, and heat exchangers for blast preheat had been invented nearly 100 years previously. In a sense, the developer of pyritic smelting, John M. Hollway, had already predicted the likely shape of the flash furnace in 1879. He envisioned a vertical shaft furnace, with a hopper of sulfide ore on top feeding the process. The heat from the flue gases is re-used to preheat the blast air, and, 'If desired, the products could be run direct into suitable reverberatory furnaces, where, after the [matte] had subsided, the slag would be run off while yet in a molten state, and in which the oxidation of the [matte] could be completed' (quoted in Sticht, 1898, p.406).

Because of a multitude of existing patents describing autogenous smelting, Outokumpu in fact had trouble patenting its invention in the United States (Särkikoski, 1999). What was finally patented was the process direction, from top to bottom in a vertical chamber (Bryk and Ryselin, 1950). The invention may not have been an exact copy of a previously existing design (Habashi, 1993, 1998), but it is difficult to see it as a great conceptual or radical leap upon existing state of the art. The largest leap of faith taken by Outokumpu was the decision to build the first pilot plant: the primary metals industry was and is naturally cautious, and investing in 1946 Finland must have been a risky bet at best.

Meanwhile in Canada, the nickel giant Inco had restarted the research that had been rudely interrupted by the war. Based on Norman's work suggesting oxygen as the solution to the heat balance problem (Norman, 1936), the research immediately focused on oxygen flash smelting of sulfide concentrates (Queneau and Marcuson, 1996). A major reason for this focus was Inco's near-world monopoly in nickel; oxygen pyrometallurgy could be more easily adapted to nickel smelting (Brundenius, 2003), although Outokumpu's method would be eventually modified for nickel as well (Särkikoski, 1999; Kuisma, 1985). Suitability for nickel production was obviously attractive to a company that had held 90 per cent of the non-Communist world's nickel supplies before the war (Pederson, 2002).

Outokumpu had also considered the possibility of oxygen, but the unavailability of oxygen generators in post-war Europe and the cost of electricity needed to run them had dissuaded them from following the same path (Särkikoski, 1999; Kuisma, 1985). Inco, however, had access to nearly unlimited cheap electricity from Niagara Falls and, as one of the true giants in the metals industry, almost inexhaustible coffers as well. At the war's end in 1945, its assets were valued (in 2010 dollars) at about US$1350 million, its sales stood at US$1480 million, and its net income was a very healthy US$300 million (Pederson, 2002). Compared to Outokumpu, whose total sales in 1945 were some US$103.5 million and net profit approximately US$15.3 million in 2010 dollars (Kuisma, 1985,

p. 442), Inco was indeed a colossus. Despite its resources, Inco's R&D was hampered by long delays caused by the post-war supply problems and high prices stipulated by the dominant supplier of oxygen systems, Linde (Queneau and Marcuson, 1996). Nevertheless, cooperation with another, smaller supplier (Air Liquide) resulted in pilot plant operations starting in January 1947 (a month before Outokumpu's) and commercial operation of new-built furnace starting on January 2, 1952 (Inco, 1955; Queneau and Marcuson, 1996).

By that time, Outokumpu had over two years of commercial-scale, although by no means trouble-free, experience in flash smelting, but with a significantly different process. Outokumpu's process originally focused on saving as much extraneous fuels as possible with 'true' autogenous smelting; Inco's flash furnace utilized electricity-generated tonnage oxygen primarily to greatly increase the productivity and only secondarily to decrease the fuel costs of contemporary reverberatory furnaces (Bryk et al., 1958; Gordon et al., 1954; Inco, 1955; Saddington et al., 1966). Interestingly, as Figure 11.3 shows, the Inco method was even from the start more energy efficient than Outokumpu's method; the difference was that Outokumpu's original method did not need expensive and, in war-ravaged Europe, scarce oxygen generators. Instead, extra energy

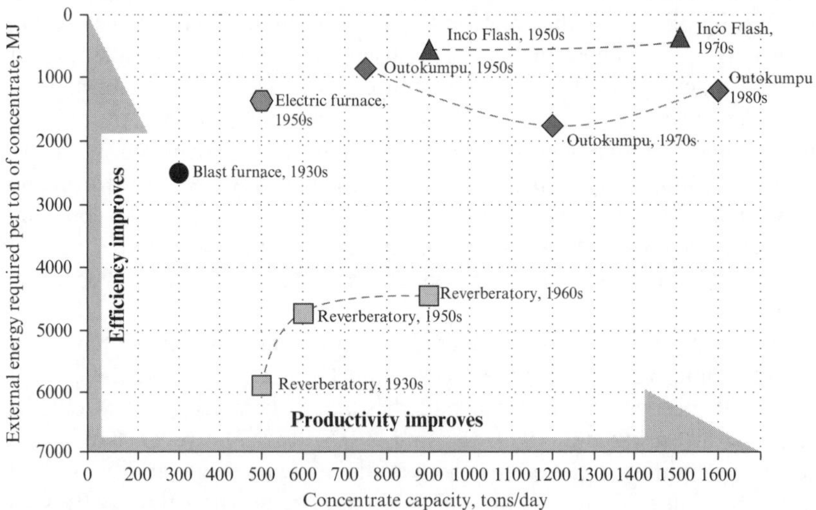

Sources: Data from Biswas and Davenport (1976), Davenport et al. (2002) and Newton and Wilson (1942).

Figure 11.3 The development of copper smelting furnaces in the twentieth century

was introduced first through heat exchangers, and after these proved troublesome in practice, from simple fuel oil burners.

What is common to both methods, however, is that they were hardly great conceptual leaps diverging significantly from existing practice. They were the almost logical outcomes – pinnacles, if you will – of a century of developments in various complementary fields, including metallurgy, chemistry, thermodynamics, cryogenics and the like. Although it could be claimed that the inventions look logical and inevitable only in hindsight, the contemporary sources give a strong impression that the inventions, even down to their specific details, were anticipated decades before developments in complementary technologies such as froth flotation and oxygen generators made their commercial realization practical. Clearly, several copper smelters had the opportunity, the means, and perhaps even the motive to invent the flash furnace; to take just one example, the Anaconda Copper Company, which sponsored Laist and Cooper's experiments, was already close to the solution by the 1930s, before experimentation was put on hold first because of the Great Depression and then because of the war (Newton and Wilson, 1942; Queneau and Marcuson, 1996). Commercialization required large investments (in 1947, building the commercial-scale Outokumpu flash furnace was estimated to cost US$30 million in 2010 dollars, representing some 12 per cent of the firm's revenues at the time) and the mining industry was notoriously conservative as long as traditional solutions served them well enough (Kuisma, 1985). In the aftermath of the Great Depression and again after the war, the problem for copper producers in general was how to cope with overcapacity, not how to increase the capacity by new furnaces.

Therefore, while crediting the *inventions* to contingent historical factors would overlook the long history of copper metallurgy, the inducive factors, such as the annexation of Finnish hydropower plants by the Soviet Union, or the availability of cheap hydropower and oxygen generators in Canada, help explain why precisely Outokumpu and Inco eventually became the *inventors* of flash smelting. This seems to have had less to do with any specific capabilities or resources of these two companies (albeit Inco's strong financial position certainly helped it to experiment and Outokumpu's interest in worldwide developments were surely conducive to innovation as well). Instead, the events seem to have more to do with historical and local contingencies. Interestingly, whereas scarce electricity was the key constraint driving Outokumpu's design decisions, it could be said that abundant electricity was one of the key factors influencing Inco's choice of oxygen-based process – which turned out to be the better of the two process choices.

Table 11.1 Outokumpu and Inco in comparison

	Outokumpu	Inco
Sales in 1945	$103.5 million (2010 USD)	$1480 million (2010 USD)
Net profit in 1945	$15.3 million (2010 USD)	$300 million (2010 USD)
Primary product	Copper	Nickel
Market share in primary product	≈ 1–2%	≈ 90%
R&D experience	Some during the war	Extensive, from 1906
Ownership	State-owned (but run as a commercial venture)	Private, traded in NYSE
Prior smelting technology	Electric	Mostly reverberatory + electric
Flash smelter developed primarily to:	Produce copper without external fuels	Improve productivity
External energy required per ton of concentrate	≈ 900 MJ	≈ 700 MJ

DISCUSSION AND CONCLUSIONS: THE NATURE OF INGENUITY UNDER RESOURCE CONSTRAINTS

Our study gives credence to the argument that radical innovation, induced by necessity, must benefit from existing innovation increments that are already invented and distributed in the innovation ecosystem, so that they are usable for the innovator company such as Outokumpu. The world-wide experimentation contributed to technological innovations that were available for the well-informed when under duress. Outokumpu had at its service a number of highly skilled experts who knew about technological developments and were able to apply them once the energy shortage forced the circumstances. However, it took the energy crisis after the Second World War in Finland for the engineers to recombine these elements (Arthur, 2009; Fleming, 2001; Hargadon and Sutton, 1997; Murray and O'Mahony, 2007) into a new technological architecture for copper smelting.

Although the details of the further development and the eventual 'triumph' of Outokumpu furnace are beyond the scope of this paper, it should be noted that the system is currently the single most common smelter furnace in copper smelting, accounting for 30–50 per cent of the total copper smelting capacity in the world (Moskalyk and Alfantazi, 2003), and is used for other metals (e.g. nickel) as well. The invention and the efforts to license it to other smelters also spawned a successful mining technology enterprise Outotec, which was spun out from its parent company in 2006. The widespread use of flash smelting may be largely credited to tightening environmental regulations, particularly sulfur dioxide emission limits, which were easier to achieve with a flash furnace, and to the 1973 energy crisis, which made competing electric furnaces in particular expensive to operate (Biswas and Davenport, 1976; Davenport et al., 2002; Särkikoski, 1999).

However, it is also instructive to note that the reasons why the Outokumpu and not the Inco method 'won' this race have much more to do with the different incentive structures and licensing strategies pursued by these two very different companies (Brundenius, 2003), rather than with the technical advantages of one technology over another. In fact, Inco's method remained in many ways technologically superior. As late as in 1976, a textbook of copper metallurgy (Biswas and Davenport, 1976) listed five significant advantages the Inco method possessed over Outokumpu's method. The authors found the success of Outokumpu's method 'somewhat surprising,' because 'it appears that the Inco process is the better from both a technical and economic point of view' (p. 170).

The efficiency and productivity advantages were diminished, but not eliminated, only when Outokumpu finally adopted oxygen injection in the early 1970s (Kojo and Storch, 2006). Brundenius (2003, pp. 24–39) argues convincingly that the reason why Inco did not license its technology was to protect its extremely lucrative advantage in nickel smelting, for which its method was particularly suitable. After all, technology sales were not big business: in 1956, Outokumpu's first licensor paid just US$1.3 million in 2010 dollars for a complete smelter design (Särkikoski, 1999). Hence, the risk of losing control of the technology may not have been a gamble worth taking for a company that was at the same time raking in revenues of some US$3700 million per year, in 2010 dollars, chiefly from sales of nickel (Pederson, 2002). As the saying went, Inco was not just a nickel company, Inco *was* nickel; by-products of nickel and technology sales were at best sideline business. Personal differences between Inco and Outokumpu management may also have played a part. Petri Bryk, an accomplished metallurgical engineer and the first inventor of flash furnace, was named Eero Mäkinen's successor as the CEO upon the latter's sudden death in 1953, and until the mid-1960s, selling 'his' flash furnaces seemed to be a sort of a pet project – even a hobby – that he indulged in (Kuisma, 1985). He may also have been motivated by monetary considerations: he received a substantial personal income from license fees, to the extent of making him the top earner in Helsinki, Finland's capital, in the early 1970s (Kuisma, 1985). In contrast, Inco's top management, though astute, was arguably less incentivized to selling 'its' technology to outsiders (Pederson, 2002).

What is striking about Outokumpu's constraint-induced innovation is how close it was to practical use in any case: had Outokumpu not developed flash smelting and had Inco retained its own patent for its own use, it is more than likely that another copper manufacturer would have invented a furnace functionally similar to Outokumpu's within the next decade at the latest. If this had happened, it seems likely that the end result would have been much closer to Inco's patent or the eventual oxygen-enriched Outokumpu furnace of the early 1970s: oxygen injection provided so many advantages that its adoption was all but inevitable.

In conclusion, based on our historical study of copper smelting over the last 200 years, ingenuity is indeed needed under resource constraints. However, this ingenuity appears to be about combinatorial capability, being able to efficiently track the existing distributed innovation increments in the ecosystem and recombine them into a working, even if at first experimental, solution (see also Baker and Nelson, 2005; Garud and Karnoe, 2003). The right moment for such new architecture building may be induced by a shock or other severe resource constraint. We extrapolate that absent the innovation increments or components accessible in the ecosystem, resource constraints do not seem to produce innovation but more likely result in poverty or the deterioration of economic circumstances. Resource constraints thus seem to be capable of both helping and hindering innovative efforts, just as Hoegl et al. (2008) and Weiss et al. (2011) have noted in the context of financial constraints. Necessity stimulates innovative efforts but results in ingenuity only when the opportunity for recombining innovation increments exists in the historical technological context, which, as described in this paper, may be global in nature. However, necessities may induce innovation in unexpected places – like Finland. In short, constraints seem to more capable of bringing about novel *innovators*, rather than novel *inventions*; necessity is a mother of inventors, not inventions.

ACKNOWLEDGEMENTS

The authors would like to acknowledge the invaluable help of Prof. Pekka Taskinen in sorting out the history of copper smelting, and Julia Kasmire and Igor Nikolic for fruitful correspondence and research cooperation that greatly helped developing the ideas presented here. All errors and omissions remain the responsibility of the authors. The preparation of this manuscript was assisted by the generous grant from Jenny and Antti Wihuri Foundation.

NOTE

1. It is interesting to note that neither the contemporary accounts written by the inventors (e.g. Bryk, 1952; Bryk et al., 1958) nor the patent (Bryk and Ryselin, 1950) mention the Laist-Cooper experiments. It is possible that these experiments were not known to Bryk et al., although the definitive history of Outokumpu flash furnace, written with access to Outokumpu's archives, does mention in passing that the pilot plant built in 1947 was modeled after the Laist-Cooper furnace (Särkikoski, 1999, p. 264, note 116). As the experiments were reported in one of the leading periodicals of the day, which was certainly read in Outokumpu, and the nearly contemporary account written by the inventors (Bryk et al., 1958) refers to T.E. Norman's 1936 study, which in turn quotes Laist and Cooper, other reasons – perhaps simple forgetfulness – seem more likely. The similarities between Outokumpu design and the Laist-Cooper furnace did not escape contemporary observers, however (Inco, 1955). These similarities have been downplayed by later scholars (Habashi, 1993, 1998) and it is definitely possible that the invention was entirely independent of Laist and Cooper's work, as the principles were well known and the parameters of the problem greatly limited the scope of practical solutions.

REFERENCES

Allen, R.C. (1983). Collective invention. *Journal of Economic Behavior & Organization*, **4**, 1–24.

Amabile, T.M. (1996). *Creativity in Context*. Boulder, CO: Westview Press.

Arthur, B.W. (2009). *The Nature of Technology: What It Is and How It Evolves*. New York: Free Press.

Baker, T. and Nelson, R.E. (2005). Creating something from nothing: resource construction through entrepreneurial bricolage. *Administrative Science Quarterly*, **50**(3), 329–66.

Biswas, A.K. and Davenport, W.G. (1976). *Extractive Metallurgy of Copper* (1st edn). Oxford, New York: Pergamon.

Bradshaw, G. (1992). The airplane and the logic of invention. In R. Giere (ed.), *Cognitive Models of Science* (pp. 239–50). Minneapolis: University of Minnesota Press.

Bridgman, H.L. (1897). Process of Reducing Ores. US Patent 578,912. United States Patent Office.

Brundenius, C. (ed.) (2003). *Technological Change and the Environmental Imperative: Challenges to the Copper Industry*. Cheltenham, UK and Northampton, MA, USA: Edward Elgar.

Bryk, P.B. (1952). Autogenes Schmelzen sulfidischer Kupfererze. *Radex Rundshau*, (1), 7–19.

Bryk, P.B. and Ryselin, J.W. (1950). Method for Smelting Sulfide Bearing Raw Materials. US Patent 2,506,557. United States Patent Office.

Bryk, P., Ryselin, J., Honkasalo, J. and Malmström, R. (1958). Flash smelting copper concentrates. *Journal of Metals*, June, 395–400.

Christensen, C. and Rosenbloom, R. (1995). Explaining the attacker's advantage: technological paradigms, organizational dynamics, and the value network. *Research Policy*, **24**, 233–57.

Cooper, J.P. and Laist, F. (1933). An experimental combination of shaft roasting and reverberatory smelting. *AIME Transactions*, **106**, 104–10.

Damanpour, F. (1991). Organizational innovation: a meta-analysis of effects of determinants and moderators. *Academy of Management Journal*, **34**(3), 555–90.

Davenport, W.G., King, M., Schlesinger, M. and Biswas, A.K. (2002). *Extractive Metallurgy of Copper* (4th edn). Oxford: Pergamon.

Davis, F.W. (1923). *The Use of Oxygen or Oxygenated Air in Metallurgical and Allied Processes. Report of Investigations No. 2502*. Washington, DC: US Bureau of Mines.

Fleming, L. (2001). Recombinant uncertainty in technological search. *Management Science*, **47**(1), 117–32.

Foster, R.N. (1985). Timing technological transitions. *Technology in Society*, **7**, 127–41.

Freeman, H. (1932). Process of Smelting Finely Divided Sulphide Ores. US Patent 1,888,164. United States Patent Office.

Gale, W.K.V. (1967). *British Iron and Steel Industry*. Newton Abbot, UK: David and Charles.

Garud, R. and Karnoe, P. (2003). Bricolage versus breakthrough: distributed and embedded agency in technology entrepreneurship. *Research Policy*, **32**, 277–300.

Gibbert, M., Hoegl, M. and Välikangas, L. (2007). In praise of resource constraints. *MIT Sloan Management Review*, **48**(3), 14–17.

Gibbert, M. and Scranton, P. (2009). Constraints as sources of radical innovation? Insights from jet propulsion development. *Management & Organizational History*, **4**(4), 1–15.

Gibbert, M. and Välikangas, L. (2004). Boundaries and innovation: special issue introduction by the guest editors. *Long Range Planning*, **37**, 495–504. doi:10.1016/j.lrp.2004.09.004

Goldenberg, J., Lehmann, D.R. and Mazursky, D. (2001). The idea itself and the circumstances of its emergence as predictors of new product success. *Management Science*, **47**(1), 69–84.

Gordon, J.R., Norman, G.H.C., Queneau, P.E., Sproule, W.K. and Young, C.E. (1954). Autogenous Smelting of Sulfides. US Patent 2,668,107. United States Patent Office.

Greenwood, H.C. (1919). *Industrial Gases*. New York: D. Van Nostrand Company.

Habashi, F. (1993). History of metallurgy in Finland: the Outokumpu story. *CIM Bulletin*, **86**(975), 57–63.

Habashi, F. (1998). The origin of flash smelting. *CIM Bulletin*, **91**(1020), 83–4.

Haglund, T.R. (1940). Roasting Process. US Patent 2,209,331. United States Patent Office.

Hargadon, A. and Sutton, R.I. (1997). Technology brokering and innovation in a product development firm. *Administrative Science Quarterly*, **42**, 716–49.

Henderson, R.M. and Clark, K.B. (1990). Architectural innovation: the reconfiguration of existing product technologies and the failure of established firms. *Administrative Science Quarterly*, **35**(1), 9–30.

Hoegl, M., Gibbert, M. and Mazursky, D. (2008). Financial constraints in innovation projects: when is less more? *Research Policy*, **37**, 1382–91.

Hoegl, M., Weiss, M. and Gibbert, M. (2010). The influence of material resources on innovation project outcomes. *2010 IEEE International Conference on Management of Innovation and Technology (ICMIT)* (pp. 450–54). Singapore.

Hughes, T.P. (1983). *Networks of Power: Electrification in Western Society, 1880–1930*. Baltimore, MD: Johns Hopkins University Press.

Inco, staff of (1955). Oxygen flash smelting swings into commercial operation. *Journal of Metals*, June, 742–50.

Katila, R. and Shane, S. (2005). When does lack of resources make new firms innovative? *Academy of Management Journal*, **48**(5), 814–829.

King, M.G. (2007). The evolution of technology for extractive metallurgy over the last 50 years – is the best yet to come? *Journal of Metals*, **59**(2), 21–27.

Klepinger, J.H., Krejci, M.W. and Kuzell, C.R. (1915). Process of Smelting Ores. US Patent 1,164,653. United States Patent Office.

Kojo, I.V. and Storch, H. (2006). Copper production with Outokumpu flash smelting: an update. In F. Kongoli and R.G. Reddy (eds), *Sohn International Symposium: Advanced Processing of Metals and Materials Volume 8 – International Symposium on Sulfide Smelting* (pp. 225–38). The Minerals, Metals & Materials Society.

Korhonen, J.M. (2013). Innovation effects: ersatz, or lasting improvements? The jet engine turbine case study. Presented at the 35th DRUID Celebration Conference, 17–19 June 2013.

Kuisma, M. (1985). *Outokumpu 1910–1985: Kuparikaivoksesta suuryhtiöksi*. Forssa: Outokumpu.

Kuisma, M. (2008). Vuorineuvos Eero Mäkinen (1886–1953). *National Biographies of*

Finland. Available at: http://www.kansallisbiografia.fi/talousvaikuttajat/?iid=405 (retrieved 2 October 2012).

Lampel, J., Jha, P.P. and Bhalla, A. (2012). Test-driving the future: how design competitions are changing innovation. *Academy of Management Perspectives*, **26**(2), 71–85.

Mäkinen, E. (1933). *Keksintöjen kirja: Vuoriteollisuus ja metallien valmistus*. Porvoo and Helsinki: Werner Söderström Osakeyhtiö.

Mone, M.A., McKinley, W. and Barker III, V.L. (1998). Organizational decline and innovation: a contingency framework. *Academy of Management Review*, **23**(1), 115–132.

Montalvo, C. (2008). General wisdom concerning the factors affecting the adoption of cleaner technologies: a survey 1990–2007. *Journal of Cleaner Production*, **16**(1), S7–S13. doi:10.1016/j.jclepro.2007.10.002

Moreau, C.P. and Dahl, D.W. (2005). Designing the solution: the impact of constraints on consumers' creativity. *Journal of Consumer Research*, **32**(1), 13–22.

Moskalyk, R. and Alfantazi, A. (2003). Review of copper pyrometallurgical practice: today and tomorrow. *Minerals Engineering*, **16**(10), 893–919.

Murray, F. and O'Mahony, S. (2007). Exploring the foundations of cumulative innovation: implications for organization science. *Organization Science*, **18**(6), 1006–21.

Nemet, G.F. (2009). Demand-pull, technology-push, and government-led incentives for non-incremental technical change. *Research Policy*, **38**, 700–9.

Newton, J. and Wilson, C.L. (1942). *Metallurgy of Copper*. New York: Wiley.

Nohria, N. and Gulati, R. (1996). Is slack good or bad for innovation? *Academy of Management Journal*, **39**, 1245–64.

Nonaka, I. and Kenney, M. (1991). Towards a new theory of innovation management: a case study comparing Canon, Inc. and Apple Computer, Inc. *Journal of Engineering and Technology Management*, **8**, 67–83.

Norman, T.E. (1936). Autogenous smelting of copper concentrates with oxygen-enriched air. *Engineering & Mining Journal*, **137**(10), 499–502, 526–67.

Pederson, J.P. (ed.) (2002). *International Directory of Company Histories, Vol. 45*. Farmington Hills, MI: St. James Press.

Peters, E.D.J. (1898a). The smelting of copper. In E.D.J. Peters (ed.), *Modern Copper Smelting* (pp. 224–35). New York: The Scientific Publishing Company.

Peters, E.D.J. (1898b). Pyritic smelting. In E.D.J. Peters (ed.), *Modern Copper Smelting* (pp. 372–95). New York: The Scientific Publishing Company.

Popp, D. (2006). Innovation in climate policy models: implementing lessons from the economics of R&D. *Energy Economics*, **28**(5–6), 596–609.

Popp, D., Newell, R.G. and Jaffe, A.B. (2010). Energy, the environment, and technological change. In B.H. Hall and N. Rosenberg (eds), *Handbook of the Economics of Innovation, Volume 2* (1st edn, pp. 873–937). Amsterdam: Elsevier BV.

Queneau, P.E. and Marcuson, S.W. (1996). Oxygen pyrometallurgy at Copper Cliff – a half century of progress. *Journal of Metals*, **48**(1), 14–21.

Romanelli, E. and Tushman, M.L. (1994). Organizational transformation as punctuated equilibrium: an empirical test. *Academy of Management Journal*, **37**(5), 1141–66.

Rosenberg, N. (1969). The direction of technological change: inducement mechanisms and focusing devices. *Economic Development and Cultural Change*, **18**(1), 1–24.

Rosenkopf, L. and Nerkar, A. (2001). Beyond local search: boundary-spanning, exploration, and impact in the optical disk industry. *Strategic Management Journal*, **22**, 287–306.

Saddington, R., Curlook, W. and Queneau, P.E. (1966). Tonnage oxygen for nickel and copper smelting at Copper Cliff. *Journal of Metals*, April, 440–452.

Särkikoski, T. (1999). *A Flash of Knowledge. How an Outokumpu Innovation Became a Culture*. (p. 304). Espoo and Helsinki: Outokumpu and Finnish Society for History of Technology.

Särkikoski, T. (2008). Bryk, Petri (1913–1977). *The National Biography of Finland*. Helsinki: Biografiakeskus, Suomalaisen Kirjallisuuden Seura. Available at: http://www.kansallisbiografia.fi/english/?id=1532 (retrieved 2 October 2012).

Sticht, R. (1898). Pyritic smelting – its history, principles, scope, apparatus, and practical results. In E.D.J. Peters (ed.), *Modern Copper Smelting* (pp. 396–441). New York: The Scientific Publishing Company.

Välikangas, L. and Gibbert, M. (2005). Constraint-setting strategies for escaping innovation traps. *MIT Sloan Management Review*, **46**(3), 58–66.

Ward, T.B. (2004). Cognition, creativity and entrepreneurship. *Journal of Business Venturing*, **19**, 173–188.

Weiss, M., Hoegl, M. and Gibbert, M. (2011). Making virtue of necessity: the role of team climate for innovation in resource-constrained innovation projects. *Journal of Product Innovation Management*, **28**(s1), 196–207.

Zeisberg, F.C. (1937). Ore Roasting. US Patent 2,086,201. United States Patent Office.

12 Stimulating organizational ingenuity with design methods

Judy Matthews

INTRODUCTION

Organizations invest in ways to stimulate new ideas for new products and services for the benefit of the organization, engaging in tournaments and competitions to generate new ideas or to combine existing ideas in new ways for new products and services (Terweisch and Uhlrich, 2009). Specifically, some large companies have developed platforms for posting intractable problems to tap into the ideas and problem solving abilities of a broader range of people (Huston and Sakkab, 2006; Morgan and Wang, 2010), and to develop new and elegant solutions often in an open innovation approach (Chesbrough, 2003).

The notion of ingenuity is often applied to individuals who create innovative solutions in situations of constraint, where ingenuity in the form of elegant solutions can be understood as one form of resourcefulness (Young, 2011). However, the notion of organizational ingenuity locates ingenuity more centrally to an organization's strategic decision making and implementation, embedding ingenuity into the company's culture.

Studies of organizations displaying ingenuity indicate a range of possibilities from extreme ingenuity (Baker and Nelson, 2005) to less dramatic but substantial changes (Thomke, 2003), sometimes in an experimental phase or as part of a move towards a new and distinct identity for ongoing innovation. Organizations that demonstrate ingenuity tend to clearly differentiate their products and services, capture our imagination and have the potential to shape the new directions for future delivery. Examples of organizational ingenuity seem to be influenced by the CEO's challenge and by senior management who encourage their staff to look for new and better ways to achieve new results.

The purpose of this chapter is to increase our knowledge and understanding of some of the processes that can lead to organizational ingenuity. We identify some of the processes that arise from situations where a design attitude and design methods are used with a view to identifying strategies for encouraging ingenuity in organizations.

The contribution we are seeking to make is a better understanding of design methods and their potential to generate new ways of working, and their implications for organizational ingenuity.

The chapter is structured as follows: first, we describe a spectrum of examples of organizational ingenuity and propose that organizations may engage in experiments that create the possibility of ingenuity being pursued at the organizational level. Second, we describe ingenuity within an organization and illustrate with an example of a company who explores new methods of developing ideas for new products and services. Third, we propose that organizational ingenuity can be stimulated by senior management where initiatives to develop, apply and disseminate through an organization need careful encouragement and sponsorship to embed within new product/service system. Finally, we suggest directions for a research agenda.

Organizational performance has been largely understood as the outcome of rational planned processes or emergent strategies. Firms gain sustained competitive advantage through relational architecture, reputation, innovation, and strategic assets (Kay, 1995). Some companies have explored processes to generate more imaginative possibilities to create novel business products and services. Firms engage in research and development (R&D) with established processes for search, idea generation, development and innovation, and capture financial benefits through patents or licensing. Firms look inside the firms for new ways of generating solutions using internal processes such as competitions and often collaborate with other companies with complementary assets in alliances or joint ventures. Stimulating firms to think in even more creative ways is not a simple process and one that requires both a planned and a spontaneous approach.

Organizations apply strategic thinking and planning processes to chart and shape their business activities and performance (Eden and Akermann, 1998: Heracleous, 1998: Heracleous and Jacobs, 2008; Hodgkinson and Healey, 2008). Some research has found that designers and designerly ways of working (Cross, 2006) can assist organizations to achieve better business performance (Lafly and Charan, 2008) beyond developing new products (Cox, 2005; DTI, 2005). Examples include Proctor and Gamble's (P&G) game changer process (Lafley and Charan, 2008), Umpqua Bank's processes (Freeze, 2006) and Bank of America's 'Keep the Change' program (*Bloomberg's Business Week*, 2006). While many of the positive outcomes of using design in business are known, the actual processes used by organizations to stimulate developing creative solutions in highly competitive markets may not be explicitly discussed. GE Medical's exploration of new products for developing economies (Govindarajan and Trimble, 2012) shows new product possibilities based on understanding the needs of the customers. Design methodologies and practices enable these possibilities (Brown, 2008).

The research gap that we attempt to address is how firms use design methods to stimulate and create new business possibilities in new markets with very distinct social and cultural contexts. We examine an example of one technology based company that has used their current deep knowledge and expertise or 'deep smarts' (Leonard and Swap, 2005) as a stimulus for exploring new ways of working in an unfamiliar context.

Capturing the life experiences of potential customers is used extensively in some new product development, but is largely novel to business practices and developing new services. This research links with the notion of discovering creativity in necessity and highlights the potential benefits of design methodologies to create new possibilities for better accessibility of the company's products to new clients. The overall theme of design processes, using design concepts and design technologies encourages exploration of the ways and means of developing new ideas for business with better outcomes.

ADDRESSING ORGANIZATIONAL CHALLENGES

Much of the previous research on ingenuity, or the process of applying ideas to solve problems or meet challenges, has been found in studies of creativity in organizations, often strongly focused on creativity at an individual level (George, 2007; Runco, 2004; Sternberg, 2006), with the exception of Hargadon and Bechky (2006) who investigated collective creativity in professional service firms. Where authors have described creativity in organizations (Amabile, 1996; Zhou and Shalley, 2008), the majority of perspectives that use constraints as a stimulus for creativity are found in the design and business literature (Stokes, 2006). For example, Stokes (2006) quotes Csikszentmihalyi (1996) and Simonton (1999) and argues 'creativity occurs when someone does something new that is also useful, generative or influential'. In this context 'useful' means solves a problem; 'generative' illustrates where one new thing leads to other ideas; and 'influential' indicates how we change the way we look at, listen to, think about or do new things' (Stokes, 2006: 1).

An exception to the use of ingenuity as a form of creativity was Flanagan, who described ingenuity as inventing or discovering a solution to a problem and 'the demonstration of a quality of genius in solving it in an unusually neat, clever or surprising way' (Flanagan, 1963: 92). Flanagan suggested that ideas, devices and procedures are ingenious if they meet three criteria for ingenuity: a practically useful solution to a real problem, an unusually clever solution and a novel solution in the sense of providing a surprisingly good solution to the special problems of the situation (Flanagan, 1963). More recently Young (2011), building on Flanagan's definition, described ingenuity as sharing similar concepts to creativity and creative problem solving such as novelty, usefulness and solutions but adding frugality (or using existing materials) and surprise.

In many organizations, creativity has a distinctive place, such as in developing new products and services, carrying out R&D, and business turnaround processes. Over time, rigidities form constraints and may limit possibilities for future approaches and strategic thinking is often carried out within these confines.

Recent research has shown that organizations benefit from creativity in multiple locations within the firms, not only in the R&D lab, but also in skunk works (Obstfeld, 2012). Firms that move outside such confines have applied design processes to imagine new possibilities, to create new directions and build new competencies for action. Organizational literature provides examples of some of these situations, such as P&G (Lafly and Charan, 2008). Some firms employ design methodologies to develop new ways of delivering their products or services to the market or the community (Bessant and Maher, 2009; Brown, 2008; Kelley, 2001; Lafley and Charan, 2008). We discuss design methods that stimulate organizational ingenuity, where constraints are used to create possibilities for new ways of working, in the context of a company that is a market leader.

Capturing the life experiences of potential customers is used extensively in some new product development, but is largely novel to business practices and developing new services. This research links with the notion of discovering creativity in necessity and highlights the potential benefits of design methodologies to create new possibilities for better accessibility of the company's products for new clients. The overall theme of design processes, using design concepts and design technologies encourages exploration of the ways and means of developing new ideas for business with better outcomes.

This chapter describes how a team from a large medical device company, when faced with a challenge to develop new customers in fast growing international markets, carried out the exploration of the needs of new clients in the largely unexplored market space of a developing country. This team used design methods and processes to identify the latent needs of new customers in situations of major economic, geographical, cultural and financial constraints. The encapsulation of the life experiences of potential customers is used extensively in some new product development, but is largely novel in business practices and in processes of developing new product/service combinations. This research links with the notion of discovering creativity in necessity and highlights the potential benefits of design methodologies to create new possibilities for better accessibility of the company's products to new clients.

USING DESIGN METHODS TO CAPTURE CUSTOMER INSIGHTS

The Case

A diverse, interdisciplinary and cross organizational research team with six members with expertise spanning innovation, design, management, marketing, sociology and health sciences was formed and engaged in research for 12 months with a medical device firm. Three research team members were experienced industrial designers, two of whom had worked with multiple organizations for extensive periods of time and one designer who had previous experience developing medical devices. The team was unified through a developed understanding of human centered design and the importance of co-design with users.

A leading medical device manufacturer was approached by the research team to participate in a research project to explore alternative approaches to the design of health services. This global company is the current market leader within its sector with a small specialized product offering and was aware of significant opportunities to grow within their current and emerging markets. This technology-led company had a significant science and engineering development team, and market research.

The research project was carried out over a nine-month period and experimented with a new approach to explore potential new product offerings that address the issues of cost and accessibility and alternative futures.

The researchers undertook the study using an action based research framework, continuously capturing and reflecting on the participant observations from the design method deployed. The findings were regularly discussed in the research team and presented to senior company R&D executives. The solutions described are examples of the types of outcomes that can be conceived using the design led innovation approach.

Data collection for this exploratory research was undertaken on location in two different countries, in multiple sites in urban and rural locations by a member of the research team and two company employees. A minimum of ten interviews, observations and videos of local conditions followed by interviews with medical practitioners, hospital administrators and medical suppliers were carried out across diverse geographical locations in each country. Data collection, analysis and interpretation were carried out in five stages. Each stage involved the construction of new materials, generating new insights through active engagement by the research team, with the data from potential customers, their families and communities, and relevant stakeholders such as local diagnosticians. Further new insights were also developed through presentation of these findings to the company's R&D team and the company's executive team.

Multiple methods of analysis were used to develop the materials for research analysis and discussion to generate detailed understanding of the lived experience in different contexts and to capture the individual, family and community aspirations and the nuances of daily life (Yin, 2003). The customer insights are a combination or the addition of problem-specific observations and personal and professional experience and hence include both subjective and objective knowledge generated from the gathered data. This design synthesis can be understood as an abductive sensemaking process (Kolko, 2009) of manipulating, organizing, running and filtering data in the context of design problem reframing; concept mapping and insight combination – prioritizing, judging and forging connections (Beckman and Barry, 2009). This approach is presented as the Design Cycle in Figure 12.1.

Each research stage is described within the context of the project case study. The stages, methodologies, technologies and outputs and the

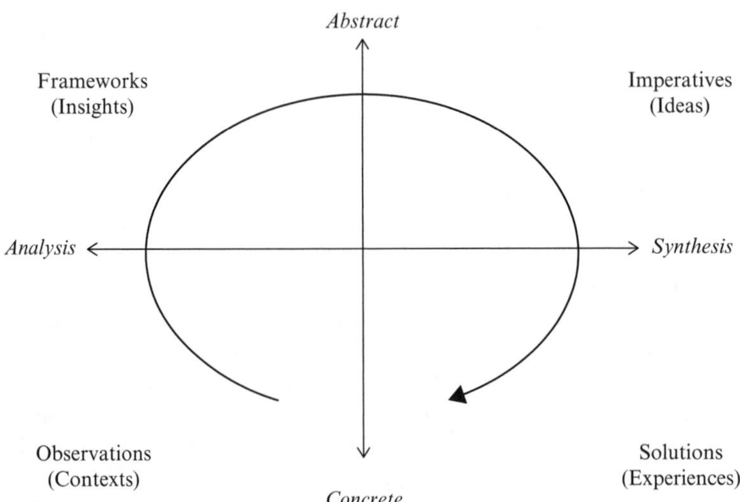

Source: Based on Beckman and Barry (2009).

Figure 12.1 Design thinking cycle

Table 12.1 Summary of research processes and outcomes

Stages	Description	Methodologies and technologies	Outputs
Stage 1	Understanding the Social Cultural Context	Semi-structured interviews with visual prompt cards; ethnographic methods: Personas	Multiple personas which captured insights about diverse contexts and needs
Stage 2	Moving from Product Interactions to Temporal Experimental Journeys	Journey of patient: diagnosis preparation; develop language	Multiple experiences of personas over life journey; multiple stakeholders and resources involved; gaps in availability of resources; value propositions for new services
Stage 3	Identifying the Latent User Needs for New Services	Combines experiential journey maps and personas Role plays; graphical representation	Graphical representations; digital services opportunities
Stage 4	Transforming Latent User Needs into Scenarios	Persona development and journey mapping developed to form 'Fragmented Connections' – essentially five sub-scenarios that present potential solutions for a potential future	Narratives for each sub-scenario
Stage 5	Communicating the results to R&D Team, Senior Management and selected industry representatives	Transform these five scenarios to two-minute vignettes for final deliverable for project	Video vignettes for new possibilities

processes used to develop material for personas, storyboards and narratives are summarized and presented in Table 12.1. Each stage of the research involved developing narratives of the potential clients, and narratives for the organization. Each narrative can be a metaphor for the challenges, possibilities and framing for new ways of working. Bartel and

Garud (2009) describe the use of narratives for organizational innovation. Stories are important to the design process. The stories collected during the field observation and interview process were shaped during framing (Beckman and Barry, 2009). Stories can create emotional connection with the challenges that potential customers face in benefitting from accessibility to the product/service.

The information captured was translated into video vignettes that were then used as narratives with company representatives and industry representatives to test the validity of new products and services, and systems for the company. Designers use processes of testing the validity of new possibilities with multiple informed stakeholders rather than reliability testing with past products to test their propositions (Martin, 2005, 2009). In essence the new notions for products and services were proposed as possible scenarios for new product service systems in a number of combinations, presented to the industry experts as new possibilities and these presentations were used to capture industry insights to further refine and shape the final products.

FINDINGS

New ideas for appropriate and potential services were developed and are being implemented by the company. The use of design methods to capture the experiences of potential customers and to present these visual and oral narratives as personal stories to the company also engaged an emotional connection for the company (Beckman and Barry, 2009).

The characteristics of the research team were to develop openness to new ideas and ways of working across the organization and its multiple locations. Many stakeholders used the discussion to generate suggestions for new ways of working and new types of technology to assist increased accessibility and quality of services, with ready acceptance of creating possibilities for new ways of working. Due to constraints of space, the processes of selecting the players, building the team, creating collaborative engaged members at a micro-level are not addressed in this paper. Ingenuity in a project team with the assistance of a research team has created new ideas for services in a science and technological product based organization. The approach used in this case study has the potential to also create new possibilities for the larger organization. The challenge now is to use these design processes in the strategic approach of the company.

Other examples of organizational ingenuity have occurred where the top leadership team has sponsored a new approach, such as in Apple and P&G (Lafly and Charan, 2008) where design principles and customer insights are used to imagine and create new possibilities. Leadership from the CEO tends to drive the organization to engage in the experimentation of new ways of working to generate new outcomes.

In contrast, in the study reported here, approval was provided for this experimental project to capture customer insights for new product service systems, but the challenge of taking this design methods approach from the project or program level as strategic thinking or strategic planning to a company-wide approach is yet to be undertaken. Here we suggest that strategic thinking and strategic planning occur iteratively over time where there is a continual quest for novel and creative strategies that can be born in the minds of strategists or can emerge from the grassroots desirability and feasibility and plan for realization (Heracleous, 1998). Strategic thinking can be understood as double-loop learning (Heracleous, 1998) and strategy is often crafted through embodied metaphors (Heracleous and Jacobs, 2008) and design methods have also been applied successfully (Monnavarian et al., 2011).

Three aspects of design, namely user understanding and empathy, concept visualization and strategic business design, are proposed as steps leading to new ways of business development (Fraser, 2008: 67–8). For example, first, reframe the business through the eyes of the user – the whole person, not just what they do but how they feel about it and how their needs surrounding the activity link to other parts of their life in terms of other activities, other people and other cues to their needs. Second, generate new possibilities for meeting human needs in an imaginative way through concept visualization, ideation and multiple-prototyping (Coughlan et al., 2007). Third, use a firm's activity system and leverage from the existing system to a future system and align strategic concepts with a future reality. Such processes can be useful for taking learning from a project and portfolio organizational level to a central role in a company's strategy.

LINKING EMERGING OPPORTUNITIES DEVELOPED THROUGH INTERNAL R&D

This study articulates one approach to stimulating ingenuity through the engagement of an external experienced designer (who had extensive prior experience with the company) and a multi-disciplinary research team and the experimentation and implementation of research methods in developing narratives tested through vignettes in company forums for new services. Through reflection on the internal working of a research project group and the engagement of the team in developing new approaches, we have a clearer notion of design processes and their application to creating new possibilities for client access as well as some detailed processes of visual storytelling and narrative, and the possibilities for application of design processes at the strategic level of the company.

LIMITATIONS

The research presented here reports an example of one company experimenting with new methods to stimulate new ideas and solutions for product/service systems by investing in in-country explorations that generate deeper understandings of the life situations of future potential customers and their system constraints. One limitation of this research is that although the design research process was carried out in multiple sites, urban and rural in two large culturally diverse countries, it was implemented with only one company. There are limits to generalizing from this one case but we argue the nature of this approach may have application for stimulating ingenuity in other companies who are exploring possibilities in new markets.

CONTRIBUTIONS AND FUTURE RESEARCH

Building on existing literature we identified an approach that describes the use of design methods for problem solving in situations of constraint. We found that design methodologies have particular potential to provide companies with insights into new market developments in emerging economies. Finally we suggest that organizations that capture learning from using ingenuity and new ways of working using design technologies in learning about future customers are more likely to benefit from using ingenuity for organizational benefits.

We contend organizational ingenuity is an important phenomenon in changing economic and organizational contexts, and as such requires a more structured approach to further explore the forms, precedents and impact of ingenuity in relation to the size of the companies and the varied nature of this phenomenon. To progress such research we propose that some of the questions that need to be addressed are: What factors encourage firms to explore possibilities in new ways to develop elegant solutions? What are the pre-conditions that make such organizational ingenuity more likely? How do design processes stimulate ingenuity and what is required to move from individual and team ingenuity to organizational ingenuity?

REFERENCES

Amabile, Teresa M. (1996), *Creativity in Context*. Boulder, CO: Westview Press.

Baker, T. and R.E. Nelson (2005), Creating something from nothing: resource construction through entrepreneurial bricolage, *Administrative Science Quarterly*, **50**, 329–66.

Bartel, C.A. and R. Garud (2009), The role of narratives in sustaining organizational innovation, *Organization Science*, **20**(1), 107–17.

Beckman, S.L. and M. Barry (2009), Design and innovation through storytelling, *International Journal of Innovation Science*, **1**(4), 151–60.

Bessant, J. and L. Maher (2009), Developing radical service innovations in healthcare: the role of design methods. *International Journal of Innovation Management*, **13**(4), 1–14.

Bloomberg's Business Week (2006), Case study: Bank of America, June 18.

Brown, Tim (2008), Design thinking, *Harvard Business Review*, June, 85–92.

Chesbrough, Henry (2003), *Open Innovation: The New Imperative for Creating and Profiting from Technology*. Boston, MA: Harvard Business School Publishing.

Coughlan, P., J. Fulton-Suri and K. Canales (2007), Prototypes as (design) tools for behavioral and organizational change: a design-based approach to help organizations change behaviors, *The Journal of Applied Behavioral Science*, **43**(1), 122–34.

Cox, G. (2005), *Cox Review of Creativity in Business*. London: HM Treasury United Kingdom.

Cross, Nigel (2006), *Designerly Ways of Knowing*. Berlin: Springer.

Csikszentmihalyi, Mihalyi (1996), *Creativity: Flow and the Psychology of Invention*. New York: Harper Collins.

DTI (2005), *Creativity, Design and Business*. DTI Economics Paper No. 15. London: Department of Trade and Industry, UK.

Eden, Colin and Fran Akermann (1998), *Making Strategy: The Journey of Strategic Management*, London: Sage.

Flanagan, J.C. (1963), The definition and measurement of ingenuity. In Taylor, C.W. and F. Barron (eds) *Scientific Creativity: Its Recognition and Development*, pp. 89–98. New York: John Wiley & Sons.

Fraser, H.M.A. (2008), The practice of breakthrough strategies by design, *Journal of Business Strategy*, **28**(4), 66–74.

Freeze, Karin (2006), *Umpqua Bank: Managing the Culture and Implementing the Brand*. Design Management Institute Case Study. Boston, MA: Harvard Business School Publishing.

George, Jennifer M. (2007), Creativity in organizations. *The Academy of Management Annals*, **1**(1), 439–47.

Govindarajan, Vijay and Chris Trimble (2012), *Reverse Innovation: Create Far From Home, Win Everywhere*. Boston, MA: Harvard Business Press Books.

Hargadon, A.B. and B. Bechky (2006), When collections of creatives become creative collectives: a field study of problem solving at work, *Organization Science*, **17**, 484–500.

Heracleous, L. (1998), Strategic thinking or strategic planning? *Long Range Planning*, **31**(3), 481–7.

Heracleous, L. and C.D. Jacobs (2008), Crafting strategy: the role of embodied metaphors, *Long Range Planning*, **41**, 309–25.

Hodgkinson, G.P. and Marl P. Healey (2008), Toward a (pragmatic) science of strategic intervention: design propositions for scenario planning, *Organization Studies*, **29**(3), 435–57.

Huston, L. and Sakkab, N. (2006), Connect and develop: inside P&G's new model for innovation, *Harvard Business Review*, March, 58–66.

Kay, John (1995), *Why Firms Succeed*. Oxford: Oxford University Press.

Kelley, Dave (2001), *The Art of Innovation: Lessons in Creativity from IDEO*. New York: Doubleday.

Kolko, J. (2009), Abductive thinking and sensemaking: the drivers of design synthesis, *Design Issues*, **26**(1), 15–28.

Lafley, A.G. and Ram Charan (2008), *The Game-Changer: How You Can Drive Revenue and Profit Growth with Innovation*. New York: Random House.
Leonard, Dorothy and Walter Swap (2005), *Deep Smarts: How to Cultivate and Transfer Enduring Business Wisdom*. Boston, MA: Harvard Business School Press.
Martin, R. (2005), Validity versus reliability: implications for management, *Rotman Magazine*, Winter, 1–8.
Martin, Roger (2009), *The Design of Business: Why Design Thinking is the Next Competitive Advantage*. Boston, MA: Harvard Business Press.
Monnavarian, A., Farmani, G. and H. Yajam (2011), Strategic thinking in Benetton, *Business Strategy Series*, **12**(2), 63–72.
Morgan, J. and R. Wang (2010), Tournaments for ideas, *California Management Review*, **52**(2), 77–97.
Obstfeld, D. (2012), Creative projects: a less routine approach toward getting things done, *Organization Science*, **23**(6), 1571–92.
Runco, M. (2004). Creativity, *Annual Review of Psychology*, **55**, 657–87.
Simonton, Dean K. (1999), Creativity from a historiometric perspective. In Robert J. Sternberg (ed.) *Handbook of Creativity*, pp. 116–33. Cambridge: Cambridge University Press.
Sternberg, Robert J. (2006), The nature of creativity, *Creativity Research Journal*, **18**(1), 87–98.
Stokes, Patricia. D. (2006), *Creativity from Constraints: The Psychology of Breakthrough*. New York: Springer.
Terweisch, Carl and K. Uhlrich (2009), *Innovation Tournaments*. Boston, MA: Harvard Business School Press.
Thomke, S. (2003), R&D comes to services, *Harvard Business Review*, **81**(4), 71–9.
Yin, Robert K. (2003), *Case Study Research: Design and Methods* (3rd edn). Newbury Park, CA: Sage.
Young, John. (2011), *How To Be Ingenious: Comedians, Engineers and Survivalists*. RSA Projects, Royal School of Arts, UK. Available at: http://www.thersa.org/__data/assets/pdf_file/0006/395475/How-to-be-Ingenious.pdf (last accessed 28 September 2013).
Zhou, Jing and Shalley, Christina E. (2008), Expanding the scope and impact of creativity research. In Jing Zhou and Christina E. Shalley (eds) *Handbook of Organizational Creativity*, pp. 347–68. New York: Lawrence Erlbaum and Associates.

13 Ingenuity research: current problems and future prospects

Israel Drori, Benson Honig and Joseph Lampel

In this volume, we examine a concept of organizational ingenuity anchored in the following definition: *the ability to create innovative solutions within structural constraints using limited resources and imaginative problem solving*. The group of authors assembled here represent a variety of perspectives that have emerged over the course of two years beginning with the conference on organizational ingenuity at McMaster University in September 2011, subsequently during the exploration and debate of the key issues that took place during the European Group on Organizational Studies (EGOS) subtheme on ingenuity in Helsinki in 2012, and lately at a panel on 'Ingenuity and National Culture' at the Academy of Management.

In this concluding chapter, we would like to take a step back and survey progress thus far with a view towards suggesting promising research issues that have not been explored in sufficient detail, but to our mind hold the promise for interesting and important findings. The current volume is the first collection of articles that deal specifically with organizational ingenuity. (A Special Issue on the same topic is planned for Organization Studies.) When it came to selecting the papers for this book we were to some extent limited by authors who were willing to join us in exploring this area. Thus, while the scholarship presented in this volume examines organizational ingenuity from a variety of perspectives, it is by no means a comprehensive survey of a research area that is still very much at a nascent stage. The same can be said for the list of future research topics that we discuss below. They are as follows: (1) expanding ingenuity research to contexts and environments not yet examined; (2) imprinting and developing a culture of ingenuity; (3) examining ingenuity from a longitudinal perspective; (4) stand alone versus process ingenuity; (5) the contingent nature of ingenuity; (6) the 'dark side' of ingenuity; and (7) the agency/action approach.

EXPANDING INGENUITY RESEARCH TO CONTEXTS AND ENVIRONMENTS NOT YET EXAMINED

Ingenuity has important consequences for organizational life at both the micro and the macro levels. At the most aggregated levels, ingenuity impacts humanity's ability to survive on the planet. One scholar refers to the 'ingenuity gap' as an ingenuity shortfall exhibited in poor regions and countries of the world, accounting for the increasing gap between rich and poor (Homer-Dixon, 2002). Exploring the role of ingenuity from an institutional or cultural perspective, including the country level, should yield promising insights. Ingenuity is essential for understanding innovation in complex systems that are evolving, for example the World Wide Web, or are failing to evolve, for example our educational system. Research should therefore examine how ingenuity impacts the utility and growth of such complex systems. Further, partly because most scholars today are still located in advanced industrial countries, the studies contained in this book (and we would argue research in general) do not address the issues of ingenuity in developing countries. Our own observations suggest that ingenuity should be of particular importance to developing countries where resources are scarce and institutions are more often a hindrance than help to development. Research on this subject in these contexts will allow us not only to gain insights on how different conditions influence ingenuity, but is also of practical importance to countries and societies that desperately need ingenuity to solve intractable problems.

IMPRINTING AND DEVELOPING A CULTURE OF INGENUITY

A substantial body of research deals with the idea that an organization's social and cultural structure, processes and action are strongly shaped by the conditions that exist at the time of the organizational founding. In this vein, Stinchcombe's seminal work on organizations (1965) provides numerous insights on organizations' enduring mechanisms. He points out that the founding conditions that imprint a new venture have durable effects, both positive and negative, on its survival and development. This implies that the initial social and economic conditions during a firm's founding imprint its capabilities and core features. Thus, we can hypothesize that organizations are 'imprinted' with the propensity for ingenuity as an outcome of historical circumstances associated with the organization's founding and evolution (Marquis and Tilcsik, 2013).

Although the legacy of early organizational life may exert lasting effect, the changing context and environment will also constrain or encourage organizational ingenuity. Furthermore, when practices and values are perceived as worthy, their preservation serves as 'templates' for the new members that may suppress ingenuity (DiMaggio and Powell, 1983). In this vein, ingenious members of organizations may engage in disruptive practices that encourage creative solutions under resource constraints (Lampel et al., 2011). Note that ingenious solutions could take different forms and actions, including creating, maintaining or disrupting an existing order (Lawrence and Suddaby, 2006). Thus, ingenuity relates to the way the organization is reconstructing experience through designing and implementing nonconventional solutions.

Organizations are designed to meet certain objectives through structures, processes and routines. Therefore, notwithstanding managerial rhetoric to the contrary, the tendency towards inertia and efficiency will often dominate (see Scott, 2008). However, research on organizational change suggests that some organizations develop imprint norms, values and practices that encourage ingenuity (Miller, 1991). Thus, future research may look at how 'history matters' in terms of embedded values and practices that encourage ingenuity, starting with the organization's founding and reinforced by the organization's evolution during its life cycle.

EXAMINING INGENUITY FROM A LONGITUDINAL PERSPECTIVE

While several of the chapters in this book employ one or another form of longitudinal analysis, none of them focuses exclusively on the life cycle of ingenuity. What kind of organizational processes initially instigate ingenuity, what are the catalysts for enhancing organizational ingenuity, and to what extent does organizational ingenuity become bureaucratized, and effectively 'killed off'? Are there different identifiable stages of organizational ingenuity? Do the stages correspond to similar stages of organizational development? Answering these questions requires careful longitudinal work, most likely case studies, where scholars who adopt ingenuity theoretical perspectives make close and repeated observations over time.

STAND ALONE VERSUS PROCESS INGENUITY

We can stipulate that there are contexts and values which are more amenable to the emergence of organizational ingenuity than others. In this vein, ingenuity could either be a 'stand alone' or process. For example, Siqueira et al. (Chapter 9) claim that ingenious action is an ongoing process and that leaders' involvement in local communities and perception of social disparities can increase the likelihood of organizational ingenuity. The organizational ingenuity of these leaders can increase the likelihood of creative solutions with an impact at the community and national levels. Consequently, ingenuity is not an isolated outcome of a given set of factors, but an active search for solutions carried out by social actors who are continuously applying different forms of creative and innovative actions within an emerging context. Ingenuity processes may encompass diverse strategies of action, varying in their degree of innovation and creativity and based on the way actors use diverse resources. Thus, ingenious processes may redefine the abilities of social actors in expanding the boundaries of problem solving by implementing creative and innovative solutions which are based on the assessment that there is a need to overcome various constraints.

In sum, ingenuity implies creative solution processes characterized by the need to provide responses to a pressing situation within or outside the organization. Ingenuity can be manifested in organizational processes that systematically produce ingenious outcomes that deviate from conventional solutions to organizational problems. The ability of actors to deviate from the common routine or practices is not limited to a single event, but has the potential to become a systematic process. So, this suggests that future research should focus on processes that examine multiple events over time.

THE CONTINGENT NATURE OF INGENUITY

Ingenuity may at times be deliberate, but it is often also a contingent, iterative process, more closely attuned to a method incorporating effectuation (Sarasvathy, 2001). Developing a research agenda incorporating a contingent experimental method suggests the need for a phased, cyclical approach, complete with feedback paths, at critical stages of ingenuity. Taking an effectuation approach to organizational ingenuity highlights the unpredictable and emergent nature of organizational decision making and action. There is not an a priori assumption that ingenuity must occur in a linear way, starting before any related activity of the organization has begun, and there is no expectation that ingenuity related activities proceed through a sequential manner. In many cases, the ingenious activity may only be fully understood retrospectively, as part of a process that may involve feedback loops allowing for the incorporation of new knowledge based on current information available to the organization. Understanding ingenuity from an effectuation approach requires studies that examine organizations from a multi-faceted perspective, so that related issues regarding constraints can be simultaneously tracked and monitored as the ingenious idea(s) develops, diffuses, is debated, implemented, and appraised.

THE DARK SIDE OF INGENUITY

The hallmark of organizational ingenuity is the idea that ingenious action implies the need for harnessing creative resources to provide a solution to problems which are governed by the organization's formal structures and processes. To achieve an ingenious solution, organizational actors adopt the idea that there are obstacles that call for an unconventional way of mobilizing resources, usually a recombination of elements exist in their accessible environment (Baker and Nelson, 2005). Thus, ingenuity in organizations is triggered as a response to constraints which, prior to the ingenious solution, blocked the effective implementation of an activity or idea (Lampel et al., 2011). Many of the ideas associated with organizational ingenuity and its effect on the way organizational members deal with resource constraints emphasize the contextual and temporal features that are associated with the task in hand. However, ingenious solutions may solve a problem only at a certain point in time and be later deemed irrelevant and even harmful. In this vein, ingenuity can be seen as an intensive effort aimed at immediate, ad-hoc problem solving regardless of the ramifications for future processes.

The fact that ingenuity is enacted by organizational members who in many instances change the 'existing order' increases uncertainty and is grounds for potential conflicts. For example, teams may launch an ingenious action focusing on a specific problem while ignoring the implication for other teams or processes within the organization. This ingenious action could be disruptive in the sense that it is a 'double edged sword'. On the one hand, enacting effective problem solving techniques is both surprising and nonconventional. On the other hand, solving one problem may disrupt the regular practices and routines that are interdependent. Further, a nonconventional solution may result in difficulties in coordination between different units because of practice or process misalignments (Lampel and Bhalla, 2011).

Ingenuity implies a claim that certain ideas and practices are tied to innovative and creative ways of solving problems. The outcome could be an organizational or technological improvement, or a radical solution to an existing problem. In many instances, taking advantage of the outcome requires an attention to management issues or alteration of strategies and processes. It may be the case that an organization could provide an internal context for ingenious action, but would not be equipped to exploit the innovation or the creative solution to the full. Ingenuity can therefore alter existing processes: ingenious outcomes may produce adverse consequences to existing individual relations and modes of coordination.

Thus, future research may study the dark side of ingenuity – its harmful outcomes that adversely affect the organization or its stakeholders. In particular, scholars might examine the organized circumstances of ingenuity that produces harmful outcomes.

THE AGENCY/ACTION APPROACH

Recently, different streams of research have converged on the notion of 'work' as a unifying concept that provides insightful theorization on the role of agency in actively influencing processes and outcomes in organizations. This is best articulated in institutional theory – 'institutional work' is defined as: 'the purposive action of individuals and organizations aimed at creating, maintaining and disrupting institutions' (Lawrence and Suddaby, 2006: 215). Thus, agency shapes outcomes through knowledgeable actors that respond to different pressures and employ varied actions (Lawrence et al., 2009). Other streams of research use the notion of 'work' to show how managers constitute both personal and social identity for justifying their action within certain organizational and environmental contexts (Sveningsson and Alvesson, 2003; Watson, 2008). Studies on identity work indicate that various discourses reconstruct managerial identities that consequently influence action (Alvesson and Willmott, 2002).

Thus, the notion of work represents a focus on practice grounded 'on the situated actions of individuals and groups as they cope with and attempt to respond to the demands of their everyday lives' (Lawrence and Suddaby, 2006). This suggests that 'ingenuity work' as an ongoing process could be used to address the following aspects of ingenuity.

First, 'ingenuity work' can be used to examine the contemplation and awareness of the nature of the problem in need of a solution. This could be done through both self reflection and assessment by actors' members who are equipped with knowledge and skills that are conducive to certain problem solving. For example, scanning the environment with the purpose of identifying rare resources (human and others) that are amenable to creative processes is an integral part of ingenuity work.

Second, we may examine concrete action which is creative and innovative with the potential to disrupt and overcome constraints and guide strategic action (Swidler, 2001). Usually ingenious action involves nonconventional practices which allocate and enact a variety of resources with the purpose of solving a seemingly insolvable problem. In addition, ingenuity work bypasses or ignores the rules, roles, plans, procedures or routines that anchor the problem in a take it for granted or inertial form (Scott, 2008). In this vein, ingenuity is not an orderly process which has its own 'schema', but could result from processes and actions that, on the face of it, would hint at failure or luck for potential solutions.

Third, we may examine the consolidation of the new solution through endorsement and acknowledgement by those who are affected by it. The ingenious reconstruction of a new means and new actions should also fit the perception, values and practices of stakeholders. For example, in their chapter on haute cuisine, Senf, Koch and Rothmann explore the institutional context of the field of haute cuisine in Germany. The authors claim that chefs are taking an active role in shaping the context of a prestigious occupation and high quality food. By resorting to ingenuity, chefs are able to produce an innovative and creative kitchen through working on an authentic and flexible profile that expresses the mannerism and characteristics of haute cuisine.

Furthermore, ingenuity work occurs when a social actor responds to a perceived or real challenge that may hamper effectiveness or pose a threat to performing a certain task or activity.

Future research may focus on ingenuity work as a set of practices that are instrumental to the implementation of innovative and creative ideas which may contradict the common knowledge and practices, but nevertheless result in a desired outcome. Furthermore, research needs to address the disruptive nature of ingenuity. In particular, ingenuity exhibits disruptive problem solving and may in certain situations lead the organization to challenge its dominant logic by sparking new ideas for new capabilities or opportunities.

CONCLUSION

As previously stated, our goal in this chapter was not to list the many possible avenues for future research, rather, it was to point the reader in the direction of research and lines of inquiry that might prove fertile. With this in mind, we wish to conclude with a few observations that sum up our thoughts on ingenuity.

Ingenuity is working against 'odds' while encountering hurdles such as: formal rules, forms of control or organizational social and symbolic boundaries that tend to prevent creative solutions to existing problems. Ingenuity involves an active transformation of a seemingly stagnant situation and the creation of a new solution which is nonconventional and innovative.

Ingenuity occurs through forming a set of creative practices that can envision, recognize or frame new opportunities and exploit them. Ingenuity is also a learning process which involves creating new knowledge for problem solving through the mechanism of bricolage or counter-intuitive ideas and action. Furthermore, an ingenious solution utilizes new organizing principles which reconstruct and reframe the relationship between the actors and the context in which the actor is operating.

Generally, future research opportunities exist that focus on the conjunction of micro-macro ingenuity or on environmental and organizational elements with a purpose of producing a multi-perspective approach to organizational ingenuity. This research agenda would draw on organizational history and on the role of culture in promoting ingenuity.

REFERENCES

Alvesson, M. and Willmott, H. (2002), Identity regulation as organizational control: producing the appropriate individual, *Journal of Management Studies*, **39**(5), 619–44.

Baker, T. and Nelson, R.E. (2005), Creating something from nothing: resource construction through entrepreneurial bricolage, *Administrative Science Quarterly*, **50**(3), 329–66.

DiMaggio, P.J. and Powell. W.W. (1983), The iron cage revisited: institutional isomorphism and the collective rationality in organizational fields, *American Sociological Review*, **48**(2), 147–60.

Homer-Dixon, T. (2002), *The Ingenuity Gap: Facing the Economic, Environmental, and Other Challenges of an Increasingly Complex and Unpredictable Future*. London: Vintage.

Lampel, J. and Bhalla, A. (2011), Living with offshoring: the impact of offshoring on the evolution of organizational configurations. *Journal of World Business*, **46**(3), 346–58.

Lampel, J., Honig, B. and Drori, I. (2011), Discovering creativity in necessity: organizational ingenuity under institutional constraints, Call for papers for the Special Issue on 'Discovering Creativity in Necessity: Organizational Ingenuity under Institutional Constraints', *Organization Studies*, **32**(5), 715–17.

Lawrence. T. and Suddaby, R. (2006), Institutions and institutional work. In S.R. Clegg, C. Hardy, T.B. Lawrence and W.R. Nord (eds), *Handbook of Organization Studies* (pp. 215–54). London: Sage.

Lawrence, T., Suddaby, R. and Leca, B. (eds) (2009), *Institutional Work*. Cambridge: Cambridge University Press.

Marquis, C. and Tilcsik, A. (2013), Imprinting: toward a multilevel theory, *Academy of Management Annals*, **7**(1), 193–243.

Miller, D. (1991), *The Icarus Paradox: How Exceptional Companies Bring About Their Own Downfall*. New York: Harper Business.

Sarasvathy, S.D. (2001), Causation and effectuation: toward a theoretical shift from economic inevitability to entrepreneurial contingency, *Academy of Management Review*, **26**, 243–63.

Scott. R. (2008), *Institutions and Organizations: Ideas and Interests*. Thousand Oaks, CA: Sage.

Stinchcombe, A.L. (1965), Social structure and organization. In J.G. March (ed.), *Handbook of Organizations* (pp. 142–93). Chicago, IL: Rand McNally.
Sveningsson, S. and Alvesson, M. (2003), Managing managerial identities: organizational fragmentation, discourse and identity struggle, *Human Relations*, **58**(9), 1141–66.
Swidler, A. (2001), *Talk of Love*. Chicago: Chicago University Press.
Watson, T.J. (2008), Managing identity: identity work, personal predicaments and structural circumstances, *Organization*, **15**(1), 121–43.

Index